Game-Based Learning, Gamification in Education and Serious Games 2023

Game-Based Learning, Gamification in Education and Serious Games 2023

Guest Editors

Carlos Vaz de Carvalho
Hariklia Tsalapatas
Ricardo Baptista

Basel • Beijing • Wuhan • Barcelona • Belgrade • Novi Sad • Cluj • Manchester

Guest Editors

Carlos Vaz de Carvalho	Hariklia Tsalapatas	Ricardo Baptista
GILT R and D	CTLL	Research Center N2i
Porto Polytechnic	Thessaly University	Instituto Politécnico da Maia
Porto	Volos	Maia
Portugal	Greece	Portugal

Editorial Office
MDPI AG
Grosspeteranlage 5
4052 Basel, Switzerland

This is a reprint of the Special Issue, published open access by the journal *Computers* (ISSN 2073-431X), freely accessible at: www.mdpi.com/journal/computers/special_issues/0731GC5T5P.

For citation purposes, cite each article independently as indicated on the article page online and using the guide below:

Lastname, A.A.; Lastname, B.B. Article Title. *Journal Name* **Year**, *Volume Number*, Page Range.

ISBN 978-3-7258-3600-0 (Hbk)
ISBN 978-3-7258-3599-7 (PDF)
https://doi.org/10.3390/books978-3-7258-3599-7

© 2025 by the authors. Articles in this book are Open Access and distributed under the Creative Commons Attribution (CC BY) license. The book as a whole is distributed by MDPI under the terms and conditions of the Creative Commons Attribution-NonCommercial-NoDerivs (CC BY-NC-ND) license (https://creativecommons.org/licenses/by-nc-nd/4.0/).

Contents

About the Editors vii

Preface ix

Ricardo Baptista, António Coelho and Carlos Vaz de Carvalho
Training and Certification of Competences through Serious Games
Reprinted from: *Computers* **2024**, *13*, 201, https://doi.org/10.3390/computers13080201 1

Elçin Yazıcı Arıcı, Michail Kalogiannakis and Stamatios Papadakis
Preschool Children's Metaphoric Perceptions of Digital Games: A Comparison between Regions
Reprinted from: *Computers* **2023**, *12*, 138, https://doi.org/10.3390/computers12070138 30

Ruth S. Contreras-Espinosa and Jose Luis Eguia-Gomez
Evaluating Video Games as Tools for Education on Fake News and Misinformation
Reprinted from: *Computers* **2023**, *12*, 188, https://doi.org/10.3390/computers12090188 52

Aikaterini Georgiadou and Stelios Xinogalos
Prospective ICT Teachers' Perceptions on the Didactic Utility and Player Experience of a Serious Game for Safe Internet Use and Digital Intelligence Competencies
Reprinted from: *Computers* **2023**, *12*, 193, https://doi.org/10.3390/computers12100193 72

Alessia Spatafora, Markus Wagemann, Charlotte Sandoval, Manfred Leisenberg and Carlos Vaz de Carvalho
An Educational Escape Room Game to Develop Cybersecurity Skills
Reprinted from: *Computers* **2024**, *13*, 205, https://doi.org/10.3390/computers13080205 91

Ying Fang, Tong Li, Linh Huynh, Katerina Christhilf, Rod D. Roscoe and Danielle S. McNamara
Stealth Literacy Assessments via Educational Games
Reprinted from: *Computers* **2023**, *12*, 130, https://doi.org/10.3390/computers12070130 106

Alexandros Kleftodimos, Athanasios Evagelou, Stefanos Gkoutzios, Maria Matsiola, Michalis Vrigkas and Anastasia Yannacopoulou et al.
Creating Location-Based Augmented Reality Games and Immersive Experiences for Touristic Destination Marketing and Education
Reprinted from: *Computers* **2023**, *12*, 227, https://doi.org/10.3390/computers12110227 121

Lazaros Lazaridis and George F. Fragulis
Creating a Newer and Improved Procedural Content Generation (PCG) Algorithm with Minimal Human Intervention for Computer Gaming Development [†]
Reprinted from: *Computers* **2024**, *13*, 304, https://doi.org/10.3390/computers13110304 155

Rytis Maskeliūnas, Robertas Damaševičius, Tomas Blažauskas, Jakub Swacha, Ricardo Queirós and José Carlos Paiva
FGPE+: The Mobile FGPE Environment and the Pareto-Optimized Gamified Programming Exercise Selection Model—An Empirical Evaluation
Reprinted from: *Computers* **2023**, *12*, 144, https://doi.org/10.3390/computers12070144 176

Vincenzo Di Nardo, Riccardo Fino, Marco Fiore, Giovanni Mignogna, Marina Mongiello and Gaetano Simeone
Usage of Gamification Techniques in Software Engineering Education and Training: A Systematic Review
Reprinted from: *Computers* **2024**, *13*, 196, https://doi.org/10.3390/computers13080196 198

Alkinoos-Ioannis Zourmpakis, Michail Kalogiannakis and Stamatios Papadakis
Adaptive Gamification in Science Education: An Analysis of the Impact of Implementation and Adapted Game Elements on Students' Motivation
Reprinted from: *Computers* **2023**, *12*, 143, https://doi.org/10.3390/computers12070143 . . . . . . **221**

About the Editors

Carlos Vaz de Carvalho

I am currently Professor Coordenador at the Computer Eng. Dep. of the Engineering School of the Porto Polytechnic (ISEP). I am also the Director of Virtual Campus Lda, an SME dedicated to Technology-Enhanced Learning and Serious Games.

I have a PhD (2001) in Technologies and Information Systems from the University of Minho, focusing on the use of e-learning in Higher Education.

As professor, I have lectured more than 20 different courses on Algorithms, Programming, Data Structures, E-Learning and Multimedia. Currently, I am lecturing Serious Game Design and Multimedia Application Development.

I started my research career in 1988 at the INESC in the Computer Graphic Group. In 1996, I focused on Technology-Enhanced Learning while completing my PhD at the University of Minho. From 2005 to 2014, I was Scientific Coordinator of GILT R&D (Games, Interaction and Learning Technologies). I directed 8 PhD and 50 MSc theses and authored over 250 publications and communications, including more than 10 books (as author and editor). I coordinated 20 national and European projects and participated in more than 40 other projects. I worked as an expert for the European Commission and associated agencies in the scope of the Horizon Europa, Eurostars, Socrates-ODL, Minerva, E-Learning, E-Contents Plus, Lifelong Learning and Erasmus+ programmes.

On the management side, I directed the Distance Education Unit of the Porto Polytechnic from 1997 until 2000. From 2001 until 2005, I was E-Learning Director of the ISEP and served as Dean of the Computer Eng. Dep. between 2003 and 2005. Between 2011 and 2013, I was President of the Portuguese Chapter of the IEEE Education Society.

Hariklia Tsalapatas

Hariklia Tsalapatas has a PhD in Computer Engineering with a focus on the development of serious games for learning from the University of Thessaly (2015). She has an MBA from Columbia Business School, NY (2005). She has an MSc Degree in Computer Science from Rice University in Houston, TX (1994). She has a Bachelor's Degree (5 years) from the University of Patras in Greece. She is engaged as an instructor at the Electrical and Computer Engineering Department on technology-enhanced learning, game design, software engineering, and other subjects. She has worked on over 40 R&D projects since 1997, 38 of which were on behalf of UTH since 2000. Before being employed at UTH, she worked as a researcher at the Foundation of Research and Technology Hellas in Heraklion, Greece (1998-2000), and as a database engineer for Oracle Corp. in San Francisco, CA (1993–1997). She has published over 70 articles on technology-enhanced learning.

Ricardo Baptista

Ricardo Baptista is a distinguished academic and researcher specializing in digital media, multimedia technology, and computer engineering. He earned his Ph.D. in Digital Media from the University of Porto in December 2017, focusing his dissertation on "Serious Games for Training and Competence Certification". Prior to his doctorate, he completed a Master's in Multimedia Technology at the Faculty of Engineering of the University of Porto in July 2008 and a Bachelor's in Computer Systems Engineering from the University of Madeira in October 2001. Since September 2023, he has served as a Coordinating Professor at the Instituto Politécnico da Maia (IPMAIA), where he also directs the Research Center (N2i).

He is an affiliated researcher at the Institute for Systems and Computer Engineering, Technology and Science (INESC TEC) and a collaborating researcher at the Polytechnic Institute of Porto's School of Engineering (ISEP). His research interests encompass serious games, training and certification, and data visualization. He has published four articles in specialized journals and contributed three book chapters. He has participated in various research projects, including those funded by the European Union's FP7 program and the Portuguese Foundation for Science and Technology (FCT). Beyond his academic and research endeavors, he has contributed to the development of serious games aimed at enhancing learning experiences and competence certification. His interdisciplinary approach bridges technology and education, underscoring his dedication to innovative teaching and learning methodologies.

Preface

Video games are extremely immersive applications that keep players highly motivated as they strive to overcome challenges and meet predefined targets. Serious games leverage this motivational power to serve a variety of purposes in education, marketing, social awareness, healthcare, and research. The use of games in the education and training field features numerous successful implementations. In parallel, gamification has become prevalent in education as it allows for boosting the students' motivation and engagement. This Special Issue aimed to showcase and discuss novel progress in these areas, and the published articles explore topics such as new scientific methodologies, experimental outcomes, and successful real-world applications.

Batista et al. introduce the "Triadic Certification Method" for assessing learning in serious games, collecting real-time learner data to evaluate skill acquisition, demonstrated through a driving simulator case study. Contreras-Espinosa and Eguia-Gomez analyze 24 video games designed to combat misinformation, showing that well-designed digital games can enhance media literacy by helping users detect fake news. Yazici Arici et al. study preschool children's metaphorical perceptions of digital games, revealing diverse cognitive and emotional associations. Georgiadou and Xinogalos assess the educational value of "Follow the Paws", a serious game for teaching cybersecurity to young learners, with positive feedback from prospective ICT teachers. Spatafora et al. develop and test a cybersecurity-themed escape-room game aimed at SME employees, demonstrating its effectiveness in increasing cybersecurity awareness. Fang et al. explore stealth literacy assessments within serious games, finding strong correlations between in-game performance and traditional reading tests, suggesting that serious games can support literacy development. Kleftodimos et al. showcase how location-based augmented reality (AR) games can enhance tourism and cultural heritage education by integrating storytelling and digital overlays, leading to increased visitor engagement. In programming education, Maskeliunas et al. present FGPE+, a web-based environment using adaptive gamification to improve learning outcomes. Lazaridis and Fragulis propose an improved procedural content generation (PCG) algorithm for video games, generating dynamic 2D maps efficiently, enhancing replayability without requiring high computational power. Di Nardo et al. systematically review 68 studies on gamification in software engineering education, highlighting benefits in engagement while cautioning readers about potential stress factors. Zourmpakis et al. explore adaptive gamification in science education, demonstrating that personalized game elements can significantly enhance student motivation. Overall, these studies highlight the growing impact of serious games and gamification, showcasing their potential in enhancing education.

Carlos Vaz de Carvalho, Hariklia Tsalapatas, and Ricardo Baptista
Guest Editors

Article

Training and Certification of Competences through Serious Games

Ricardo Baptista [1,2,3,*], António Coelho [1,2] and Carlos Vaz de Carvalho [4]

1. Department of Informatics Engineering, Faculty of Engineering, University of Porto, 4200-465 Porto, Portugal; acoelho@fe.up.pt
2. INESC TEC—INESC Technology and Science, Rua Dr. Roberto Frias, 4200-465 Porto, Portugal
3. Instituto Politécnico da Maia, Avenida Carlos de Oliveira Campos—Castêlo da Maia, 4475-690 Maia, Portugal
4. GILT (Games, Interaction, Learning Technologies), Instituto Superior de Engenharia do Porto, 4200-072 Porto, Portugal; cmc@isep.ipp.pt
* Correspondence: ricardjose@fe.up.pt

Abstract: The potential of digital games, when transformed into Serious Games (SGs), Games for Learning (GLs), or game-based learning (GBL), is truly inspiring. These forms of games hold immense potential as effective learning tools as they have a unique ability to provide challenges that align with learning objectives and adapt to the learner's level. This adaptability empowers educators to create a flexible and customizable learning experience, crucial in acquiring knowledge, experience, and professional skills. However, the lack of a standardised design methodology for challenges that promote skill acquisition often hampers the effectiveness of games-based training. The four-step Triadic Certification Method directly responds to this challenge, although implementing it may require significant resources and expertise and adapting it to different training contexts may be challenging. This method, built on a triadic of components: competencies, mechanics, and training levels, offers a new approach for game designers to create games with embedded in-game assessment towards the certification of competencies. The model combines the competencies defined for each training plan with the challenges designed for the game on a matrix that aligns needs and levels, ensuring a comprehensive and practical learning experience. The practicality of the model is evident in its ability to balance the various components of a certification process. To validate this method, a case study was developed in the context of learning how to drive, supported by a game coupled with a realistic driving simulator. The real time collection of game and training data and its processing, based on predefined settings, learning metrics (performance) and game elements (mechanics and parameterisations), defined by both experts and game designers, makes it possible to visualise the progression of learning and to give visual and auditory feedback to the student on their behaviour. The results demonstrate that it is possible use the data generated by the player and his/her interaction with the game to certify the competencies acquired.

Keywords: serious games; competencies and skills; "in-game" assessment

Citation: Baptista, R.; Coelho, A.; Vaz de Carvalho, C. Training and Certification of Competences through Serious Games. *Computers* **2024**, *13*, 201. https://doi.org/10.3390/computers13080201

Academic Editor: Wenbing Zhao

Received: 8 July 2024
Revised: 6 August 2024
Accepted: 8 August 2024
Published: 15 August 2024

Copyright: © 2024 by the authors. Licensee MDPI, Basel, Switzerland. This article is an open access article distributed under the terms and conditions of the Creative Commons Attribution (CC BY) license (https:// creativecommons.org/licenses/by/ 4.0/).

1. Introduction

The global computer games industry holds immense learning potential. Games, now marketed to all ages and genders, are increasingly being recognised as powerful tools for learning (GBLs). They have been shown to motivate players, enhance problem-solving, and facilitate collaboration and competition. These benefits are not limited to a specific discipline or education level but apply across the board, from formal classrooms to non-formal learning processes. This recognition of the potential of game-based learning is a cause for optimism about the future of education.

There are already many studies about the effectiveness of Serious Games, both in training and other activities [1,2]. Sousa and other authors [3] consider measuring their actual effects on learning as one of the biggest challenges for accepting Serious Games as an effective educational method. Mayer [4] highlighted that few existing evidence-based approaches to assessing their contribution to learning are still available. Various frameworks have been developed relating to theories of instruction and learning to the mechanics and elements of games, one of two distinct domains: one with the design and development of Serious Games on distinct approaches like describing game-based learning scenarios [5], a scenario generation framework for mission-based virtual training from both the trainer and trainee's perspective [6], a conceptual framework to design Serious Games that have empathy as part of the learning outcomes [7], considering the flow framework that uses the dimensions of flow experience to analyse the quality of educational games [8], a framework of an evaluation-driven design which offers guidance in the evaluation process [9], a Serious Games design framework in cultural heritage with steps to follow during the whole process [10], and a conceptual framework based upon a systematic literature review of developments in student-centred digital learning [11]. The other domain is evaluation, which uses other approaches, such as the serious game design assessment framework, as a constructive structure to examine purpose-based games [12]; a holistic approach to serious game evaluation with four key areas: theoretical, technical, empirical, and external [13]; an interpretive evaluation framework that can identify the educational value in COTS games [14]; and the dimensionalisation of game-based-learning and further decomposition into factor/sub-factors based on theoretical constructs [15].

Yet, the current state of Serious Games and game-based assessments does not translate into valid certifications. Students still need to prove their knowledge through traditional evaluations. Despite games supporting learning, assessing the knowledge and skills acquired this way needs significant improvement. This research aims to develop a design and development process for Serious Games, considering that they integrate training, evaluation, and skills certification. The focus is on the potential to incorporate the evaluation process and the consequent certification of competencies within the game context, governed by specific norms that systematise student performance measurement.

Four key questions guide this research, each exploring a crucial aspect of the intersection between gaming and skills training:

1. To determine whether a significant relationship exists between game genres and the training of specific skills. By analysing different game types, we aim to uncover which genres are most effective in cultivating particular competencies, thereby enhancing the educational potential of Serious Games. Another proposition is to identify practical elements of serious game design for evaluating learning and training player skills. As such, the research also explores the possibility of a competency certification method using Serious Games, with game design considering learning objectives and certification performance metrics.
2. To investigate the elements of serious game design that best support the evaluation of learning and training players' skills based on structured competency frameworks. This proposal examines how game mechanics, narratives, and feedback systems can be optimised to engage players and assess and develop their abilities.
3. To identify a robust method for certifying competencies using games. This proposal entails creating a game design that aligns with learning objectives and incorporates performance metrics that can reliably measure and validate players' skills. By establishing clear criteria and standards, we can ensure the certification process is rigorous and credible, reassuring us about its reliability.
4. To understand the integration of Learning Analytics into the in-game assessment system to precisely measure player performance in skills training. This proposal involves leveraging data analytics to track and analyse player behaviour and outcomes, providing detailed insights into their learning progress. By doing so, we can enhance the accuracy and effectiveness of skill assessments, ensuring that players receive meaningful and actionable feedback.

This study was able to answer the various questions posed, the answers to which allowed the four-step Triadic Certification Method to be structured and supported by new tools. With the successful integration of the evaluation process into Serious Games, the results obtained made it possible to check the players' performance in the various training contexts, attesting to the validity of the training design, as well as validating the skills acquired by visualising the progression of skill acquisition through the Triadic Certification Method. This component of the method is particularly comprehensive and well-rounded, as it conventionally designs SGs for the acquisition of competencies based on a balance between the three elements: essential competencies (skills), mechanics (game), and reality (training).

This paper has four sections. The first section presents a literature review on digital games, genre taxonomies of games, Serious Games and their taxonomies, game-based learning, and learning design assessment. In the second section, we discuss the training and certification of competencies. The subsequent section focuses on the Triadic Certification Method, a comprehensive approach developed to address the identified problem, integrating training, evaluation, and skills certification within the game context. The final section describes a case study on the training and certification of competencies in teaching automobile driving and the results obtained. It serves as a proof of concept for the developed method, demonstrating its practical application and effectiveness.

2. Digital Games and Game Based-Learning (GBL)

In this part, we endeavour to elucidate the significance of games within educational contexts and their consequential impact on human behaviour, drawing from constructivist perspectives. We commence by delineating fundamental concepts such as "game" and "play". A game is a purposeful competition governed by rules wherein players strive for victory. Conversely, play encompasses many intentional activities, often undertaken for recreational or leisurely purposes.

Johan Huizinga's conceptualisation of the "magic circle" accentuates the discrete space wherein game-related activities manifest devoid of real-life repercussions, underscoring the immersive nature intrinsic to gameplay. This concept has proved pivotal in elucidating the symbiotic relationship between players and games, shaping discourse across digital and traditional gaming paradigms [16].

Scholars including Huizinga [16], Caillois [17], Juul [18], and Salen and Zimmerman [19] underscore the inherent allure of games to players, influenced by variables such as age, cognitive aptitude, and individual personality traits. Game designers leverage these factors to augment player engagement by iteratively adjusting goals, rules, challenges, and participant dynamics.

Contrary to prevalent misconceptions, games are structured environments imbued with clearly delineated objectives, adversaries, and regulatory frameworks. They thereby afford players opportunities for cognitive stimulation and skill refinement. Nonetheless, unlike real-world scenarios, games have distinct consequences for successes and failures, contributing to their intrinsic allure and divergence from reality [20].

Games epitomise rule-bound activities characterised by delimited beginnings, middles, and denouements. They present players with cognitive challenges necessitating proactive engagement. While games ostensibly simulate real-world scenarios, they deviate in outcome predictability and repercussions, thereby furnishing distinct pedagogical and experiential paradigms for players [21].

2.1. Game Genre Taxonomies

The defining characteristic of computer games lies in the interactive pattern established between the player and the game environment. Video games are categorised into genres primarily based on their patterns of gameplay interaction rather than their visual or narrative elements [21]. However, the taxonomy of game genres has been a contention, with numerous proposals in the field [22,23]. These genre classifications organise games into distinct categories defined by their underlying gameplay mechanics.

Understanding game genres is crucial for game designers as it enables them to align additional content, such as new levels or characters, with established gameplay mechanics like jumping or shooting in a platformer game. It also allows them to innovate existing ones, for instance, by introducing a new gameplay mechanic like time manipulation in a puzzle game [24].

Game genres are more than just categories—they are a unifying force. The primary factor uniting games within a genre is the similarity in the interactions facilitated between the player and the game environment. These interactions manifest through various gameplay mechanics, encompassing the actions of in-game objects and players throughout gameplay [25]. It is these recurrent actions or challenges that ultimately define the genre of a game, creating a sense of belonging and connection among players and designers alike.

While there is no universally standardised taxonomy of video game genres, the industry's recognition is a validation of their importance. The industry commonly recognises several overarching categories. These typically include action, strategy, role-playing, sports, management simulation, adventure, puzzle, and quiz genres, acknowledging the diversity and significance of each [24,26–28] (Table 1).

Table 1. Taxonomy of game genres.

Game Genre	Goals	Sub-Genre
Action	To overcome mental or physical challenges against one or more opponents by engaging in a series of actions (timing—reaction speed, in which accuracy may be emphasised). Realism is not relevant.	Beat-'em-ups Beat 'em ups, Shooter games (1st and 3rd person), Platform games
Strategy	Deploy tactics/strategies to overcome complex challenges against one or more opponents by planning a superior series of actions (physical challenges are not emphasised).	4X (eXplore, eXpand, eXploit, eXtermine), Real-time strategy games, Real-time tactics, Turn-based strategy, War games
Role-Playing	Victory is achieved through superior planning or out-thinking the opponents (physical challenges and chance take a more minor role). Distinct from action games, RPGs seldom test a player's physical skill (combat is more tactical than physical) and involve other non-action gameplay (resource management).	RPGs, MMORPGs
Sports	Similar to action games, except for the realism of movements and techniques, which are very important.	Exergames, Sports/management games
Management Simulation	To overcome economic challenges, a series of actions must be planned. Direct action upon an opponent is not emphasised. They are typically designed to be never-ending (no-win scenario). Goal example: to build a collection of objects.	Racing games/Vehicle, Virtual worlds/Pets, Life simulation/social games, Business
Adventure	To use an avatar for the exploration of an interactive story and to overcome challenges in isolation (puzzle adventure) by planning a superior series of actions (physical challenges are not emphasised).	Graphics Adventure, Puzzle adventure
Puzzle	To overcome mental challenges in isolation (not around a conflict with another opponent) by planning a superior series of actions. Games usually involve shapes, colours, or symbols that the player must directly or indirectly manipulate into a specific pattern.	Action/Arcade puzzle (timed), Reveal the picture game, Physics game.
Quiz	Gamepad controlled, mouse keyboard, Wii balance board.	

However, it is essential to note that this classification is not exhaustive, and numerous hybrid genres exist that blend elements from multiple established categories. Moreover, the continuous technological advancements within the gaming industry constantly give rise to new genres, particularly with the introduction of novel platforms or input devices. For instance, the advent of Nintendo's Wii console spurred the emergence of "physical" games like Wii Fit, exemplifying the industry's ongoing evolution and diversification of genres.

2.2. Serious Games

Serious Games (SGs) are not just a trend in corporate settings and research communities; they are a transformative tool. This game definition harnesses the engaging features of video games to make learning processes not just bearable but exciting [17,18]. With SGs, players can learn while playing, a concept revolutionising education and training. These games are designed to engage players with specific topics, effectively teaching educational content or training workers to perform particular tasks. This transformative power of Serious Games inspires educators and researchers in their work.

SGs refer to digital games used for purposes other than entertainment, such as training, advertising, simulation, or education. Intentionally, these games help learning, skills acquisition, and behaviour change through a game design process that focuses on achieving learning outcomes through gameplay [4,21,29,30]. Clark C. Abt [31], a pioneer in the field, introduced the concept of Serious Games in his book. This marked a shift in the perception of games, extending their meaning beyond mere entertainment when used or embedded in a specific context. Serious Games, as he defined them, are not primarily for amusement but have an explicit and carefully thought-out educational purpose. Since the first Serious Games initiative sponsored by Woodrow Wilson in 2002 and the Serious Games Summit in 2004, there has been significant growth in game-based learning. Serious Games are interactive computer applications with a challenging goal, are fun to play and engage, incorporate some concept of scoring, and impart skills, knowledge, or attitudes that can be applied in the real world.

The concept of Serious Games still lacks a precise definition, with some authors using other terms like immersive learning simulations, digital game-based learning, gaming simulations, and "games you have to play" [32,33]. The main goal of Serious Games is to provide an interactive means for the transference of knowledge to the player. One main goal of Serious Games is to provide an interactive means for transferring knowledge to the player.

SGs educate or train the player, contain a direct means of assessing a skill or learning, and employ a game interface that provides these features. They are considered a new tool within the active learning paradigm. If game design focuses on learning outcomes, learning becomes a natural consequence of playing.

Serious Games have many applications, from government and corporate training to health, public policy, and strategic communication [34]. Despite their diverse uses, the primary focus of SGs remains their educational purpose [29,35]. They are designed to facilitate learning and training and to apply new pedagogies. Research has shown that SGs can accelerate learning, increase motivation, and support the development of higher-order cognitive thinking skills [5]. This diverse range of applications intrigues game developers and professionals in the field.

The key to the success of these games is motivation. It is a psychological process that stimulates an individual to act upon something to attain a desired effect or goal. Motivation in learning can be affected by intrinsic motivation [36], extrinsic motivation [37,38], and emotional stability. These motivational factors should be considered when designing and developing Serious Games. However, due to the broad range of individuals' emotional stability, it may take much work to address it.

Serious Games (SGs) are not limited to a single field; they have distinct classifications that cater to various purposes. These games are used in government, defence, education, corporate, and industry settings, highlighting their versatility. They address aspects such as occupational safety, skills, communications, and orientation, proving that Serious Games can be applied in various scenarios. Examples of SGs include Alcoa SafeDock, Rosser Surgery Skills w/Games, Shield of Freedom, America's Army, Darfur is Dying, and Tactical Language & Culture.

Several attempts have been made to classify Serious Games into genres or similar typologies. The criteria used to classify the games vary greatly, with the most commonly used being the educational content and field of application of Serious Games. Michael and Chen [29] name military, government, educational, corporate, healthcare, political, religious, and art games. This typology solely bases itself on the application areas of the games, showcasing the diverse range of Serious Games. For instance, in the health games, Susi et al. [39] list the subgroups of exergaming [40], health education [41], biofeedback [42] and therapy [43].

Sawyer and Smith [44] suggest a higher-resolution taxonomy that crosses game and learning types with application areas. They list advergames, games for work, or games for health as game types. The core innovation of Sawyer and Smith was to separate the designed purpose from actual application areas. Based on the promising work of Ratan and Rittefeld [45], we would propose the following label categories, which allow for the inclusion of specifically designed Serious Games as well as COTS games for "serious" purposes, as the following Table 2 presents:

Table 2. Categories for classifying serious games.

Label/Tag Category	Exemplary Labels
1. Platform	Personal Computer, Sony Play Station 3, Nintendo Wii, Mobile Phone
2. Subject Matter	World War II, sustainable development, physics, Shakespeare's works
3. Learning goals	Language skills, historical facts, environmental awareness
4. Learning principles	Rote memorisation, exploration, observational learning, trial and error, conditioning
5. Target audience	High school children, nurses, law students, the general public, preschoolers, military recruits
6. Interaction mode(s)	Multiplayer, co-tutoring, single-player, massively multiplayer, tutoring agents
7. Application area	Academic education, private use, professional training
8. Controls/interfaces	Gamepad-controlled mouse keyboard, Wii balance board.
9. Common gaming labels	Puzzle, action, role-play, simulation, card game, quiz

2.3. Game Assessment

Game assessment is a critical component of the learning process. It involves comparing the expected learning goals with the evidence obtained from the learning actor. In the context of digital game-based learning (DGBL), this evaluation is typically conducted through traditional methods such as questionnaires, interviews, log file analysis, or observation of experience [46]. However, new technologies and media are emerging to support more advanced evaluation tools, addressing the current gaps in evaluation.

Over the past decades, research has aimed to develop new approaches that support the evaluation paradigm in game-based learning. This analysis is divided into two contexts: measuring learning and incorporating the evaluation component in the game development.

2.3.1. Assessment of Digital Learning

The assessment of the game's learning experience is primarily carried out outside the game's context, using external tools such as Acumen Team Skills Assessment and Profiles Team Analysis [47]. These tools attest to the results obtained in higher education, management, and personal and team skills development.

The literature refers to three robust theoretical frameworks that underpin the evaluation of learning in Serious Games: RETAIN (Relevance, Embedding, Transfer, Adaptation, Immersion, and Naturalization) [48], Kirkpatrick's levels of evaluation [49], and the CRESST learning model [50]. These frameworks provide a solid foundation for developing and evaluating learning games, ensuring they effectively incorporate educational content. This thorough evaluation process of Serious Games gives educators and researchers confidence in the effectiveness of this tool.

First, the RETAIN framework supports learning games' development and evaluates how they contain and incorporate educational content. It aims to identify the best combination between the various game elements, associating them with the genre taxonomy. This framework's relevance to this work is related to the analysis of the conceptualisation of the relationship between the components of games—the elements and genres of games—and the objectives and levels of learning competencies. The educational potential of games depends on the coherence between different elements using distinct levels of learning conceptualisation and assessment. These models join the curriculum and motivational aspects of their design. The models of learning conceptualisation include Bloom's taxonomy, which corresponds to a taxonomy of educational objectives, dividing the learning into three main objectives: to generate skills, to develop competencies, and to transfer knowledge in three distinct domains: cognitive, behaviour, and aptitude; and Gagné's Events of Instruction [33,51], which propose motivational events as positively influencing the achievement of the expected results. Involvement with experience does not derive from a hierarchy of events but rather from the assumption that what goes on inside and outside the game learning experience are lines of and many elements in a single event can be combined or interconnected. This approach, in particular, is a structuring example in terms of the design of the structure of events in the gaming experience for this study, which focuses on skills learning. The closer it is to reality, the more involvement it brings to the transfer of knowledge after learning. Finally, Keller's ARCS Model [52] states that motivation in student learning corresponds to a systematic process represented by four steps, within which motivation can be achieved or promoted. These models aim to maximise the potential of educational situations by choosing the most appropriate combination of factors to incorporate into the game's development.

The second model for conceptualising evaluation is Kirkpatrick's four levels of assessment: Reaction, Learning, Behaviour, and Results. This model presents a hierarchy of levels for evaluating learning or training programmes. The transition between levels plays a vital role in the evaluation process, adding value to the information collected. However, the process becomes more complex and time-consuming. Each evaluation level produces expected results, but this methodology must be applied for a correct analysis [49]. This framework is widely used in training evaluation due to its structure, aligning with the learning aptitudes and competencies cycle.

The CRESST framework by Baker and Mayer [50] comprises five fundamental cognitive requirements for learning: content comprehension, problem-solving, self-regulation, communication, and collaboration/teamwork. This model focuses on the actions considered for each group, which will be tested to validate the expected results.

In summary, the three frameworks presented reflect the domains of learning evaluation but still need to improve their convergence with technologies. Evaluation in game-based learning (GBL) must consider a hierarchical set of needs (Bloom Taxonomy), expected results (the four levels of Kirkpatrick's assessment), and cognitive requirements (CRESST model). The learning environment must always be considered, as it must promote student-friendly involvement. For successful learning, the experience should increase students' motivation in the game process through tactics (Keller's ARCS model) and/or events (Gagné's Nine Events of Instruction model), focusing on results.

2.3.2. Evaluation Design in Serious Games

The evaluation design in Serious Games involves several conceptual frameworks that apply various evaluation models integrated into the game development process.

The first is the Evidence-Centred Assessment Design (ECD) framework [53–56], which posits that evaluation results from evidence analysis functioning in a triangular interaction. The keys to a balanced triangular interaction are cognition (theory and information about how students learn), observation (student task performance can demonstrate their learning), and interpretation (used to draw inferences from observations).

Another evaluation framework is the structure of De Freitas and Oliver's four dimensions [4,57,58]: context, student specification, modes of representation, and pedagogical principles. These dimensions allow for the evaluation of game-based learning and simulations and should be considered interactive, each containing key characteristics. Harteveld and other authors [59–61] propose another model that contends that the design fundamentals of any game have a serious purpose.

This model is in continuity with the structures presented previously, where the balance between the forces exerted by the three intervening areas (pedagogy, game elements, and reality) in the design or use of educational games is key. This model has three nuclear pillars: Play Space, Meaning, and Reality. Another structure developed for designing and developing educational games is the "Design, Play, and Experience Framework" (DPE) [62–64]. Its main objective is to describe the relationship between the designer and the player as a mediated experience to achieve the expected results through the game. This model is supported by three pillars, each of which contributes to the game's design according to the phase/level of the game and the type of player.

When the "DPE Framework" is applied to Serious Games, it expands with another set of layers related to the specificity of the design of these types of games (learning, storytelling, gameplay, and user experience), which are transversal to the three components of the structure (Design, Play, and Experience).

Two more models in the context of Serious Games design are the "Experimental Gaming Model" and the EFM Model (effective learning environment, flow experience, and motivation). These models emphasise the experience in the game context, focusing on goals such as motivation, the player's learning experience, and other emotional or affective aspects.

According to several researchers [65–67], the "Experimental Gaming Model" presents the learning process as circular, based on constructing cognitive schemes through activities within the game environment. The direct interactions between players and their experiences with the environment create a circular learning mechanism that includes all the necessary steps to ensure the success and achievement of the objectives. The principal elements of an educational game should be contained in the scenario that will define the learning objectives. Feedback is crucial in providing insight into the acquired knowledge and evaluating the player's performance. According to Song and Zhang [68] and Hussein [69], the EFM model suggests clever design practises to inspire motivation and help learners genuinely learn from the game. It proposes ideas for developing games with effective learning environments where students develop increased motivation during the experience flow. An effective learning environment supports seven basic requirements by presenting specific tasks with clear objectives and appropriate challenges while achieving a high degree of interaction and feedback. The model includes two distinct levels: a group of nine components of the flow of experience, subdivided into three categories (conditional, experience, and results), and another group of strategies with four essential components (relevance, trust, satisfaction, and attention) to stimulate motivation.

2.3.3. Final Considerations on the Design and Evaluation of Learning with Games

Despite the existing research and diverse approaches to applying evaluation in game design and development, the use of assessment in game-based learning has yet to gain full recognition for its role in the success of this learning approach.

The referenced models aim to associate the assessment process with students' gains in game-based learning. They highlight how various elements of games contribute to practical evaluation within the games themselves. Key aspects of the game, such as the interface, play environment, narrative, mechanics, and student motivation and involvement, are crucial in this process.

Another critical aspect of the evaluation process is its stakeholders' respective roles. Evaluators face new challenges as they are expected to collaborate and assist in the evaluation process and in understanding and identifying the entertainment elements of the game.

Evaluating knowledge acquisition and transfer through games focuses on the need to hierarchise learning to achieve expected results. However, learning is not seen merely as a sequence of goals achieved through a gaming experience. Though still performed traditionally (i.e., summative tests), the evaluation now considers convergence models. Motivation, a key element in the decision to carry out activities and corresponding knowledge acquisition, is highlighted in various models.

According to Conole [70,71] and Wills and others researchers [72], there is a need for greater convergence between the role of technology and its impact on evaluation. In game-based learning, new models must be developed that allow us to explore and take advantage of the success of the gaming experience. These issues are very important for future research, especially in the methodologies for designing Serious Games.

3. Training, Competences, and Certification

Training and education are the same in definition but slightly different in context. Both are actions associated with acquiring competencies (knowledge, skills, and attitudes). However, education plays a crucial role in workplace learning, where learning is developed, preferably in the workplace, to improve the performance of employees. They gain practise with tools, equipment, and other elements that can be used daily. Complementing this is another concept: certification. An initial definition of this concept is the validation of the skills that the individual has achieved after the training programme, and this issue will be discussed later in this chapter.

Training is associated with many contexts and is a planned learning experience that ensures permanent change in individual knowledge, attitudes, or skills. The meaning of this learning corresponds to improving the individual's performance to achieve a certain level of knowledge or skill through the organised transmission of information and/or guided instructions.

Author Michael Armstrong [73] reinforces the idea of performance associated with training by stating that it is a systematic development of the knowledge, skills, and attitudes required of an individual to perform a particular task or job. Similarly, Edwin Flippo [74] states that training increases an employee's knowledge and skills to perform a specific or particular job. Finally, the author Aswathappa [75] defines the concept as improving skills and attitudes, where training contributes to updating old skills and developing new ones.

Training, as a systematic process, must be directed in such a way as to achieve the expected benefits. We can characterise a training system (programme) in four phases: (1) assessment of training needs; (2) design of training programmes; (3) implementation of the training programme; and (4) evaluation of the training programme.

We can conclude that skills training reflects a programme whose structured approach corresponds to an individual's training needs to achieve specific results. When the training is completed, the evaluation is carried out on-site, at work, or in the context in which the task is carried out to verify whether or not the acquisition of desired knowledge, skills, or attitudes is necessary.

Building on this understanding, certification is a voluntary process that precedes the on-site verification of competencies (assessment). It is a powerful tool for professionally recognising knowledge, skills, and other practises [76]. According to the authors Byrne, Valentine, and Carter [77], the act of certification, when aimed at validating a more advanced level of knowledge and practise, is a formal procedure that allows an individual or an accredited/authorised entity to assess, verify, and attest, in writing and by issuing a certificate, to the attributes, characteristics, quality, and/or other aspects related to the status of individuals or organisations, procedures, or pre-processes, which are following established requirements or standards [78,79].

Acquiring a certification signifies that the individual's competencies and attitudes gained endure. This enduring nature, coupled with benefits such as personal development, career progression, financial reward, professional recognition, and perceived empowerment [80], underscores the value and security of the investment in training and certification.

Finally, training is an organised activity that imparts information or instructions to improve performance or help individuals attain the required knowledge and skills. In training, it is essential to distinguish between competency and competence. The former refers to an individual's ability to make deliberate choices from a repertoire of behaviours in specific professional contexts. The latter is context-dependent and involves integrating knowledge, skills, judgement, and attitudes. Competences are "domain-specific cognitive dispositions that are required to cope with certain situations or tasks successfully and acquired by learning processes" [81].

Different types of competencies can be considered in organisations or specific fields. First, we have personal competencies, representing the core knowledge, skills, and attitudes each person should have for superior performance. Next, functional competencies are related to specific technical knowledge within a particular area or profession. Finally, task competencies are implicit and associated with specific role functions.

Regarding skills, "skill" is often preferred over "competence" in the training and work environment. Hard skills encompass specific technical abilities or solid factual knowledge required for a job, such as machine operation, programming languages, and safety standards. These skills are typically trainable and easy to observe, quantify, and measure. On the other hand, soft skills (also known as "people skills") are more subjective and are associated with personal attributes and character. Soft skills are essential for applying technical skills in the workplace, including communication, teamwork, problem-solving, and time management. These skills are more challenging to observe and quantify.

3.1. Reference Structures for Skills and Competences

This section aims to present various frameworks used to prepare and recognise the skills that are being learned. The skills matrix is a tool for assessing the skills needed to achieve maximum impact and locating where these skills can be found. In this way, the skills framework is a structure that establishes and defines each skill (such as problem-solving or people management) required of people who work in or are part of an organization.

This matrix/framework can take several approaches, most notably when recruiting employees by aptitude standard and performance appraisal or identifying the aptitudes required to perform an activity in any given role. A matrix can, therefore, be considered an inventory of skills categorised by level, with a given required/chosen level of skills. This matrix results in what can be learned (skills) and the quantification (points) required to acquire and improve a skill.

3.1.1. European Qualifications Framework

The recognition of qualifications in Europe, more specifically in the European Union, is carried out through the European Qualifications Framework (EQFs) (https://wwwcdn.dges.gov.pt/sites/default/files/brochure_eqf_en.pdf (accessed on 7 July 2024)), which acts as a standard reference system to link all national qualifications systems. For each of the eight levels defined, there is a set of indicators that specify the expected learning outcomes corresponding to the qualifications of that level in any qualifications system, covering several education levels (primary, secondary and higher education, vocational training) [82] as well as the processes of recognition, validation, and certification of competences obtained, whether by non-formal or informal means [83].

This approach is based on learning outcomes, with eight reference levels defining what is necessary and sufficient for each student to know, understand, and be able to achieve after completing the learning process. These criteria are defined in terms of knowledge as the result of assimilating information during learning; aptitude, as the ability to apply knowledge and know-how to complete tasks and solve problems; and attitude, as the proven ability to use knowledge; skills; and personal, social, and methodological competences in a work or study context and professional and personal development [84].

3.1.2. Lominger's Competency Models and Education Competencies: A Comprehensive Approach

A competency model is a comprehensive framework that outlines the behaviours employees must exhibit to achieve success in their roles or perform specific tasks effectively. Unlike job descriptions, which enumerate the tasks and responsibilities associated with a particular position, competency models delve deeper by elucidating how employees should perform their duties. While job descriptions provide a list of tasks and functions required for a role, competency models identify the requisite behaviours, skills, and knowledge essential for executing those tasks proficiently.

Lominger's sixty-seven competencies have emerged as a universal standard for achieving task success. Known as the Leadership Architect Competencies, this assessment tool enables us to compile a comprehensive list of competencies by combining existing models. The goal is to encapsulate the essential skills for success in various contexts [85]. This competency model represents "a collection of competencies associated with successful performance" [86]. To apply it specifically to education and training, the same authors collaborated with Microsoft to create a similar approach known as Education Competencies or the Educational Competency Wheel [87]. This tool encompasses various attributes, behaviours, knowledge areas, and abilities for effective job performance.

The competency table, as depicted in Table 3, comprises six core skill sets and personality characteristics. These include individual excellence (IE), organisational skills (OrSs), courage (C), results (Rs), strategic skills (SSs), and operating skills (OpSs). While these categories initially draw from Lominger's standard set of 44 competencies, they can be extended beyond education to other domains, such as competency training. The competency wheel offers additional resources to identify core competencies critical for an organization's success [88]. These resources include clear definitions, proficiency levels, sample interview questions, and activities aimed at skill development, all geared toward helping organisations achieve their goals.

The six qualities or success factors can be categorised into two main types: hard and soft skills. Hard skills are teachable abilities or skill sets that lend themselves to quantification. In contrast, soft skills are more subjective and challenging to measure. Among the core skill sets, we can consider individual excellence, courage, results, and strategic skills to be soft skills. These enable effective collaboration, direct communication, goal-oriented action, and pursuit of longer-term objectives. On the other hand, operating skills and organisational skills fall into the hard skills category. They encompass the practical skills for daily task management, relationship building, and effective communication across diverse organizational contexts. With this restricted and adequate number of core competencies, the mapping aligns with the technical and personal competence needs based on the expected results.

Comprising 37 competencies referred to as success factors, this set of categories, while aligning with the Lominger matrix, is not limited to formal education. Its versatility extends to areas like skills training and shares striking similarities with other performance standards, such as the Baldrige Education Criteria for Performance Excellence, defined by the International Society for Technology in Education and the National Standards for Educational Technology [89]. Another important aspect is that this competency standard is recognised by UNESCO, whose general competencies are ICT-related competencies [84,90] and application competencies [88].

To conclude this section about the competency standards, it is essential to differentiate between certification and qualification as they carry distinct meanings. According to the European Qualifications Framework (EQFs), qualification represents the formal outcome of an assessment. In contrast, certification involves a validation process conducted by a competent body to determine whether an individual has achieved specific learning outcomes according to a predefined standard.

Numerous international and national standards govern professional certification. Notably, the ISO/IEC 17024 standard, developed by the International Organization for Standardization (ISO) and the International Electrotechnical Commission (IEC), specifies requirements for certification bodies [91]. These standards apply independently of any specific area of expertise. The European Community has also adopted ISO/IEC 17024. In the United States, the National Organization for Competency Assurance (NOCA) has established standards and an accreditation process for certification programmes since the late 1970s.

Table 3. Educational Competency Wheel [87].

Educational Success Factors					
Individual Excellence (IE)	Organisational Skills (OrSs)	Courage (C)	Results (Rs)	Strategic Skills (SSs)	Operating Skills (OpSs)
Building Effective Teams (IE1)	Comfort Around Authority (OrS1)	Assessing Talent (C1)	Action Oriented (R1)	Creativity (SS1)	Developing Others (OpS1)
Compassion (IE2) Organisational)	Organisational Agility (OrS2)	Conflict Management (C2)	Drive For Results (R2)	Dealing with Ambiguity (SS2))	Directing Others (OpS2)
Customer Focus (IE3)	Presentation Skills (OrS3)	Managerial Courage (C3)		Decision Quality and Problem Solving (SS3)	Managing and Measuring Work (OpS3)
Humour (IE4)	Written Communications (OrS4)			Functional / Technical Skills (SS4)	Managing Through Processes Systems (OpS4)
Integrity and Trust (IE5)				Intellectual Acumen (SS5)	Organising (OpS5)
Interpersonal Skills (IE6)				Learning on the Fly (SS6)	Planning (OpS6)
Listening (IE7)				Strategic Agility and Innovation Management (SS7)	Priority Setting (OpS7)
Managing Relationships (IE8)				Technical Learning (SS8)	Time Management (OpS8)
Managing Vision and Purpose (IE9)					Timely Decision-Making (OpS9)
Managing Vision & Purpose (IE9)					
Motivating Others (IE10)					
Negotiating (IE11)					
Personal Learning and Development (IE12)					
Valuing Diversity (IE13)					

A professional certification effort, a journey of empowerment, involves three relatively independent dimensions. Firstly, the professional role characterisation includes defining the specific professional role to be certified. Secondly, the list of required abilities and skills to identify the abilities and skills necessary for professionals in that role. Finally, the description of the certification process outlines the certification process and its organizational aspects, all designed to empower professionals in their respective roles.

ISO/IEC 17024 is a crucial standard for individual certification. It is a benchmark for recognising certification bodies and their national and international certification schemes. This standard plays a pivotal role in defining the certification process, encompassing all activities through which a certification body establishes that a person fulfils specified competence requirements.

The competence certification system, a beacon of recognition, is a powerful tool that enables professionals working in the labour market to gain recognition based on their qualifications. By achieving specific parameters, professionals demonstrate their competency, empowering them to showcase their skills and knowledge. For instance, the EQFs assesses whether an individual has acquired learning outcomes aligned with relevant standards. The validation process involves four phases: identification, where dialogue is used to identify an individual's experiences; visibility, which consists of making these experiences visible through documentation; formal assessment, which evaluates the experiences formally; and recognition, which leads to certification for partial or complete qualifications. This process focuses on assessing the skills and knowledge demonstrated by learners in specific tasks, ensuring their competency applies to real-world scenarios.

4. Triadic Certification Method

Creating virtual environments conducive to learning through games represents a significant milestone. These environments guarantee success in acquiring knowledge and experience for players and students, driven by motivational and engaging elements. Previous literature reviews have highlighted the characteristics and strategies contributing to successful game learning.

While individuals can be trained to achieve expected results in various situations and contexts, assessing skills remains challenging. Specifically, further progress is required to certify the knowledge and skills acquired during the learning processes conducted through SGs.

This research addresses the development of new methods to maximise the benefits of successful learning through games. To achieve convergence, we integrate certification of competency training. Unlike assessing learning, this approach validates the knowledge and skills acquired for professional functions or activities.

The Triadic Certification Method (TCM) aims to incorporate certification into SG development (from conception to design and implementation). This method involves four steps that influence game design, ensuring elements necessary for certification success. Communication between key stakeholders—the trainer/instructor and the designer—guarantees fundamental decisions regarding SG functionality.

The method, a testament to its versatility, is not restricted to game taxonomy and applies universally regardless of game type; it applies to any training context and adapts to diverse training scenarios. Skills acquired during training levels align with proficiency levels, reflecting the learning state. Certification occurs only when all defined competencies are successfully trained, reassuring professionals of its comprehensive applicability.

To address the research question about integrating certification into game development, we must embed the context of training and competency certification within the game design process. This inclusion necessitates rethinking the entire SG development chain. Additionally, we introduce a new team member—the instructor/coach—who defines skills and competencies. The instructor actively contributes to specifying contextual elements, such as characteristics, missions, specific objectives, and expected learning outcomes.

4.1. Relationship between Game Taxonomy and Competencies Development Survey

This research significantly defines the correlation matrix between game taxonomy and competencies. The effectiveness of training-based games, especially Serious Games, hinges on their ability to provide challenges that facilitate the acquisition of knowledge, experience, and professional aptitudes. However, there is no ideal design methodology to support this process.

A critical piece of information for game designers aiming to adapt mechanics for practical certification through SGs is the game genre. To address this, this study analyses standard options used in challenges based on a set of competencies. The evaluation draws from various game taxonomies, including those proposed by Adams and Dormans [92], Adams and Rollings [26], ESA [93], Bateman and Boon [94], Stahl [95], and Wolf [28]. Since no standard or universally accepted taxonomy exists, the researchers define their taxonomy as consisting of 8 categories subdivided into 22 subcategories.

Table 4 provides a quantitative overview of the analysed games, categorised by genre and subgenre. The genre with the highest number of Serious Games analysed is simulation, followed by puzzle and adventure games.

Table 4. Quantitative summary by genre and subgenre.

Genre	Subgenre	Subtotal	Total
Action	"Beat-'em-ups" "Beat 'em ups"	2	8
	1st/ 3rd person game	1	
	Platform games	5	
Strategy	4X (eXplore, eXpand, eXamine, eXtermine)	1	8
	Real-time strategy	1	
	Real-time tactics	1	
	Turn-based strategy	4	
	War games	1	
Role-Playing	Action RPGs	2	4
	MMORPGs	2	
Sport	Exergames	2	3
	Sports/management games	1	
Management simulation	Racing/vehicle games	6	41
	Virtual worlds/fantasy/pets	24	
	Business	8	
	Social games and life simulations	3	
Adventure	Graphics adventures	17	19
	Puzzle adventures	2	
Puzzle	Arcade/Action puzzle (timed)	14	22
	Physics games	4	
	Hidden images games	2	
	Traditional games	2	
Quiz		11	11

The chosen competency model, between the previous reference structures, is educational competencies, developed by Microsoft. This set of competencies aligns with current references and is considered essential for future success in performing various functions.

The research analyses 116 Serious Games from different sites and open repositories available in [96]. For each game, they collect information such as description, classification, domain areas, game genre, topics, audience, and type of realism. By analysing available data and, when possible, playing the games, the researchers identify the specific competencies involved.

The study's results provided a cross-reference of the genre categories with a set of competencies in a matrix, allowing us to identify some areas with significant intersections to achieve learning outcomes [97–99].

Many genres' potential to support aptitude learning is significant. It contributes in the same way to developing various game design strategies. The contribution of this study results in the mixture/combination of genres or the reinforcement of challenges to reach skills such as Decision Quality and Problem-Solving (SS3) and Technical Learning (SS8); Organisation (OpS5) and Timely Decision-Making (OpS9); and both results category competencies, Action-Oriented (R1) and Move for Results (R2), which can be synchronised in different strategies to achieve better student performance.

In summary, this research bridges the gap between game design, competencies, and certification, emphasising the importance of aligning game mechanics with desired learning outcomes. Developers can create more effective Serious Games for training and certification by understanding the interplay between game genres and competencies.

4.2. Design of Triadic Certification Method

The Triadic Certification Method (TCM) represents an important advancement in SGs, particularly concerning competency and skills certification. The TCM enables performance measurement during training missions by directly integrating training guidelines into SG design.

The TCM architecture comprises four steps, each contributing to evaluating skills acquisition and certification. Unlike traditional post-game questionnaires [99], the TCM assesses player performance within the game itself. This approach provides clear guidelines to the development team, especially designers, ensuring competencies are seamlessly woven into the game construction alongside an evaluation map.

The following figure, Figure 1, shows the method design:

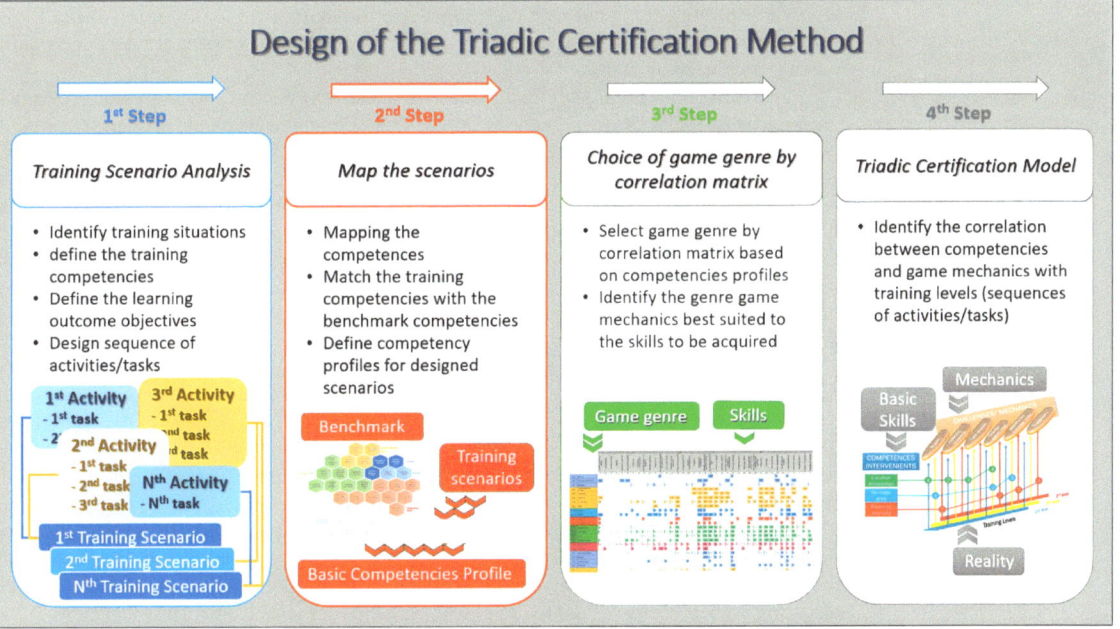

Figure 1. Workflow of Triadic Certification with used tools and defined goals.

In the first step, the analysis/diagnosis of the training context involves collaboration between the trainer/instructor and the development team. The training needs and competencies required for specific learning objectives are defined. Identifying situations, scenarios, and learning outcomes allows for detailed training planning, focusing on knowledge, skills, and aptitudes.

The TCM also leverages two methodologies: The Mission Essential Task List (METL) [100, 101], which hierarchically lists essential tasks and activities, and the CRAWL-WALK-RUN Approach [102–104], which defines task sequences to promote progressive learning. Constructing a reference table for training scenarios associates training stages with lists of essential tasks, ensuring successful training by achieving expected performance levels. To visualise the scenarios, a concept similar to the use case diagrams (through the Unified Modelling Language—UML [105]) can be used to reference training scenario actors.

The second step of the method involves mapping educational competencies to align with the list of tasks and activities determined in the preceding stage, with the trainer or instructor remaining the central figure. During this phase, the focus shifts to defining the fundamental skills required by each target group profile, which will be developed through training. This definition is derived from aligning training competencies with educational competencies [87]. This mapping exercise contextualises the specific training environment with a standardised reference point, ensuring that skills acquisition can be appropriately assessed across different scenarios.

Initiating this stage of the method involves utilising a chosen reference matrix as a competency model to identify the essential competencies necessary for performing or training in a specific role across various contexts such as employment, occupation, organisation, or industry. The aim is to construct a behavioural description representative of the function to be performed based on the definition of competencies associated with each occupational role, as suggested by Fogg [106].

Once the competencies for a specific task or position have been identified through mapping from the reference matrix, the next step involves determining the most suitable actions within the game to achieve the learning objectives, known as game mechanics. Upon completing this step and defining the basic skills profile, attention shifts to the subsequent step: selecting the genre of the SG based on the correlation between game mechanics and basic skills.

The third step entails choosing the SG genre that best aligns with the previously defined basic skills profile. To accomplish this, extensive research on various Serious Games is conducted to comprehend the contributions of different gaming genres towards competency acquisition. While identifying the optimal game mechanics for skill acquisition can be challenging, analysing game genres aims to uncover patterns of competencies that specific game genres effectively encompass.

By conducting a high-level analysis of the mechanics in various Serious Games, designers can determine the most suitable game mechanics for acquiring specific skills. This methodology aspect falls under the designer's responsibility, providing a guiding framework for game design while allowing room for creativity.

In certain instances, the choice of genre may not be straightforward but rather a combination of genres, where insights from the correlation matrix combine mechanics and challenges from various genres to train the desired skills effectively. A key conclusion drawn from this correlation matrix is the importance of leveraging past successful experiences with Serious Games to inform future development, thereby facilitating correct and efficient implementation.

The fourth and final step of the method involves integrating previous design contributions into the new game. This module aims to adapt the serious game design for skills training while maintaining autonomy in operation and configuration, contingent upon receiving values/elements related to player performance within established mechanics and challenges. Additionally, this step finalises the development of the Triadic Certification Method (TCMd).

The TCM serves as a communication tool among stakeholders, aiming to standardise game design for competency acquisition by balancing three components: identified skills and competencies (basic skills), mechanics and challenges based on game type (mechanics), and training levels (reality). Through the "in-game" certification method facilitated by the TCM, design contributions are defined, certified, and validated, ensuring that games effectively foster learning. Figure 2 is the Certification Triadic Model for training local tour guides, which comprises three reference axes: vertical, horizontal, and oblique. Each axis assumes a specific function to achieve defined competencies [107].

Starting with the vertical axis, it encompasses competencies aligned with various proficiency levels (basic, intermediate, advanced, expert), distributed across a mechanics framework. The progression of skills throughout the training sessions is marked by assigning the achieved proficiency level.

The horizontal axis represents the mechanics of skill acquisition, applied transversally across competencies. Learning progresses linearly through the accumulation of successful tasks, with higher competency levels indicating previous success.

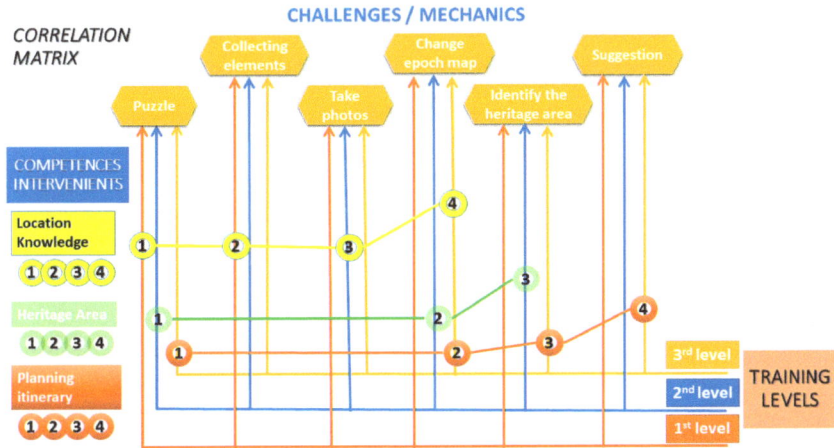

Figure 2. Certification Triadic Model for training local tour guides.

The oblique axis refers to sessions or training levels, illustrating the progression of competencies from basic to expert levels. Each subsequent level builds upon the results of the previous level, fostering a cumulative advancement in proficiency profiles.

While the model emphasises progressive learning, it acknowledges that certain competencies may depend on a single mechanic or that a single training level may incorporate multiple mechanics. It also recognises the possibility of continuity or discontinuity in learning objectives between different training levels, with some outcomes remaining consistent across multiple proficiency profiles.

After presenting the model's guidelines, we propose demonstrating its application as a customisable process since the starting point is always the context in which the skills are trained (reality). The Triadic Certification Method shown in Figure 2 refers to the training of local tour guides. Starting from this specific reality, it is a priority to understand the elements of certification for this activity, such as which tourist region and tourist resources contribute to a variety of tourist experiences: cultural, gastronomic, and traditional points of interest, whether it is one location or a group of locations. The certification of competencies for this professional is based on a successful profile that requires mastery of various areas of knowledge and skills, such as the geography of tourism, history, and cultural and architectural heritage, as well as culture and traditions of the regions, various types of communication (oral, written and active listening), group facilitator, mastery of several languages, and planning and organising tourist routes and circuits [99].

Following the TCM, various steps were defined (training scenarios, identifying the educational competencies to support certification, and finally, identifying the correlation of the expected competencies with the game genre and its most appropriate mechanics). Bearing in mind that scenarios for exploring the tourist region have been defined, the following competencies are defined:

- Planning (OpS6), Organisation (OpS5), and Time Management (OpS8).
- Written Communications (OrS4) and Presentation Skills (OrS3).
- Action Orientated (R1).

- Decision Quality and Problem Solving (SS3), Technical Learning (SS8), and Strategic Agility and Innovation Management (SS7).

These competencies correspond to various tasks related to the acquisition and demonstration of knowledge in the following aspects:
- Monuments;
- Cultural and architectural heritage (centuries-old and specific stories and traditions);
- Understanding the people's traditions as a key to organising and planning different thematic and tourist itineraries for different target audiences.

Considering the third step, the most suitable game genre combines both adventure genres: graphic and puzzle.

The game aims to use an avatar to explore an interactive story, which together takes on mental challenges in mini-games (puzzles and challenges) about tourist resources. Three training levels were defined to demonstrate the Triadic Certification Method's construction: navigation, knowledge, and recommendation about a tourist region. With more detail on each training level, on the navigation level, the player will use a map of the tourist region, monitored by a GPS device whose route taken between two points of interest (POIs) will help validate route choices, time, and distance travelled. This level also ensures the physical recognition of routes and their POIs. As for the second level, knowledge acquisition occurs by identifying the POI and answering questionnaires and other challenges in various on-site situations. The questions will be trivia about random locations, destinations, or other more specific contexts relating to current or past events. The third and final level, the recommendation, has a double meaning: the extra motivation to share opinions and ratings on the spot of the various resources encountered and the collection of other helpful information from other participants that will be fundamental to the planning component of the thematic and other more specific itineraries that will have to be trained. The final classification of the route planning depends on the knowledge already acquired of the POIs included in the route, as well as the recommendations made to them.

However, we must bear in mind that the levels are cumulative, and it is necessary to collect several navigation routes between crucial tourist spots and the respective POIs found to ensure that all the other knowledge acquisition actions happen.

In conclusion, the Triadic Model offers a comprehensive and customisable framework for understanding and implementing skills training through Serious Games in different areas. It emphasises the integration of competencies, mechanics, and training levels to facilitate effective learning outcomes. In summary, the TCM bridges game design, competencies, and certification, emphasising the importance of aligning game mechanics with desired learning outcomes. By involving players in the design process, the TCM supports effective Serious Games for training and certification purposes.

5. Case Study of Driver's Licences—Comprehensive Training Method for Light Vehicle Driving

To test our research hypothesis, we develop gaming applications, specifically prototypes, to support the case study of competence training for obtaining a driver's licence. This case study focuses on road safety, which remains an enduring priority. Utilising games as a learning tool offers scalability of results, cost reduction, and solid consolidation of learning.

Acquiring driving skills is considered complex and dynamic because it involves various psychological processes on the driver's part. This complexity can be broken down into three acquisition stages: information gathering, information processing, and action. New Serious Game-based learning tools can enhance the quality of driving skills acquisition and training to promote safer and more responsible drivers.

5.1. Case Study Context and Implementation in DRIS

The motivations for studying the acquisition of driver's licences are twofold. Firstly, analysis of OECD (Young Drivers—The Road to Safety: Conference of Ministers of Transport (ECMT): OCDE 2006; https://www.oecd-ilibrary.org/transport/young-drivers_97

89282113356-en (accessed on 7 July 2024)) data revealed that road accidents were a major cause of death among individuals aged 15 to 24. Deaths and serious injuries resulting from road accidents pose a significant public health issue, with young drivers being major contributors. Seeking validation solutions for driving tasks could gradually reduce this ongoing catastrophe.

Secondly, we tested this case study in the context of automotive driving learning using the virtual automobile simulator (DriS) in the Traffic Analysis Laboratory of the Civil Engineering Department at the Faculty of Engineering of the University of Porto [108,109]. The virtual simulator was configured with a training environment to monitor students' real-time performance evolution and corresponding validation according to predefined plans/assignments and learning objectives. The simulation room has an image projection system with a projector and a screen. The driving position consists of a real vehicle (customised Volvo 440 turbo). Figure 3 shows images of the simulation room and its driving position. The vehicle's integrated instrumentation includes sensors for actuating the pedals (clutch, brake, and accelerator). The car also has instrumentation for reading the gear engaged, direction indicators, the position of the ignition key, and all the light controls.

Figure 3. Picture of the simulation room and the driving position.

Utilising DriS, our objective was to test the application of the concept of validating vehicle control and mastery skills (operational) and adaptation to constant changes in the road environment (tactical) based on the learning support matrix of Driving: GDE—Goals for Driver Education [110–112]. This matrix hierarchically defines the driving task, emphasising individual driver characteristics impacting driving, including experience, attitudes, skills, motivations, decisions, and behaviours. This matrix allows for defining educational objectives and performance indicators in driver training as a tool for defining the skills necessary to become a safe driver. Understanding learning guidelines is vital, as they indicate that some areas must be learned before others may progress and that the development of different components has varying timings.

Feedback is another crucial aspect of learning to drive, informing driving practises at higher- and lower-order factors. In the first case, feedback acts as a regulator and behavioural motivator, while in the second case, it actively engages the trainee throughout the driving task, connecting to necessary automatisms and procedures. The amount of feedback perceived in the driving task at this lower level is greater, suggesting that low-level skills are learned faster than high-level skills. A prototype was implemented for the training certification system module to validate the acquisition of a driving licence. Developed for the Windows Operating System, the module aims to integrate with any serious game, providing a set of metrics representing the necessary mechanics for real-time training. The module's coupling is intended to be generic and established through communication via "sockets" in TCP/IP networks. The programming module is designed to be compatible with other application cases through the definition of distinct projects associated with a separate database file. This option ensures independence and portability for each application case.

The tool allows each participant to independently interact with three distinct work areas in the training process: trainer, who defines training competencies; game designer, who defines game elements and mechanics associated with training; and certification, as the training outcome component, which analyses student performance data and provides feedback on the training plan. The feedback system is a critical component in this implementation. It enables monitoring of tasks performed by students during implementation through visual and auditory means.

5.2. TCM Design

The implementation of the TCM commences with the analysis of training scenarios. To conduct this competence diagnosis, it was imperative to consult the legal code of the road and driving instructional manuals, which organise various theoretical themes of road safety, traffic rules, and traffic signals. Additionally, technical files developed by the IMT (Portuguese public entity regulating mobility and transport) and testimonies from professional instructors were utilised to understand the procedures for initiating the practise of driving a Category B motor vehicle.

Driving instruction in Portugal follows a matrix structure allowing for the organization of various levels of learning for drivers. The legislative and regulatory responsibility for driving education lies with the Portuguese State through the Institute of Mobility and Transport (IMT, IP). Based on the gathered information about the practical learning of driving tasks, frameworks of learning phases were developed, aligning with the Portuguese driving education system.

In the initial step of this Triadic Method, we conduct a meticulous analysis of the tasks and competencies essential for driving light vehicles. Our objective is to design distinct training levels that ensure thorough mastery of these competencies, ultimately preparing drivers for real-world scenarios.

We have identified three core competencies crucial for effective driver training: speed adaptation and vehicle control, complete vehicle mastery, and traffic situation resolution. Each competency is broken down into specific learning topics, which are then linked to targeted tasks, ensuring a comprehensive learning experience.

Our training strategy is founded on the proven "Crawl-Walk-Run" methodology. In the "Crawl" phase, trainees begin with fundamental tasks, building a solid foundation of basic skills. In the "Walk" phase, the difficulty and realism of tasks gradually increase, promoting steady progress. Finally, in the "Run" phase, trainees achieve high-level performance through advanced practise, simulating real-world driving conditions.

Using the Mission Essential Task List (METL) methodology, we create a structured hierarchy of tasks for each training mission. Training scenarios are designed to increase in complexity progressively, allowing learners to build on their knowledge and skills systematically. For this case study, we focus on three key competencies: speed control, navigating crossings and intersections, and manoeuvring through wide curves.

To instil automatic responses in trainees, we include specific tasks such as safe vehicle startup and stopping, speed changes emphasising the coordination of gearbox and pedals, light driving on straight and curved tracks, and defensive driving decision-making based on road signs. A variety of routes are created to incorporate different driving situations, connecting various activities through specific route signalling, ensuring comprehensive learning.

The routes (T1 to T5) include right and left curves with appropriate signalling. We simulate real traffic conditions by placing other vehicles in the trainee's path and at intersections, enhancing the realism of the training. Driver performance is assessed based on compliance with signage and successful task execution along the routes, covering all competencies cumulatively.

The interconnected routes allow for continuous competency training. During evaluation, these connections ensure that the training paths align seamlessly with the assessment scenarios, promoting a cohesive learning experience. The sequence of training paths, detailed in Figure 4, showcases the interconnection of routes and competencies:

Competency	T1	T2	T3	T4	T5	T6	T7	T8
Speed limit control	1t	2t	3t		4t		1a	2a
Stopping at the STOP sign	1t		2t					1a
Right of way		1t					1a	
In-road vehicle control	1t	2t	3t	4t			1a	2a

Figure 4. Table of interconnection between routes and competences, where each competence has a colour and corresponding training and assessment context.

This comprehensive training framework ensures drivers progressively acquire essential driving skills, leading to superior performance in real-world situations. Our meticulous approach guarantees that every driver is thoroughly prepared, confident, and safe on the road. Embrace this training method to master the art of driving light vehicles and transform your driving experience today.

The second step in our Triadic Method involves competency mapping, focusing on four specific competencies: speed limit control, in-road vehicle control, and approach to crossings and junctions, distinguished by STOP signalling and signalling with and without right of way. By aligning these competencies with our educational competency matrix, we identified that they align best with strategic skills (SSs) and operational skills (OpSs).

From our reference matrix (Figure 5), the following competencies emerged as critical:

- Strategic skills (SSs): Decision Quality and Problem Solving (SS3) and Functional/Technical Skills (SS4)
- Operational skills (OpSs): Planning (OpS6), Priority Setting (OpS7), and Timely decision-making (OpS9)
- Results (Rs): Action Oriented (R1)

In the third step, we selected the game genre based on a correlation matrix between game genres and competence benchmarks. This step ensures the skills identified are seamlessly integrated into the game design, facilitating effective training. Our analysis revealed that action, strategy, and simulation genres had the highest success rates. Thus, our case study involves a hybrid of simulation games with vehicles (such as rally or heavy-vehicle driving) combined with action and strategy elements. This blend allows us to leverage individual mechanics effectively, with lower levels engaging in action mechanics and higher levels incorporating strategic elements.

In the fourth step, we implemented the game using this combination of mechanics, aligned with the chosen game taxonomy. This approach enabled us to pinpoint specific challenges that correlate with the game's objectives. For each selected genre, we identified mechanics and actions crucial for training the anticipated competencies:

- Spatial perception: enhancing the ability to navigate through the game environment to develop a spatial relationship essential for reaching destinations.
- Points: providing feedback on progress within the scenario, enhancing visualisation and goal tracking.
- Levels: introducing new sets of challenges in different scenarios to demonstrate progression.
- Detailed simulation actions: including acceleration, deceleration with pedals and gearbox, braking, coordinating the vehicle within the lane, and stopping the car.

Figure 6 illustrates the mapping of competencies within the training plan, where each horizontal axis corresponds to a different path (e.g., Path 1, Path 2). The alignment between skills and paths is established through the mechanics or challenges implemented in the game.

			IE							OrS		SS						OpS						R		C
			IE1	IE2	IE6	IE7	IE8	IE9	IE13	OrS1	OrS4	SS1	SS2	SS3	SS4	SS5	SS9	OpS1	OpS5	OpS6	OpS7	OpS8	OpS9	R1	R2	C3
Action	Beat' em ups	2	0%	0%	0%	0%	0%	0%	0%	0%	0%	0%	0%	0%	100%	0%	50%	0%	0%	0%	0%	50%	0%	50%	0%	0%	25%
Action	shooter games (1st and 3rd person)	1	100%	0%	100%	0%	0%	0%	0%	0%	0%	0%	100%	100%	0%	0%	0%	0%	100%	0%	0%	100%	0%	0%	100%	0%	17%
Action	platform games	5	0%	0%	0%	20%	0%	0%	20%	0%	20%	20%	0%	40%	0%	0%	40%	0%	20%	0%	0%	0%	40%	60%	20%	20%	23%
Strategy	4X (eXplore, eXpand, eXploit, eXterminate)	1	0%	0%	0%	0%	0%	0%	0%	0%	0%	100%	100%	100%	100%	100%	0%	0%	100%	100%	100%	0%	100%	0%	100%	0%	83%
Strategy	real-time strategy	1	0%	0%	0%	0%	0%	0%	0%	0%	0%	100%	100%	100%	0%	100%	0%	0%	100%	100%	100%	0%	100%	0%	100%	0%	67%
Strategy	real-time tactics	1	100%	0%	0%	0%	0%	0%	0%	0%	0%	100%	0%	100%	100%	100%	0%	0%	100%	0%	100%	0%	100%	100%	0%	0%	83%
Strategy	turn – based strategy	4	50%	0%	0%	25%	50%	25%	25%	25%	0%	25%	75%	100%	75%	75%	50%	0%	100%	75%	75%	50%	50%	50%	50%	50%	71%
Strategy	wargames	1	0%	0%	0%	0%	0%	0%	0%	0%	0%	100%	100%	100%	100%	0%	0%	0%	100%	100%	100%	0%	0%	100%	0%	100%	83%
Role-Playing	Action RPGs	2	0%	0%	0%	50%	0%	0%	50%	50%	0%	0%	100%	50%	0%	0%	0%	0%	100%	50%	0%	0%	50%	100%	0%	50%	42%
Role-Playing	MMORPGs	2	100%	0%	50%	0%	50%	0%	50%	0%	0%	0%	100%	100%	50%	50%	0%	50%	0%	50%	0%	50%	100%	100%	100%	67%
Sports	Exergames	2	50%	0%	0%	0%	50%	0%	0%	0%	0%	0%	100%	50%	0%	0%	50%	100%	100%	50%	50%	0%	50%	50%	100%	58%
Sports	Sport /management games	1	50%	0%	0%	0%	0%	0%	0%	50%	0%	0%	50%	50%	50%	50%	50%	0%	0%	0%	0%	50%	0%	50%	0%	25%
Management Simulation	Racing games / vehicles	6	33%	0%	0%	17%	0%	0%	0%	33%	0%	0%	100%	83%	100%	67%	100%	0%	67%	83%	83%	0%	83%	50%	50%	0%	80%
Management Simulation	virtual worlds/Pets	24	33%	0%	8%	21%	29%	29%	21%	29%	13%	21%	67%	92%	83%	46%	79%	0%	75%	67%	75%	33%	71%	25%	67%	46%	69%
Management Simulation	Business	8	0%	0%	0%	0%	0%	13%	0%	0%	25%	13%	50%	88%	63%	75%	88%	0%	63%	88%	63%	25%	75%	13%	100%	75%	65%
Management Simulation	Life simulation/Social Games	3	33%	0%	67%	67%	33%	0%	0%	33%	0%	0%	67%	100%	100%	100%	67%	0%	67%	100%	100%	67%	33%	67%	33%	100%	83%
Adventure	Graphics adventure	17	18%	0%	24%	24%	6%	12%	35%	6%	0%	41%	88%	47%	59%	53%	12%	71%	41%	29%	29%	41%	29%	53%	41%	46%
Adventure	puzzle adventure	2	0%	50%	0%	50%	0%	0%	50%	0%	50%	0%	100%	0%	0%	50%	0%	50%	0%	50%	0%	50%	50%	50%	0%	42%
Puzzle	Action/Arcade puzzle (timed)	14	0%	0%	7%	0%	0%	0%	0%	14%	14%	21%	93%	21%	57%	79%	0%	14%	14%	0%	7%	57%	21%	43%	21%	34%
Puzzle	Physics game	4	0%	0%	0%	0%	0%	0%	0%	0%	0%	0%	100%	25%	50%	50%	0%	0%	0%	0%	50%	50%	25%	0%	38%
Puzzle	Reveal the picture game	2	0%	0%	0%	0%	0%	0%	50%	50%	50%	0%	0%	50%	0%	0%	0%	0%	0%	0%	50%	0%	0%	0%	17%
Puzzle	traditional game	2	0%	0%	0%	0%	0%	0%	0%	0%	0%	0%	50%	0%	50%	100%	0%	50%	50%	0%	50%	50%	50%	50%	42%
Quizz		11	0%	0%	0%	9%	0%	0%	9%	0%	27%	0%	0%	45%	18%	0%	100%	0%	0%	9%	0%	0%	0%	36%	9%	0%	18%

Figure 5. Summary grid highlighting the competencies identified for the training scenario.

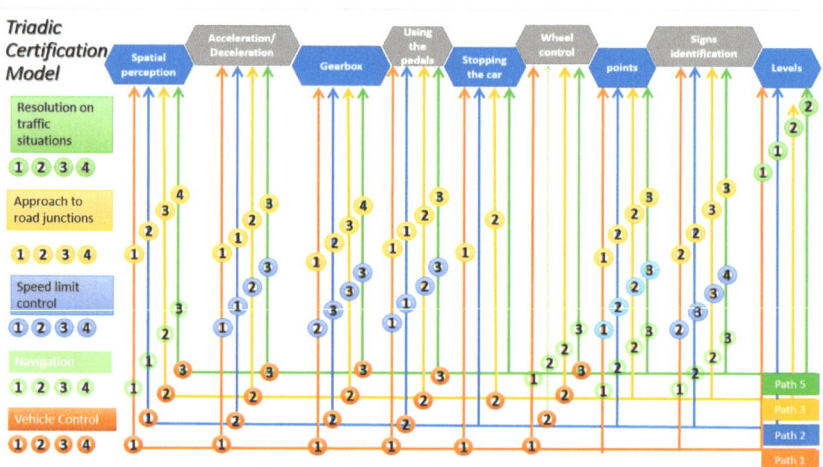

Figure 6. Mapping of the training (paths) with the level of competence acquired through the mechanics identified by the game taxonomy.

Each defined competency has a scalable profile (1—basic, 2—intermediate, 3—advanced, 4—expert) associated with it, classifying the use of mechanics in the task. This profile classification serves a dual purpose: categorising all mechanics and competency alignments with a learning level and contextualising the student during training tasks to achieve learning outcomes.

Each training path incorporates one or more skills that can be learned or trained sequentially. This mapping allows us to track the evolution of competency along the scale, contingent on the successful performance of the mechanics. This structure frames the learning of competencies within each path, enabling students to accumulate evidence of their learning by successfully performing tasks through mechanics. Post-training, the learning outcome is determined by combining successful evidence from various mechanics, creating a profile, and then assigning the achieved degree of competence.

By adopting this method, we ensure a structured, progressive, and practical approach to mastering light vehicle driving. This method prepares drivers comprehensively and instils confidence and ensures safety on the road, ultimately transforming the driving experience.

Mapping of competencies along the competency training plan allows for alignment between skills and paths through mechanics or challenges implemented in the game. Each training path incorporates one or more skills that can be learned or trained sequentially.

If the mechanics' performance is successful, the evolution of competency along the scale is verified. This structure enables framing the learning of competencies within each path, accumulating evidence of learning through successful task performance. After training, the learning result is achieved by combining successful evidence from various mechanics, creating a profile, and assigning the achieved degree of competence.

This comprehensive methodology ensures a systematic approach to competency training in driving education, facilitating effective learning outcomes for drivers.

5.3. Analysis of Results

Following the completion of the tests, 50 volunteers participated, comprising 38 men and 12 women aged 18 to 65. Unfortunately, four participants withdrew prematurely due to simulation-induced nausea. The remaining 46 participants were categorised into two groups based on whether they held a driver's licence: 31 participants held licences, while 15 did not, as delineated in Table 5.

Table 5. Description of participants.

Category	Groups	Frequency	Percentage
Age	18–23	26	56.5%
	24–30	12	26.2%
	31–40	4	8.7%
	41–50	3	6.5%
	52–65	1	2.2%
Gender	Female	11	23.9%
	Male	35	76.1%
Driver's Licence	Yes	31	67.4%
	No	15	32.6%

All participants underwent a comprehensive familiarisation process with the simulator throughout the experiment. This involved understanding acceleration, deceleration, steering wheel control, and traffic signs. Even those without a driver's licence were fully briefed on traffic signs and their meanings. The data collected from the experiment provided a meticulous analysis of each competency, showcasing the individual learning progression under the well-structured training plan.

Although the study presented in this document has a more methodological focus, the instrument validation must be addressed. To analyse and process the data collected with the user tests, we considered various statistical instruments before and after the descriptive analysis of the results, which we do not present here. Firstly, we check the internal consistency of the data using Cronbach's alpha test, with two separate analyses:

one for performance by competence and the other for overall performance. Secondly, the sample distribution was validated using the frequency and standard distribution curve, and lastly, the correlation between the variables through Pearson's coefficient (r).

We considered qualitative variables such as gender, driving ability, and age to assess participant performance during training (paths 1, 2, 3, and 5) and evaluation (paths 7 and 8). Performance was analysed based on both components' success in training and evaluation and the transition from training failure to evaluation success. Graphical data analysis allowed us to evaluate overall performance differences and individual progress.

The analysis of speed competence revealed high failure rates globally and in specific segments. Only 18 out of 46 participants succeeded, possibly due to the training design and the imposed 40 km/h speed limit. The repetitive nature of the training routes likely caused disinterest and monotony, contributing to the low success rate. In contrast, the control of the car within the lane showed a positive learning evolution, with a 10% improvement despite a reduced growth margin. This positive trend was further reinforced by the fact that most participants (37 out of 41) had completed training and evaluation.

There was an improvement in the competence of approaching crossings and junctions with signalling. The success rates increased from 9 participants in training to 23 in evaluation, marking a remarkable 255% improvement. However, the slight difference between training and both moments (7 participants) indicated that many struggled with mandatory STOP signalling.

The right-of-way competence also saw positive results, with 31 out of 46 participants (67%) achieving success. The desired behaviour improved significantly, with 17 participants showing positive evolution between training and evaluation. Globally, the performance evaluation indicated that 86% of successful participants were qualified to drive, validating our training plan. Only seven participants (six males and one female, with four under 23) successfully acquired all four competencies. The low overall success rate could be attributed to the repetitive nature of the training routes and the 40 km/h speed limit, which caused frustration and demotivation.

The CRAWL-WALK-RUN method, which involves sequential and repetitive task performance, yielded limited success in skills acquisition. Despite this, the high success rate among participants with driver's licences (86%) confirms the validity of our training method for designing and validating competencies.

In conclusion, despite some challenges, our comprehensive training method effectively prepares participants for real-world driving scenarios. It ensures a structured, progressive, and practical approach to mastering light vehicle driving. This method instils confidence, ensures safety, and transforms the driving experience.

6. Conclusions

In recent years, Serious Games have emerged as a compelling alternative for acquiring, training, and certifying skills because they provide a more engaging and meaningful learning experience. By incorporating rules, behavioural simulations, and feedback mechanisms, Serious Games create an environment where learners can make mistakes without real-life consequences and receive instant feedback. However, Serious Games must be designed using appropriate training validation and certification methodologies. The gameplay element is crucial for progression and successful learning outcomes.

This research was conducted with a rigorous and innovative approach to answer whether using player performance and interaction with the game to certify acquired competencies is possible. A comprehensive literature review confirmed the relevance of this question, leading to a focused exploration of integrating evaluation processes and certification within game-based learning, following specific norms for measuring student performance.

The conclusions of this research underscore the unique features of the Triadic Certification Method, distinguishing it from the existing literature and providing it with strong academic value. The integration of the correlation matrix between competencies and game taxonomy is a pioneering contribution, enabling game designers to craft Serious Games that both entertain and foster effective and specific learning of competencies. This breakthrough sets a benchmark that can be applied to diverse training contexts, potentially replicating best practises in developing new Serious Games.

In addition, the study comprehensively identifies the aspects of game design that favour competency-based assessment and training, bringing them in line with the demands of the 21st century. Structuring the four-step Triadic Certification Method provides a systematic approach to measuring and certifying player performance, using learning metrics that go beyond the traditional ones and incorporate elements such as agility, attention, cooperation, and precision.

Integrating competence matrices into serious game design offers practical and context-specific training scenarios. This aspect empowers designers to tailor game mechanics for skill acquisition, enhancing Serious Games' effectiveness. The introduction of the Triadic Certification Method, which balances basic skills, game mechanics, and reality, stands out for its ability to keep those involved in the training and certification process independent while promoting effective communication between them. This visual model enables situational awareness of the progression of skills acquisition over the various training levels, ensuring a robust and visually comprehensible methodology.

Another important result was the integration of learning metrics into in-game evaluation systems. Implementing performance validation based on scores derived from successful activities within a training plan could assess the effectiveness of knowledge and skill acquisition and offer a viable alternative to traditional validation methods such as surveys and observations. The ability to quantify training success based on in-game results reinforces the effectiveness of the Triadic Certification Method, opening up new possibilities for assessing and certifying competencies in innovative learning environments.

Overall, this research validates using Serious Games in skills certification and introduces pioneering tools and methodologies. These innovations have the potential to transform educational and training practises in various areas, offering new and effective ways to assess and certify competencies.

Author Contributions: Conceptualization, R.B.; methodology, R.B., A.C. and C.V.d.C.; software, R.B.; validation, R.B., A.C. and C.V.d.C.; investigation, R.B.; resources, R.B.; data curation, R.B.; writing—original draft preparation, R.B.; writing—review and editing, A.C. and C.V.d.C.; supervision, A.C. and C.V.d.C. All authors have read and agreed to the published version of the manuscript.

Funding: This research received no external funding.

Data Availability Statement: The data presented in this study are openly available in Ricardo Baptista Ph.D. Thesis (https://hdl.handle.net/10216/110820 (accessed on 7 July 2024)).

Conflicts of Interest: The authors declare no conflicts of interest.

References

1. Uskov, A.; Sekar, B. Serious games, gamification and game engines to support framework activities in engineering: Case studies, analysis, classifications and outcomes. In Proceedings of the IEEE International Conference on Electro/Information Technology, Milwaukee, WI, USA, 5–7 June 2014; pp. 618–623.
2. Boller, S.; Kapp, K. *Play to Learn*; ATD Press: Alexandria, VA, USA, 2017.
3. Sousa, L.; Figueiredo, M.; Monteiro, J.; José, B.; Rodrigues, J.; Cardoso, P. Developments of Serious Games in Education. In *Handbook of Research on Human-Computer Interfaces, Developments, and Applications*; IGI Global: Hershey, PA, USA, 2016; pp. 392–419.
4. Mayer, R.E. *Computer Games for Learning: An Evidence-Based Approach*; MIT Press: Cambridge, MA, USA, 2014.
5. de Freitas, S.; Jarvis, S. A framework for developing serious games to meet learner needs. In Proceedings of the Interservice/Industry Training, Simulation and Education Conference, Orlando, FL, USA, 1 December 2006.
6. Luo, L.; Yin, H.; Cai, W.; Lees, M.; Zhou, S. Interactive scenario generation for mission-based virtual training: Interactive scenario generation. *Comput. Animat. Virtual Worlds* **2013**, *24*, 345–354. [CrossRef]
7. Huang, W.-H.D.; Tettegah, S.Y. Cognitive Load and Empathy in Serious Games: A Conceptual Framework. In *Gamification for Human Factors Integration: Social, Education, and Psychological Issues*; Jonathan, B., Ed.; IGI Global: Hershey, PA, USA, 2014; pp. 17–30.

8. Kiili, K.; Lainema, T.; de Freitas, S.; Arnab, S. Flow framework for analysing the quality of educational games. *Entertain. Comput.* **2014**, *5*, 367–377. [CrossRef]
9. Emmerich, K.; Bockholt, M. Serious Games Evaluation: Processes, Models, and Concepts. In *Entertainment Computing and Serious Games: International GI-Dagstuhl Seminar 15283*; Dörner, R., Göbel, S., Kickmeier-Rust, M., Masuch, M., Zweig, K., Eds.; Springer International Publishing: Dagstuhl, Germany, 2015; pp. 265–283.
10. Andreoli, R.; Corolla, A.; Faggiano, A.; Malandrino, D.; Pirozzi, D.; Ranaldi, M.; Santangelo, G.; Scarano, V. A Framework to Design, Develop, and Evaluate Immersive and Collaborative Serious Games in Cultural Heritage. *J. Comput. Cult. Herit.* **2017**, *11*, 1–22. [CrossRef]
11. Coleman, T.E.; Money, A.G. Student-centered digital game-based learning: A conceptual framework and survey of state of the art. *High Educ.* **2020**, *79*, 415–457. [CrossRef]
12. Mitgutsch, K.; Alvarado, N. Purposeful by design? A Serious Game design assessment framework. In Proceedings of the FDG '12: International Conference on the Foundations of Digital Games, Raleigh, NC, USA, 29 May 2012; pp. 121–128.
13. Wilson, D.W.; Jenkins, J.; Twyman, N.; Jensen, M.; Valacich, J.; Dunbar, N.; Wilson, S.; Miller, C.; Adame, B.; Lee, Y.-H.; et al. Serious games: An evaluation framework and case study. In Proceedings of the 2016 49th Hawaii International Conference on System Sciences (HICSS), Washington, DC, USA, 5–8 January 2016; Sprague, R.H., Bui, T.X., Eds.; IEEE Computer Society: Washington, DC, USA, 2016.
14. Ulrich, F.; Helms, N.H. Creating Evaluation Profiles for Games Designed to Be Fun: An Interpretive Framework for Serious Game Mechanics. *Simul. Gaming* **2017**, *48*, 695–714. [CrossRef]
15. Tahir, R.; Wang, A.I. Codifying game-based learning: The league framework for evaluation. In Proceedings of the 12th European Conference on Game Based Learning—ECGBL 2018, Sophia Antipolis, France, 4–5 October 2018; pp. 4–5.
16. Huizinga, J.; Hull, R.C. *Homo Ludens. A study of the Play-Element in Culture*, 1st ed.; Beacon Press Boston: Boston, MA, USA, 2014.
17. Caillois, R.; Halperin, E.P. The structure and classification of games. *Diogenes* **1955**, *3*, 62–75. [CrossRef]
18. Juul, J. The repeatedly lost art of studying games. *Game Stud.* **2001**, *1*. Available online: https://www.gamestudies.org/0101/juul-review/?ref=driverlayer.com (accessed on 30 May 2024).
19. Salen, K.; Zimmerman, E. *Rules of Play: Game Design Fundamentals*; The MIT Press: Cambridge, MA, USA, 2004.
20. de Carvalho, C.V.; Batista, R. Work in progress-learning through role-play games. In Proceedings of the 38th Annual Frontiers in Education Conference, Saratoga Springs, NY, USA, 22 October 2008; p. T3C-7.
21. Bergson, B. *Developing Serious Games; Game Development Series*, 1st ed.; Charles River Media, Inc.: Needham, MA, USA, 2006.
22. Apperley, T.H. Genre and game studies: Toward a critical approach to video game genres. *Simul. Gaming* **2006**, *37*, 6–23. [CrossRef]
23. Breuer, J.S.; Bente, G. Why so serious? On the relation of serious games and learning. *Eludamos J. Comput. Game Cult.* **2010**, *4*, 7–24. [CrossRef] [PubMed]
24. Adams, E. *Fundamentals of Game Design*; Pearson Education: London, UK, 2014.
25. Pinelle, D.; Wong, N.; Stach, T. Using genres to customize usability evaluations of video games. In Proceedings of the 2008 Conference on Future Play Research, Play, Share-Future Play '08, New York, NY, USA, 3 November 2008; pp. 129–136.
26. Adams, E.; Dormans, J. *Game Mechanics: Advanced Game Design*, 1st ed.; New Riders: San Francisco, IN, USA, 2012.
27. Grace, L. Game Type and Game Genre. Available online: http://aii.lgracegames.com/documents/Game_types_and_genres.pdf (accessed on 30 May 2024).
28. Wolf, M.J. *The Medium of the Video Game*, 1st ed.; University of Texas Press: Austin, TX, USA, 2002.
29. Michael, D.R.; Chen, S. *Serious Games: Games That Educate, Train, and Inform*; Muska Lipman/Premier: Cincinnati, OH, USA, 2005.
30. Tadayon, R.; Amresh, A.; Burleson, W. Socially relevant simulation games: A design study. In Proceedings of the 19th ACM International Conference on Multimedia, Scottsdale, AZ, USA, 28 November 2011; pp. 941–944.
31. Abt, C.C. *Serious Games*; University Press of America: Lanham, MA, USA, 1987.
32. Marlow, C.M. Games Learning in Landscape Architecture. In Proceedings of the 10th International Conference on Information Technologies in Landscape Architecture-Digital Landscape Architecture, Valletta, Malta, 5–7 June 2024; pp. 129–136.
33. Becker, K. How are games educational? Learning theories embodied in games. In Proceedings of the DiGRA 2005 Conference: Changing Views: Worlds in Play, Vancouver, BC, Canada, 1 January 2005.
34. Zyda, M. From visual simulation to virtual reality to games. *Computer* **2005**, *38*, 25–32. [CrossRef]
35. Wiberg, C.; Jegers, K. Satisfaction and learnability in edutainment: A usability study of the knowledge game 'Laser Challenge' at the Nobel e-museum. In Proceedings of the HCI International—10th International Conference on Human Computer Interaction, Crete, Greece, 22 June 2003.
36. Wigfield, A.; Guthrie, J.T.; Tonks, S.; Perencevich, K.C. Children's motivation for reading: Domain specificity and instructional influences. *J. Educ. Res.* **2004**, *97*, 299–310. [CrossRef]
37. Deci, E.L.; Koestner, R.; Ryan, R.M. A meta-analytic review of experiments examining the effects of extrinsic rewards on intrinsic motivation. *Psychol. Bull.* **1999**, *125*, 627. [CrossRef]
38. Lepper, M.R.; Greene, D.; Nisbett, R.E. Undermining children's intrinsic interest with extrinsic reward: A test of the "over justification" hypothesis. *Psychol. Bull.* **1973**, *28*, 129. [CrossRef]
39. Susi, T.; Johannesson, M.; Backlund, P. Serious Games: An Overview. 2007. Available online: https://web.archive.org/web/20240327104441/https://www.diva-portal.org/smash/get/diva2:2416/fulltext01.pdf (accessed on 30 May 2024).

40. Stratton, G.; Ridgers, N. Energy expenditure in adolescents playing new generation computer games. *Br. J. Sport Med.* **2008**, *42*, 592–594.
41. Lieberman, D.A. Interactive video games for health promotion: Effects on knowledge, self-efficacy, social support, and health. In *Health Promotion and Interactive Technology: Theoretical Applications and Future Directions*; Street, R.L., Gold, W.R., Manning, T.R., Eds.; Routledge: New York, NY, USA, 2013; pp. 103–120.
42. Forstner, B.; Szegletes, L.; Angeli, R.; Fekete, A. A general framework for innovative mobile biofeedback based educational games. In Proceedings of the 2013 IEEE 4th International Conference on Cognitive Infocommunications (CogInfoCom), Budapest, Hungary, 2 December 2013; pp. 775–778.
43. Griffiths, M. The therapeutic use of videogames in childhood and adolescence. *Clin. Child Psychol. Psychiatry* **2003**, *8*, 547–554. [CrossRef]
44. McCallum, S.; Boletsis, C. A taxonomy of serious games for dementia. In *Games for Health: Proceedings of the 3rd European Conference on Gaming and Playful Interaction in Health Care*; Springer Science & Business Media: Berlin, Germany, 2013; pp. 219–232.
45. Ratan, R.A.; Ritterfeld, U. Classifying serious games. In *Serious Games*, 1st ed.; Routledge: London, UK, 2009; pp. 32–46.
46. Mohamed, H.; Jaafar, A. Development and potential analysis of heuristic evaluation for educational computer game (PHEG). In Proceedings of the 5th International Conference on Computer Sciences and Convergence Information Technology, Seoul, Republic of Korea, 30 November 2010; pp. 222–227.
47. Boyle, L.; Hancock, F.; Seeney, M.; Allen, L. The Implementation of Team Based Assessment In Serious Games. In Proceedings of the 2009 Conference in Games and Virtual Worlds for Serious Applications, Marrakesh, Morocco, 23 March 2009; pp. 28–35.
48. Gunter, G.A.; Kenny, R.F.; Vick, E.H. Taking educational games seriously: Using the RETAIN model to design endogenous fantasy into standalone educational games. *Educ. Technol. Res. Dev.* **2008**, *56*, 511–537. [CrossRef]
49. Kirkpatrick, D. Great ideas revisited. *Train. Dev.* **1996**, *50*, 54–59.
50. Baker, E.L.; Mayer, R.E. Computer-based assessment of problem solving. *Comput. Hum. Behav.* **1999**, *15*, 269–282. [CrossRef]
51. Wilcox, D. Design and implementation of educational games: Theoretical and practical perspectives. In *Design and Implementation of Educational Games: Theoretical and Practical Perspectives*; Zemliansky, P., Wilcox, D., Eds.; IGI Global: Hershey, PA, USA, 2007.
52. Roodt, S.; Joubert, P. Evaluating Serious Games in Higher Education: A Theory-based Evaluation of IBMs Innov8. In Proceedings of the 3rd European Conference on Games Based Learning, Graz, Austria, 12–13 October 2009; pp. 332–339.
53. Mislevy, R.J.; Steinberg, L.S.; Almond, R.G. Focus article: On the structure of educational assessments. *Meas. Interdiscip. Res. Perspect.* **2003**, *1*, 3–62. [CrossRef]
54. Becker, K.; Parker, J. Serious Instructional Design: ID for digital simulations and games. In Proceedings of the Society for Information Technology Teacher Education International Conference, Austin, TX, USA, 25 March 2012; pp. 2480–2485.
55. Arieli-Attali, M.; Ward, S.; Thomas, J.; Deonovic, B.; Von Davier, A.A. The Expanded Evidence-Centered Design (e-ECD) for Learning and Assessment Systems: A Framework for Incorporating Learning Goals and Processes Within Assessment Design. *Front. Psychol.* **2019**, *10*, 853. [CrossRef] [PubMed]
56. Muftuoglu, C.T.; Sahin, M. Framework of Assessment Design Based on Evidence-Centered Design for Assessment Analytics. In *Assessment Analytics in Education: Designs, Methods and Solutions*; Springer International Publishing: Cham, Switzerland, 2024; pp. 157–172.
57. De Freitas, S.; Liarokapis, F. Serious games: A new paradigm for education? In *Serious Games and Edutainment Applications*; Ma, M., Oikonomou, A., Jain, L., Eds.; Springer: London, UK, 2011; pp. 9–23. Available online: https://link.springer.com/chapter/10.1007/978-1-4471-2161-9_2#citeas (accessed on 7 July 2024).
58. Van Staalduinen, J.P.; De Freitas, S. A game-based learning framework: Linking game design and learning outcomes. In *Learning to Play: Exploring the Future of Education with Video Games*; International Academic Publishers: Hershey, PA, USA, 2011; Volume 53, pp. 29–45.
59. Harteveld, C. *Triadic Game Design: Balancing Reality, Meaning and Play*; Springer Science Business Media: Berlin, Germany, 2011.
60. Kellerhals, U.; Burgess, N.; Wetzel, R. Let it Bee: A Case Study of Applying Triadic Game Design for Designing Virtual Reality Training Games for Beekeepers. In Proceedings of the 16th International Conference on the Foundations of Digital Games, Montreal, QC, Canada, 3 July 2021; pp. 1–12.
61. Troiano, G.M.; Schouten, D.; Cassidy, M.; Tucker-Raymond, E.; Puttick, G.; Harteveld, C. All good things come in threes: Assessing student-designed games via triadic game design. In Proceedings of the 16th International Conference on the Foundations of Digital Games, Bugibba, Malta, 15 September 2020; pp. 1–4.
62. Winn, B.M. The Design, Play, and Experience Framework. In *Handbook of Research on Effective Electronic Gaming in Education*; Ferdig, R.E., Ed.; IGI Global: Hershey, PA, USA, 2009; pp. 1010–1024.
63. Mellecker, R.; Lyons, E.J.; Baranowski, T. Disentangling fun and enjoyment in exergames using an expanded design, play, experience framework: A narrative review. *Games Health Res. Dev. Clin. Appl.* **2013**, *2*, 142–149. [CrossRef] [PubMed]
64. Romero, M.; Ouellet, H.; Sawchuk, K. Expanding the game design play and experience framework for game-based life-long learning (GD-LLL-PE). In *Game-Based Learning across the Lifespan: Cross-Generational and Age-Oriented Topics*; Springer: Berlin/Heidelberg, Germany, 2017; pp. 1–11.
65. Kiili, K. Content creation challenges and flow experience in educational games: The IT-Emperor case. *Internet High. Educ.* **2005**, *8*, 183–198. [CrossRef]

66. Zapušek, M.; Rugelj, J. Learning programming with serious games. *EAI Endorsed Trans. Game-Based Learn.* **2013**, *13*. Available online: https://eudl.eu/doi/10.4108/trans.gbl.01-06.2013.e6 (accessed on 7 July 2024). [CrossRef]
67. Cetto, A.; Netter, M.; Pernul, G.; Richthammer, C.; Riesner, M.; Roth, C.; Sänger, J. Friend inspector: A serious game to enhance privacy awareness in social networks. *arXiv* **2014**, arXiv:1402.5878.
68. Song, M.; Zhang, S. EFM: A model for educational game design. In Proceedings of the Technologies for E-Learning and Digital Entertainment: Third International Conference, Edutainment 2008, Nanjing, China, 25 June 2008; Springer: Berlin/Heidelberg, Germany, 2008; pp. 509–517.
69. Abd Elsattar, H.K.H. Designing for game-based learning model: The effective integration of flow experience and game elements to support learning. In Proceedings of the 14th International Conference on Computer Graphics, Imaging and Visualization, Marrakesh, Morocco, 22 May 2017; pp. 34–43.
70. Conole, G. Assessment as a Catalyst for Innovation. Available online: https://eprints.soton.ac.uk/9735/ (accessed on 7 July 2024).
71. Conole, G.; Warburton, B. A review of computer-assisted assessment. *ALT-J* **2005**, *13*, 17–31. [CrossRef]
72. Wills, G.B.; Bailey, C.P.; Davis, H.C.; Gilbert, L.; Howard, Y.; Jeyes, S.; Millard, D.E.; Price, J.; Sclater, N.; Sherratt, R. An e-learning framework for assessment (FREMA). *Assess. Eval. High. Educ.* **2009**, *34*, 273–292. [CrossRef]
73. Armstrong, M. *A Handbook of Human Resource Management Practice*, 10th ed.; Kogan Page: London, UK, 2006.
74. Flippo, E.B. *Personal Management*; Tata McGraw Hill: New York, NY, USA, 1992.
75. Aswathappa, K. *Human Resource and Personnel Management*; Tata McGraw-Hill Education: New York, NY, USA, 2005.
76. Flarey, D.L. Is Certification the Current Gold Standard? *JONA'S Healthc. Law Ethics Regul.* **2000**, *2*, 43.
77. Byrne, M.; Valentine, W.; Carter, S. The Value of Certification—A Research Journey. *AORN J.* **2004**, *79*, 825–835. [CrossRef]
78. Briggs, L.A.; Brown, H.; Kesten, K.; Heath, J. Certification a benchmark for critical care nursing excellence. *Crit. Care Nurse* **2006**, *26*, 47–53. [CrossRef] [PubMed]
79. Kaplow, R. The value of certification. *AACN Adv. Crit. Care* **2011**, *22*, 25–32. [CrossRef]
80. Aswathappa, K.; Atif, A.; Richards, D.; Bilgin, A.; Marrone, M.; Nyberg, A. The value of certification. In Proceedings of the 6th International Conference of Education, Research and Innovation, Seville, Spain, 18 November 2013; p. 34.
81. Koeppen, K.; Hartig, J.; Klieme, E.; Leutner, D. Current issues in competence modeling and assessment. *Z. Psycholo-Gie/J. Psychol.* **2009**, *216*, 61–73. [CrossRef]
82. Coles, M. *Qualifications Frameworks in Europe: Platforms for Qualifications, Integration and Reform*; EU, Education and Culture DG: Brussels, Belgium, 2007.
83. Méhaut, P.; Winch, C. The European Qualification Framework: Skills, Competences or Knowledge? *Eur. Educ. Res. J.* **2012**, *11*, 369–381. [CrossRef]
84. Ferrari, A. *Digital Competence in Practice: An Analysis of Frameworks*; Publications Office of the European Union: Luxembourg, 2012; Volume 10, p. 82116.
85. Lombardo, M.M.; Eichinger, R.W. *The Leadership Machine*; Lominger Limited: Minneapolis, MN, USA, 2006.
86. Garman, A.N.; Johnson, M.P. Leadership competencies: An introduction. *J. Healthc. Manag.* **2006**, *51*, 13–17. [CrossRef]
87. Education Competency Wheel. Available online: https://download.microsoft.com/download/3/4/7/3477e49d-315d-4ee7-a8ca-ff653a4455d6/Competency_Wheel.pdf (accessed on 30 May 2024).
88. Murthy, S. Academagogical Framework for Effective University Education-Promoting Millennial Centric Learning in Global Knowledge Society. In Proceedings of the 2011 IEEE International Conference on Technology for Education, Washington, DC, USA, 14 July 2011; pp. 289–290.
89. Fletcher, G.H. Building a Better CTO: CoSN Creates a New Framework of Skills That Draws a Greater Range of Responsibility for the 21st Century Technology Leader. *The Free Library*. 1 January 2010. Available online: https://thejournal.com/articles/2010/01/08/building-a-better-cto.aspx (accessed on 14 May 2024).
90. Hine, P. *UNESCO ICT Competency Framework for Teachers*; United Nations Educational: Paris, France, 2011.
91. Hubregtse, M.; Moerbeek, S.; Veldkamp, B.P.; Eggen, T. Testing competences worldwide in large numbers: Complying with ISO/IEC 17024: 2012 standard for competence testing. In Proceedings of the Computer Assisted Assessment. Research into E-Assessment: 18th International Conference, Zeist, The Netherlands, 22–23 June 2015; pp. 27–39.
92. Adams, E.; Dormans, J. *Fundamentals of Game Design*, 1st ed.; Game Design and Development Series; Prentice-Hall, Inc.: Hoboken, NJ, USA, 2006.
93. 2015 Essential Facts about the Computer Video Game Industry. Available online: https://www.theesa.com/wp-content/uploads/2024/02/2015-EF-FINAL.pdf (accessed on 30 May 2024).
94. Bateman, C.; Boon, R. *21st Century Game Design*; Game Development Series; Charles River Media, Inc.: Massachusetts, MA, USA, 2005.
95. Stahl, T. Video Game Genres. The History of Computing Project 2005. Available online: www.thocp.net/software/games/reference/genres.htm (accessed on 2 June 2014).
96. Baptista, R. Jogos Sérios Para Treino E Certificação de Competências. Ph.D. Thesis, University of Porto, Porto, Portugal, 21 December 2017.
97. Baptista, R.; Coelho, A.; de Carvalho, C.V. Relationship Between Game Categories and Skills Development: Contributions for Serious Game Design. In Proceedings of the ECGBL2015-9th European Conference on Games Based Learning, Steinkjer, Norway, 8 October 2015; p. 34.

98. Baptista, R.; Coelho, A.; de Carvalho, C.V. Relation Between Game Genres and Competences for In-Game Certification. In Proceedings of the Serious Games, Interaction, and Simulation: 5th International Conference, SGAMES 2015, Novedrate, Italy, 16–18 September 2015; Springer International Publishing: New York, NY, USA, 2015; pp. 28–35.
99. Baptista, R.; Coelho, A.; de Carvalho, C. Methodology for In-Game Certification in Serious Games. In Proceedings of the EDULEARN13, Barcelona, Spain, 1 July 2013; pp. 3143–3152.
100. Tritten, J.J. Joint mission-essential tasks, Joint Vision 2010, core competencies, and global engagement: Short versus long view. *Air Space Power J.* **2004**, *11*, 32.
101. Grimaila, M.R.; Mills, R.F.; Haas, M.W.; Kelly, D.J. Mission Assurance: Issues and Challenges. In Proceedings of the 2010 International Conference on Security & Management, Las Vegas, NV, USA, 12–15 July 2010; pp. 651–657.
102. Filipović, J. Update of the crawl, walk, run methodology framework. *Facta Univ. Ser. Econ. Organ.* **2019**, *16*, 229–238. [CrossRef]
103. Myers, D. Computer games genres. *Play. Cult.* **1990**, *3*, 286–301.
104. Munro, I.; Mavin, T.J. Crawl-walk-run. In Proceedings of the 10th International Symposium of the Australian Aviation Psychology Association, Sydney, Australia, 10 November 2012.
105. Jacobson, I.; Booch, G.; Rumbaugh, J. *The Unified Software Development Process*; Addison-Wesley Educationa: Boston, MA, USA, 2011.
106. Fogg, C.D. *Implementing Your Strategic Plan: How to Turn"Intent" into Effective Action for Sustainable Change*; AMACOM/American Management Association: New York, NY, USA, 1999.
107. Baptista, R.; Nóbrega, R.; Coelho, A.; de Carvalho, C.V. Location-based tourism in-game certification. In Proceedings of the INTED2015, Madrid, Spain, 2 March 2015; pp. 3602–3610.
108. Campos, C.; Leitão, J.M.; Coelho, A.F. Building Virtual Roads from Computer made Projects. In Proceedings of the HCI International 2015-Posters' Extended Abstracts: International Conference, HCI International 2015, Los Angeles, LA, USA, 2 July 2015; pp. 163–169.
109. Campos, C.J.; Pinto, H.F.; Leitão, J.M.; Pereira, J.P.; Coelho, A.F.; Rodrigues, C.M. Building Virtual Driving Environments From Computer-Made Projects. In *Interface Support for Creativity, Productivity, and Expression in Computer Graphics*; IGI Global: Hershey, PA, USA, 2019; pp. 306–320.
110. Engström, I.; Gregersen, N.P.; Hernetkoski, K.; Keskinen, E.; Nyberg, A. Young Novice Drivers, Driver Education and Training: Literature Review. Available online: https://www.diva-portal.org/smash/get/diva2:675234/FULLTEXT01.pdf (accessed on 30 May 2024).
111. Keskinen, E. What is GDE all about and what it is not. In Proceedings of the GDE-Model as a Guide in Driver Training and Testing, Umeå, Sweden, 7–8 May 2007; pp. 3–13.
112. Peräaho, M.; Keskinen, E.; Hatakka, M. Driver competence in a hierarchical perspective: Implications for driver education. *Rep. Swed. Road Adm.* **2003**, 1–52. Available online: https://media.freeola.com/other/23050/howtoteachthegde-matrix_1.pdf (accessed on 30 May 2024).

Disclaimer/Publisher's Note: The statements, opinions and data contained in all publications are solely those of the individual author(s) and contributor(s) and not of MDPI and/or the editor(s). MDPI and/or the editor(s) disclaim responsibility for any injury to people or property resulting from any ideas, methods, instructions or products referred to in the content.

Article

Preschool Children's Metaphoric Perceptions of Digital Games: A Comparison between Regions

Elçin Yazıcı Arıcı [1], Michail Kalogiannakis [2,*] and Stamatios Papadakis [3]

[1] Department of Preschool Education, Düzce University, 81700 Duzce, Turkey; elcinyazici@duzce.edu.tr
[2] Department of Special Education, University of Thessaly, 38221 Volos, Greece
[3] Department of Preschool Education, University of Crete, 74100 Rethymnon, Greece; stpapadaksi@uoc.gr
* Correspondence: mkalogian@hotmail.com or mkalogian@uoc.gr

Abstract: Preschoolers now play digital games on touch screens, e-toys and electronic learning systems. Although digital games have an important place in children's lives, there needs to be more information about the meanings they attach to games. In this context, the research aims to determine the perceptions of preschool children studying in different regions of Turkey regarding digital games with the help of metaphors. Four hundred twenty-one preschool children studying in seven regions of Turkey participated in the research. The data were collected through the "Digital Game Metaphor Form" to determine children's perceptions of digital games and through "Drawing and Visualization", which comprises the symbolic pictures children draw of their feelings and thoughts. Phenomenology, a qualitative research model, was used in this study. The data were analyzed using the content analysis method. When the data were evaluated, the children had produced 421 metaphors collected in the following seven categories: "Nature Images, Technology Images, Fantasy/Supernatural Images, Education Images, Affective/Motivational Images, Struggle Images, and Value Images". When evaluated based on regions, the Black Sea Region ranked first in the "Fantasy/Supernatural Images and Affective/Motivational Images" categories. In contrast, the Central Anatolia Region ranked first in the "Technology Images and Education Images" categories, and the Marmara Region ranked first in the "Nature Images and Value Images" categories. In addition, it was determined that the Southeast Anatolia Region ranks first in the "Struggle Images" category.

Keywords: preschool education; metaphor; digital game

Citation: Yazıcı Arıcı, E.; Kalogiannakis, M.; Papadakis, S. Preschool Children's Metaphoric Perceptions of Digital Games: A Comparison between Regions. *Computers* **2023**, *12*, 138. https://doi.org/10.3390/computers12070138

Academic Editors: Carlos Vaz de Carvalho, Hariklia Tsalapatas and Ricardo Baptista

Received: 14 June 2023
Revised: 1 July 2023
Accepted: 3 July 2023
Published: 11 July 2023

Copyright: © 2023 by the authors. Licensee MDPI, Basel, Switzerland. This article is an open access article distributed under the terms and conditions of the Creative Commons Attribution (CC BY) license (https://creativecommons.org/licenses/by/4.0/).

1. Introduction

Today's children have a more comprehensive range of modern digital technologies, such as tablets and smartphones, than previous generations [1]. Modern digital technology has changed children's play platforms and how children interact with materials. In this change process, contemporary children benefit from digital developments called "digital play" [2]. Digital developments and the affordability of mobile devices have made digital games an increasingly common phenomenon among children [3]. For this reason, the "digital game" concept is seen as a new game form [4].

Digital games are entertainment and media played on digital tools and have learning opportunities [5]. Nevski and Siibak [6] defined digital play as the actions children perform on the touch screen, while Kinzie and Joseph [7] defined it as fun and exciting actions where the rules are followed in line with a goal. In short, digital play is expressed as an area where children use digital technologies [8]. Based on these definitions, it is seen that digital games include learning through exploration and inquiry, as well as learning through play in a broader context. In particular, a digital game can be played individually or in groups, with or without adult support, with devices such as computers or tablets [9].

For this reason, children have become able to use digital technology devices anytime and anywhere [10]. This shows that the digital age has reached the early childhood age, and its usage by children has become widespread. At the same time, it stated in studies that 30% of children aged six years and under play digital games [11]. In a study by the Hong Kong Ministry of Health, 70.3% of children aged 4–14 years played video and computer games five days a week on average [10]. In addition, it is stated that 86% of children aged 5–6 years in Russia can use digital tools [12]. From this point of view, it is noteworthy that children's games in this period have gradually changed from traditional to digital games [3].

Children no longer play face-to-face but with their friends on screens; they have begun socializing in virtual environments rather than in real life. When it comes to playing games, digital games come to mind first, and children prefer to play with digital tools, not with their friends [13]. Studies also indicate that digital technologies limit children's activities with the individuals around them [14,15]. On the other hand, it is stated that children who play these games and are in the developmental period face many negative consequences, including worsened vision and hearing, delay in development, aggressive behaviour, addiction, inability to socialize, decrease in creative activities, deficiencies in language development, and emotional problems [12]. In short, digital games are criticized for causing problems such as children's inactive lifestyles, sleep disorders, and a lack of physical activity [14]. However, it was pointed out that digital games, like all other game types, are not low-level game modes and provide more opportunities for children to play [3]. It is an active component of children's playing and learning [16].

When the positive and negative aspects of digital games are evaluated in general, the significant point depends on the design of digital games. Digital games should have meaning for children and be designed to suit their abilities and desire to play and explore [17] because the rules, challenges, and feedback of digital games can be rewarding or frustrating, build or destroy self-confidence, help to develop desirable or undesirable skills or create engaging experiences. Children learn in a well-designed digital game because they need to, and motivation is built into it. In other words, if digital games are designed to serve children's interests and abilities, their desire to play and explore, and their inner learning needs, children will be more likely to develop and strengthen their curiosity and attention [11,18]. In short, digital games should provide rich, fun, and interactive experiences by supporting children's learning, cognitive development, skill development, social interaction, physical activities, and health behaviours [19]. In this way, children will gain competence in self-management, competence, and a desire to learn. Children's early years lay the groundwork for their lifelong development, and therefore, it is crucial to know how to design games that will serve them well and to choose games that are designed for them [11,20].

From this point of view, digital games that support holistic development should not be compared with traditional game formats and should not be differentiated from them. Digital games should be considered an educational activity that allows children to interact with a digital environment and explore the world [9]. Digital games support effective learning and activate children's interests [21]. They help children reach their academic expectations. They connect in-school and out-of-school learning experiences. They improve children's social skills and benefit their problem-solving skills [14,18]. They also support the development of creativity and imagination, the basic building blocks for children's future emotional and cognitive development and academic skills. In this way, children master various skills and knowledge [11,19,20].

The context in which children play should be re-evaluated, as children's games have now become an area that includes technology and the latest tools. Based on the developmentally appropriate practice philosophy, it is essential to consider how children's experiences change during this process and how current play strategies affect their learning [22]. For this reason, the relationship between play and technology should be addressed in the early childhood curricula, and the existing gap should be closed. Therefore, it is crucial to properly select, use, integrate, and evaluate digital games in order to develop children's learning and discovery skills. In this process, digital games should be used as pedagogical tools that serve the basic features of "learning through play" and "playing through learning" by combining games and technology. The point here is how to escalate the positive outcomes of these new media in a way that enriches children's play experiences [9,23].

Digital games have become indispensable for children growing up in the technology age; therefore, children's games have been extensively studied in the last decade [24]. These studies generally focus on how long digital games are played [25], advantages [5,19,26,27], disadvantages [28], and types of digital games [1]. However, adult opinions (educators, software, parents) are included in these studies [2,3,12,21,24]. Although these studies provide essential information, we know that young children growing up in the age of technology are playing games and spending more time with the increasingly common digital tools. Topics related to children's understanding of digital games still need to be adequately researched. The present research is one of the original studies that reflect children's perceptions of digital games since it is a candidate to be the first in the field as a metaphorical study on how children conceptualize and perceive digital games in seven regions of Turkey and at least three cities from each region. A metaphor is an essential tool that directly compares a concept in one field to another unrelated field. At the same time, metaphors are seen as a fundamental element of human cognition, shaping how we think and reason about the world. They provide a new way of thinking about general concepts. Metaphors are complex cognitive mechanisms that affect thinking, learning, and reasoning [29,30].

Therefore, children must acquire new knowledge. Specifically, reflective learning mechanisms can help children create new explanations and analogies and imagine alternative possibilities. As a result, these mechanisms can expand children's conceptual repertoire to create new ideas and solutions. Because metaphors bring new perspectives to general knowledge, they are fundamental mechanisms for childhood learning and creating conceptual change. Metaphors and metaphorical thinking can contribute to preschoolers' remarkable conceptual innovation and learning abilities [29]. With the recent increase in the number of studies on metaphors, this phenomenon is a robust mental mapping and modelling tool used for understanding and structuring children's worlds and was reported to be effective. Metaphors are significant, especially in acquiring complex concepts and terms and concretizing and visualizing abstract concepts. From this point of view, this research aims to examine the views of preschool children studying in different regions of Turkey on digital play through metaphors. In line with this purpose, sub-objectives are given below.

- What are the metaphors of preschool children regarding digital play in terms of education in different regions?
- Considering the standard features of these metaphors, under which conceptual categories are the metaphors included?

2. Materials and Methods

2.1. Research Model

In phenomenology, the researcher tries to reveal perceptions about a phenomenon. The researcher tries to understand the participants' world and describe their perceptions and reactions [31].

2.2. Working Group

The study group of this research consisted of 48–66-month-old children attending preschool education institutions in 7 regions of Turkey (Black Sea, Mediterranean, Marmara, Aegean, Eastern Anatolia, Southeastern Anatolia, and Central Anatolia) in the 2022–2023 academic year—this study group was collected using the convenience sampling method [32]. The study group of this research included a total of 421 children who were from the Black Sea Region (*n*:62), Mediterranean Region (*n*:50), Marmara Region (*n*:66), Aegean Region (*n*:57), Eastern Anatolia Region (*n*:53), Southeastern Anatolia Region (*n*:66) and Central Anatolia Region (*n*:67).

2.3. Data Collection Tool

The "Personal Information Form", "Digital Game Metaphor Form", and "Drawing and Visualization" were used as data collection tools in the research.

- Personal Information Form. This form, developed by the researchers, contains information about the children and their families. The children's classroom teachers filled out the forms.
- Digital Game Metaphor Form. The expression 'Digital game is like ... Because ... ' was used to determine children's perceptions regarding digital games. It is a data collection tool prepared by the researchers. Children were given an approximate time to complete this statement and were asked to focus on only one metaphor. This statement given by the children constitutes the data source of the research as a "document". The class teachers recorded the answers of the children.
- Drawing and Visualization. This technique is a powerful tool for gathering information from children. The aim is to reveal children's feelings and thoughts about the world through pictures [33]. In this study, children were asked to draw pictures reflecting the metaphor sentence to examine digital game-themed metaphors deeply.

2.4. Data Collection

To collect the research data, firstly, the cities representing the seven regions of Turkey were determined. At least three cities and independent kindergartens or kindergartens affiliated with the Ministry of National Education in these cities were selected from each region. Interviews were held with the teachers of the selected institutions via telephone and email. While determining the study group in the research, the principle of voluntarism was taken as a basis. The teachers were informed about the purpose of the research and the points to be considered during the application. The expression "Digital game ... like this. Because ... " was introduced for this. It was explained how the children should complete the dotted parts in the expression. This form has been presented within the framework of ethical rules. The study was conducted following ethical principles (E-78187535-050.06-290724).

The children were told they should be given time to think about their metaphors. At this stage, it was stated that the children should be given general information, that they should mention a single concept, and that the children should explain why they thought about this concept. Afterwards, it was explained to the children that they should be given A4 size paper and crayons, draw a picture describing their feelings and thoughts about their metaphor, and briefly describe the picture they drew. It was also emphasized that the interviews should be individual with each child, as they could be affected by the discourses and drawings of their friends. The papers containing these expressions, written by the classroom teachers in their handwriting, are documents and constitute the primary data source of this research. The data were transmitted to the researcher via digital media.

2.5. Analysis of Data

The data were analyzed using the content analysis method. Content analysis is one of the qualitative methods and is used to analyze and interpret data. This analysis reduces the data to concepts representing the research phenomenon [34]. Analyses were performed in five stages [35] (Figure 1).

1. <u>Coding and debugging phase.</u> The metaphors produced by the researchers regarding the concept of digital games were examined individually in the context of the regions, and a tentative list was created. In this framework, forms that did not include a metaphor or a justification did not produce any metaphors and were incompatible with a justification sentence related to the metaphor in question were elected. Then, the metaphor presented by each child was coded. In this context, 75 out of 496 forms were excluded from the research scope due to the evaluation of the metaphors for the concept of digital games. A total of 421 forms were included in the study.

2. Sample metaphor image compilation phase. After the first stage was completed, the raw data were reviewed again. Sample metaphors were selected from the expressions representing each metaphor. During the selection process, attention was paid to the metaphor's ability to express the analogy clearly, how much it emulated the analogy and their relationship. In addition, information about who produced the metaphor image is given in parentheses before the metaphor expression in question. The codes BSRC1 (Black Sea Region Child 1), MERC1 (Mediterranean Region Child 1), MRC1 (Marmara Region Child 1), ARC1 (Aegean Region Child 1), EARC1 (Eastern Anatolia Region Child 1), SARC1 (Southeast Anatolia Region Child 1), and CARC1 (Central Anatolia Region 1), which comprise a region and a number, refer to a child who attends a preschool education institution in that region. For example, the code (BSRC1) is used for the number one child in the Black Sea Region.
3. Category development phase. Each metaphor image produced by children was analyzed in terms of the subject of the metaphor and its typical features. Then, considering the codes given to the metaphors, metaphors with similar themes were included in the same category, and a total of 7 conceptual categories were created.
4. The stage of ensuring validity and reliability. The data in the research process are presented objectively, and the data analysis process is explained in detail. In addition to the code/category tables, detailed descriptions are made. For this reason, direct quotations from the children's statements are frequently included. However, the findings are presented quantitatively as well as qualitatively. Expert opinion was used for reliability. Expert and researcher category pairings were compared. Miles and Huberman's [36] formula was used for comparisons, and the reliability was determined as 0.93. Since it was thought to make the research results reliable, the application was more comprehensive than just completing the metaphor. However, data variations (triangulation) such as drawing and visualization were also used.
5. Transferring and interpreting data to a computer environment. In the last stage, all data were transferred to the computer environment. Then, the frequency of children's use of metaphors was calculated, and tables suitable for the categories were created. The data were interpreted according to the findings obtained.

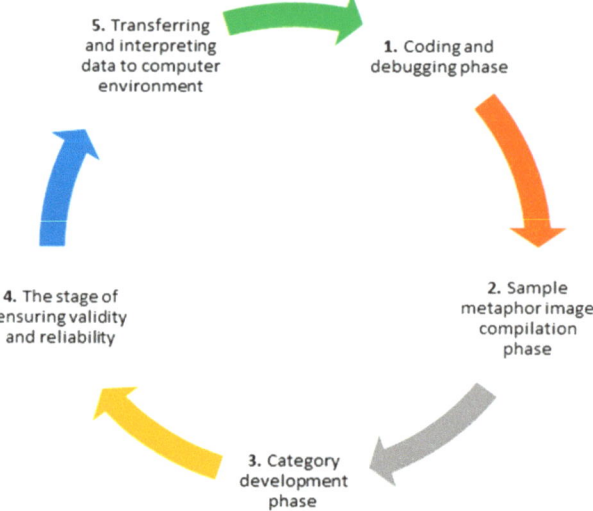

Figure 1. Stages of the data analysis process.

3. Results

In this section, the research results are presented in tables.

Table 1 shows the distribution of metaphors created by children regarding "digital games" by region.

Table 1. Distribution of metaphors produced by regions.

Regions	Metaphors	f	%
The Black Sea Region	House (2), Sky (2), Lipstick, Park, Car, Animal, Flower, Space, Fish, Fun (7), Feeling Good (2), Boring, Changing Clothes, Good Play, Bad, Dare, Surprise, Emotion, Good Thing, Funny, Minecraft (4), Roblox (3), Tom, FIFA, Monster, Dinosaur, Flying, Alien, Internet (3), Tablet (2), Phone (2), Robot, Television, Computer, Game (2), Educational Toy, Color, Picture, Race Car, Parkour, Car Game, Football, Human	62	14.72
The Mediterranean Region	Car, Balloon, Park, Dog's Nest, Rabbit, Cat, Bear, Rainbow, Watermelon, Strawberry, Snake, House, Truck, Eating, Jumping, Dentist, Minecraft (3), PUBG (3), Roblox, Princess, Ghost, Hacker, Dinosaur, Computer (3), Tablet (2), Robot (2), Phone, Memory Game (4), Drawing (2), Color (2), Card Game, Toy, Shape, Parkour, Chase, Friendship, Man	50	11.87
The Marmara Region	Car (4), Bus (3), Money (3), Balloon (2), Truck (2), Motor, Dog, House, Panda, Bear, Caravan, Horse, Tower, Ball, Diamond, Garden, Sheep, Flower, Animal, Beauty, Bathing, Making Happy, Fight, Very Good, Chocolate, Headache, Roblox (2), Mario (2), Lightning McQueen, Barbie Doll, Freezing Crew, Telephone (3), Computer, Electromagnetic Item, Chess (2), Game, Puzzle, Jigsaw, Card, Line, Coloring, Numbers, Parkour (2), Catch, Football, Basketball, Fishing, Miner, Boxer, Soldier, Human	66	15.67
The Aegean Region	Car (2), House, Ferris Wheel, Bucket, Bicycle, Seat, Pillow, Dog, Spider, Red Bull, Ship, Grandpa Moon, Lipstick, Money, Buckle, Fun (2), Driving, War, Cat Play, Eye Pain, Minecraft (2), Roblox (2), Rabbit Game (2), Wing Game, Pop It, Angela, Snake Game, Dinosaur, Unreal, Electric, Video, Toy (2), Game, Coloring, Puzzle, Doll Dress Up, Painting, Cat Game, Car Game (4), Race (2), Race Car, Parkour, Ball Game, Balloon Game, Soldier, Hairdressing, Help	57	13.53
The Eastern Anatolia Region	Eating, Pizza, Cake, Car Racing, Song, Heart, Barbie Doll (5), Pop It (2), Roblox, Elsa, Toca Boca, Dinosaur, Car (4), Airplane (3), Cake (3), Ball (2), Ice Cream, Bed, Butterfly, Key, Bracelet, Forest, Shop, World, Box, Flower, Tablet (2), Game (2), Puzzle, Figure Racing (3), Fighting, War, Shark, Child (2)	53	12.53
The Southeastern Anatolia Region	Train (2), Ball (2), Grape, Stone, Tree, Sheep, Slide, Strawberry, Cave, Rain, Cat, Field, Eraser, Machine, Oven, Telephone (7), Computer (4), Television (2), Technology, Tablet, Cat Game (3), Nail game (2), Game (2), Dress Up, Mind Game, Baby Game, Rectangle, Match (2), Racing Game, Flamingo Racing, Gold Collecting, Football, Robot War, Car Game, Horse Racing, Motor, Fun (3), Affection, Cartoon, Papchi (3), PUBG, Tom, Minecraft, Monster	66	15.67

Table 1. *Cont.*

Regions	Metaphors	f	%
The Central Anatolia Region	Car (4), Sun (2), Bus, Rainbow, Slide, Money, Cat, Ladybug, Pen, Fridge, Fun, Heart, Jumping Game, Mario, Roblox, Galaxy, Monster, Tablet (9), Computer (6), Telephone (5), Television (5), Internet, Technology, Mobile Phone, Game (3), Sauce Game (2), Shape (2), Wrestling, Tennis, Toy, Coloring, Chess, Brain Game, Race Car, Football Race, Car Game, Catch, King	67	15.91
Total		421	100

The numbers in brackets indicate how many times the metaphor is repeated.

When Table 1 is examined, it is seen that 421 metaphors were produced based on all regions. When the metaphors produced by the children are evaluated, it is found that the number of metaphors in each region is as follows: Central Anatolia Region, f = 67 (15.91%); Marmara and Southeastern Anatolia regions, f = 66 (15.67%); Black Sea Region, f = 62 (14.72%); Aegean Region, f = 57 (13.53%); Eastern Anatolia Region, f = 53 (12.53%); and the Mediterranean Region, f = 50 (11.87%).

In Table 1, the metaphors produced based on regions are given. It was suggested that they be divided into groups. They are divided into groups according to regions. Table 1 presents the general situation.

According to the analysis results, seven "Digital Games" categories were created when all regions were considered. A visual presentation of the seven categories of the metaphors obtained is presented in Figure 2.

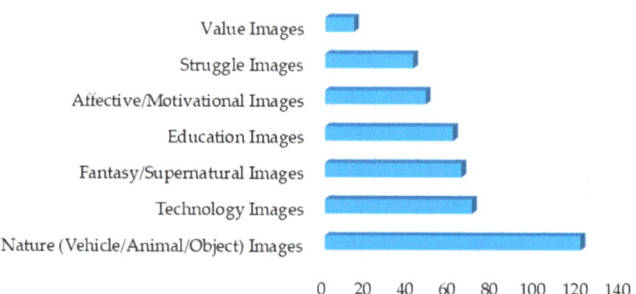

Figure 2. Categories created for digital games related to all regions.

When Figure 2 is examined, it can be seen that 28.34% (f = 121) of the answers given by children in the regions included in the study are nature (vehicle/animal/object) images, 16.62% (f = 70) are technology images, 15.43% (f = 65) are fantastic/supernatural images, 14.48% (f = 61) are educational images, 11.40% (f = 48) are affective/motivational images, 9.97% (f = 42) are struggle images, and 3.32% (f = 14) seem to be related to value images.

The responses for each category were evaluated based on regions and are presented below.

When Table 2 is examined, it is seen that 43 different metaphors related to digital games were produced in the Black Sea Region, and the metaphors produced are grouped under seven different categories according to their similarity. Among these, the Affective/Motivational Images category (f = 18 (29.03%)) is the area where the most intense metaphors were produced. In comparison, the Value Images category (f = 1 (1.61%)) is where the least number of metaphors were produced. The children's expressions for the relevant categories and examples of pictures (Figure 3) are given below.

Table 2. Metaphoric perception categories of children attending preschool education institutions in the Black Sea Region regarding "Digital Games".

Categories	Metaphors	No. of Metaphors	f	%
1. Affective/Motivational Images	Fun (7), Feeling Good (2), Boring, Changing Clothes, Good Play, Bad, Dare, Surprise, Emotion, Something Good, Funny	11	18	29.03
2. Fantasy/Supernatural Images	Minecraft (4), Roblox (3), Tom, FIFA, Monster, Dinosaur, Flying, Alien	8	13	20.96
3. Nature (Veh./Anim./Obj.) Images	Home (2), Sky (2), Lipstick, Park, Car, Animal, Flower, Space, Fish	9	11	17.74
4. Technology Images	Internet (3), Tablet (2), Phone (2), Robot, Television, Computer	6	10	16.12
5. Education Images	Game (2), Educational Toy, Color, Picture	4	5	8.06
6. Struggle Images	Race Car, Track, Car Game, Football	4	4	6.45
7. Value Images	Human	1	1	1.61
	Total	43	62	100

The numbers in brackets indicate how many times the metaphor is repeated.

BSRC12

BSRC4

Figure 3. Examples of children's drawings.

Category 1. "Affective/Motivational Images"
BSRC20—"Digital game is like courage. Because if someone attacks me, I have the feeling of protecting myself. Moreover, I think I can make armour to protect myself when I grow up."
Category 2. "Fantasy/Supernatural Images"
BSRC4—"Digital game is like a monster. Because monsters have extraordinary powers, they can do anything with these powers. You can do anything in digital games. You can even create new monsters."
Category 3. "Nature (Veh./Anim./Obj.) Images"
BSRC17—"Digital game is like flowers. Because, like flowers, it is colourful. It has as many games as you want in it."
Category 4. "Technology Images"
BSRC12—"Digital game is like a computer. Because, like him, he is swift, and we can do whatever we want."

Category 5. "Education Images"
BSRC6—"Digital games are like a painting. Because I carry paints and I take beautiful pictures. Whatever colour shows the paint, I take that colour and paint it."

Category 6. "Struggle Images"
BSRC34—"Digital game is like parkour. Because there are obstacles on the track, there are also obstacles in the digital game. Then you win the game by going through the obstacles, but if you are fast."

Category 7. "Value Images"
BSRC56—"Digital game is like a human. Because people help each other, the game also helps me and teaches me everything. I also like to help."

In Table 3, the answers given by the children in the Mediterranean Region regarding digital games are given. It is seen that the children produced 37 different metaphors, and these metaphors are separated into seven different categories. The Nature (Veh./Anim./Obj.) Images category (f = 13 (26.00%)) is the area where the most intense metaphors were produced. In comparison, the Value and Struggle Images categories (f = 2 (4.00%)) are the areas where minor metaphors were produced. The children's expressions for the relevant categories and examples of pictures (Figure 4) are given below.

Table 3. Metaphoric perception categories of children attending preschool education institutions in the Mediterranean Region regarding "Digital Games".

Categories	Metaphors	No. of Metaphors	f	%
1. Affective/Motivational Images	Eating, Jumping, Dentist	3	3	6.00
2. Fantasy/Supernatural Images	Minecraft (3), PUBG (3), Roblox, Princess, Ghost, Hacker, Dinosaur	7	11	22.00
3. Nature (Veh./Anim./Obj.) Images	Car, Balloon, Park, Dog Nest, Rabbit, Cat, Bear, Rainbow, Watermelon, Strawberry, Snake, House, Truck	13	13	26.00
4. Technology Images	Computer (3), Tablet (2), Robot (2), Phone	4	8	16.00
5. Education Images	Memory Game (4), Drawing (2), Color (2), Card Game, Toy, Shape	6	11	22.00
6. Struggle Images	Parkour, Chase	2	2	4.00
7. Value Images	Friendship, Man	2	2	4.00
	Total	37	50	100

The numbers in brackets indicate how many times the metaphor is repeated.

MERC56

MERC6

Figure 4. Examples of children's drawings.

Category 1. "Affective/Motivational Images"
MERC22—"Digital games are like eating food. Because I am happy when I play, just as I am happy when I eat."

Category 2. "Fantasy/Supernatural Images"
MERC6—"Digital game is like a princess. Because princesses are omnipotent, they can fly and become invisible at any moment. Furthermore, I am the princess of the game."
Category 3. "Nature (Veh./Anim./Obj.) Images"
MERC1—"Digital games are like a park. Because there are too many toys in the park for us to play with, I can ride any toy I want when my mom takes me there. There are also many games when playing games on the phone. I can play however I want."
Category 4. "Technology Images"
MERC12—"Digital game is like a robot. Because when you set the robots, you can get everything done. I can also adjust my game while playing the game. For example, I can choose the car and person I want."
Category 5. "Education Images"
MERC48—"Digital game is like a memory game. Because I am playing a game of finding fruits, you must find the same one when the strawberry picture appears. When found, another fruit emerges. When you find all the fruits, you move to another level."
Category 6. "Struggle Images"
MERC37—"Digital game is like a chase. They are trying to catch me because I am running in the game. If I do not get caught, I win the game. Of course, I have strategies for that."
Category 7. "Value Images"
MERC56—"Digital game is like a friend. Because he is always playing games with us, I think it helps us to make us happy."

When Table 4 is examined, it is seen that 51 different metaphors were produced by the children in the Marmara Region regarding digital games. The metaphors are grouped into seven different categories according to their similarities. Among these, the Nature (Vehicle/Animal/Object) Images category (f = 28 (42.42%)) is the field where metaphors were produced the most. In comparison, the Value Images category (f = 4 (6.06%)) is where metaphors were produced the least. The children's expressions for the relevant categories and examples of pictures (Figure 5) are given below.

Table 4. Metaphoric perception categories of children attending preschool education institutions in Marmara Region regarding "Digital Games".

Categories	Metaphors	No. of Metaphors	f	%
1. Affective/Motivational Images	Beauty, Bathing, Making You Happy, Fighting, Adorable, Chocolate, Headache	7	7	10.60
2. Fantasy/Supernatural Images	Roblox (2), Mario (2), Lightning McQueen, Barbie Doll, Rafadan Tayfa	5	7	10.60
3. Nature (Veh./Anim./Obj.) Images	Car (4), Bus (3), Money (3), Truck (2), Balloon (2), Dog, House, Panda, Bear, Caravan, Horse, Motor, Tower, Ball, Diamond, Garden, Animal, Sheep, Flower	19	28	42.42
4. Technology Images	Telephone (3), Computer, Electromagnetic Equipment	3	5	7.57
5. Education Images	Chess (2), Games, Puzzles, Jigsaw, Card, Line, Coloring, Numbers	8	9	13.63
6. Struggle Images	Track (2), Catch, Football, Basketball, Fishing	5	6	9.09
7. Value Images	Miner, Boxer, Soldier, Human	4	4	6.06
	Total	51	66	100

The numbers in brackets indicate how many times the metaphor is repeated.

MRC60   MRC44

Figure 5. Examples of children's drawings.

Category 1. "Affective/Motivational Images"
MRC6—"Digital games are like making children happy. Because when a child goes to a house as a guest, the owner does not give toys to the child; the child plays with his mother's phone and is happy."
Category 2. "Fantasy/Supernatural Images"
MRC47—"Digital gaming is like Lightning McQueen. Because cars talk and fly, I wish my cars could talk too."
Category 3. "Nature (Veh./Anim./Obj.) Images"
MRC33—"Digital gaming is like a tower. Because towers are very tall, and too many people are in them. Digital games are also very long. Moreover, there are many games in it."
Category 4. "Technology Images"
MRC29—"Digital game is like an electromagnetic object. Because there are hoses connected inside, they make us play nice games. Nevertheless, we cannot see them."
Category 5. "Education Images"
MRC44—"Digital game is like chess. Because while playing chess, I always think and place the pieces in the right place. I cannot move forward if I do not put the animals in their nests in the digital game."
Category 6. "Struggle Images"
MRC29—"Digital gaming is like fishing. Because the one who catches the fish wins, there is also winning in digital games."
Category 7. "Value Images"
MRC60—"Digital game is like a miner. Because miners are always helping others, and they are your friends."

When Table 5 is examined, it is seen that 47 different metaphors related to digital games were produced in the Aegean Region, and they are grouped under seven different categories according to their similarity. Among these, the Nature (Vehicle/Animal/Object) Images category (f = 16 (28.07%)) is the area where the most intense metaphors were produced. In comparison, the Technology Images category (f = 2 (3.50%)) is where the most minor metaphors were produced. The children's expressions for the relevant categories and examples of pictures (Figure 6) are given below.

Table 5. Metaphoric perception categories of children attending preschool education institutions in the Aegean Region regarding "Digital Games".

Categories	Metaphors	No. of Metaphors	f	%
1. Affective/Motivational Images	Fun (2), Driving, War, Cat Play, Eye Pain	5	6	10.52
2. Fantasy/Supernatural Images	Minecraft (2), Roblox (2), Rabbit Game (2), Wing Game, Pop It, Angela, Snake Game, Dinosaur, Unreal	9	12	21.05
3. Nature (Veh./Anim./Obj.) Images	Car (2), House, Ferris Wheel, Bicycle, Bucket, Seat, Pillow, Dog, Spider, Red Bull, Ship, Grandpa Moon, Lipstick, Buckle, Money	15	16	28.07
4. Technology Images	Electricity, Video	2	2	3.50
5. Education Images	Toy (2), Game, Coloring, Puzzle, Doll Dress Up, Picture, Cat Game	7	8	14.03
6. Struggle Images	Car Game (4), Race (2), Race Car, Parkour, Ball Game, Balloon Game	6	10	17.54
7. Value Images	Soldier, Hairdressing, Helping	3	3	5.26
	Total	47	57	100

The numbers in brackets indicate how many times the metaphor is repeated.

ARC13   ARC9

Figure 6. Examples of children's drawings.

Category 1. "Affective/Motivational Images"
ARC25—"Digital gaming is like eye pain. Because if I play games for a long time, my eyes hurt. My mother is also angry with me. It hurts your eyes, and you will be wearing glasses soon."

Category 2. "Fantasy/Supernatural Images"
ARC50—"Digital game is like Wing game. Because they wear orange, it has blue and white stripes. Planes are fighting each other. They have extraordinary powers."

Category 3. "Nature (Veh./Anim./Obj.) Images"
ARC13—"Digital game is like a car. Because the car runs on gasoline, it works with a charge on the tablet we play games on. That is why I liken it to a car."

Category 4. "Technology Images"
ARC41—"Digital gaming is like electricity. Because there are robots and robots are powered by electricity. Electricity goes into them, and the robots can move."

Category 5. "Education Images"
ARC54—"Digital game is like a toy. Because we are playing games with both of them, my toys teach me a lot, too, and the games I play on the tablet."

Category 6. "Struggle Images"
ARC9—"Digital game is like a ball game. Because you are throwing it, you are trying to get it up and down the hole. The winner gets a prize."

Category 7. "Value Images"
ARC38—"Digital games are like helping; for instance, I learned numbers thanks to the game. Even the colours. It helps me by teaching me so many good things. Everyone asks me how do you know this."

Table 6 shows that the children in the Eastern Anatolia Region produced 35 different metaphors about digital play, and these metaphors are grouped under seven categories. Among these, the Nature (Vehicle/Animal/Object) Images category (f = 22 (41.50%)) has the most intense metaphors, while the Technology and Value Images categories (f = 2 (3.77%)) have the most minor metaphors produced. The children's expressions for the relevant categories and examples of pictures (Figure 7) are given below.

Table 6. Metaphoric perception categories of children attending preschool education institutions in Eastern Anatolia Region regarding "Digital Games".

Categories	Metaphors	No. of Metaphors	f	%
1. Affective/Motivational Images	Eating, Pizza, Cake, Racing Car, Song, Heart	6	6	11.32
2. Fantasy/Supernatural Images	Barbie Doll (5), Pop It (2), Roblox, Elsa, Toca Boca, Dinosaur	6	11	20.75
3. Nature (Veh./Anim./Obj.) Images	Car (4), Airplane (3), Cake (3), Ball (2), Ice Cream, Bed, Butterfly, Key, Bracelet, Forest, Shop, World, Box, Flower	14	22	41.50
4. Technology Images	Tablet (2)	1	2	3.77
5. Education Images	Game (2), Puzzle, Number	3	4	7.54
6. Struggle Images	Racing (3), Fighting, War, Shark	4	6	11.32
7. Value Images	Child (2)	1	2	3.77
	Total	35	53	100

The numbers in brackets indicate how many times the metaphor is repeated.

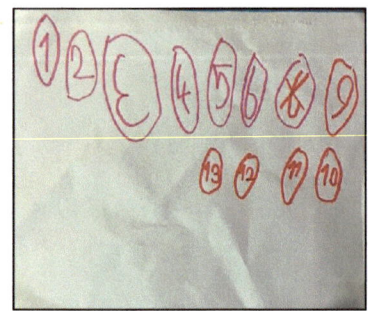

EARC3　　　　　　　　　　　　EARC41

Figure 7. Examples of children's drawings.

Category 1. "Affective/Motivational Images"
EARC15—"Digital games are like eating food. Because it is both fun and beautiful."

Category 2. "Fantasy/Supernatural Images"
EARC45—"Digital game is like a Barbie doll. Because she is talking to me, and I am talking to her. I make her eat, drink and sleep like my sister. Moreover, sometimes, she shows me magical powers."

Category 3. "Nature (Veh./Anim./Obj.) Images"

EARC3—"Digital game is like a shop. Because there are many surprise games for kids in it."
Category 4. "Technology Images"
EARC19—"Digital game is like a tablet. I can play games with a tablet because there are so many games on it. Nevertheless, I cannot play all the time."
Category 5. "Education Images"
EARC41—"Digital game is like numbers. Because we can play forever, the numbers are endless. I am counting, but it never ends."
Category 6. "Struggle Images"
EARC28—"Digital game is like an adventure. Because there are obstacles, I get gifts as I surpass them."
Category 7. "Value Images"
EARC33—"Digital game is like a child. Because children help everyone, digital games also help us. It entertains us when we are bored."

In Table 7, it is seen that 45 different metaphors related to digital games were produced in the Southeastern Anatolia Region, and they are grouped under seven different categories according to their similarity. Among these, the Nature (Vehicle/Animal/Object) Images category (f = 17 (25.75%)) is the area where metaphors were produced the most. In comparison, the Value Images category f = 1 (1.51%) is where the most minor metaphors were produced. The children's expressions for the relevant categories and examples of pictures (Figure 8) are given below.

Table 7. Metaphoric perception categories of children attending preschool education institutions in Southeastern Anatolia Region regarding "Digital Games".

Categories	Metaphors	No. of Metaphors	f	%
1. Affective/Motivational Images	Fun (3), Love, Cartoon	3	5	7.57
2. Fantasy/Supernatural Images	Papchi (3), PUBG, Tom, Minecraft, Monster	5	7	10.60
3. Nature (Veh./Anim./Obj.) Images	Ball (2), Train (2), Grape, Stone, Tree, Sheep, Slide, Strawberry, Cave, Rain, Cat, Field, Eraser, Machine, Oven	15	17	25.75
4. Technology Images	Phone (7), Computer (4), Television (2), Technology, Tablet	5	15	22.72
5. Education Images	Cat Game (3), Nail Game (2), Game (2), Dress Up, Mind Game, Baby Game, Rectangle	7	11	16.66
6. Struggle Images	Match (2), Racing Game, Flamingo Racing, Gold Collecting, Football, Robot Battle, Car Game, Horse Racing, Motorbike	9	10	15.15
7. Value Images	Spider-Man	1	1	1.51
	Total	45	66	100

The numbers in brackets indicate how many times the metaphor is repeated.

Category 1. "Affective/Motivational Images"
SARC7—"Digital gaming is like love. Because if we love someone, we always want to see them. I always want to play on a tablet too. Because I have much love."
Category 2. "Fantasy/Supernatural Images"
SARC40—"Digital gaming is like a beast. Because we do not know what happened, we do not know about monsters either. What do I do with it if it comes off my tablet while playing? I do not know."
Category 3. "Nature (Veh./Anim./Obj.) Images"
SARC13—"Digital game is like an eraser. Because it can be easily cleaned."
Category 4. "Technology Images"
SARC22—"Digital gaming is like a phone. Because it is speedy and has a lot of stuff, you can find friends while playing games. You can also talk to your friends on the phone."

Category 5. "Education Images"
SARC41—"Digital game is like nail game. Because you paint the nails how you want according to their colour, we apply nail polish on the same nail; it becomes like that."
Category 6. "Struggle Images"
SARC33—"Digital game is like a flamingo. Because we are always trying to win, we overcome obstacles. You have to be fast, of course."
Category 8. "Value Images"
SARC16—"Digital game is like Spider-Man. Because spider-men can go anywhere with their webs, he helps everyone. I think games help us too. It teaches a lot."

SARC33

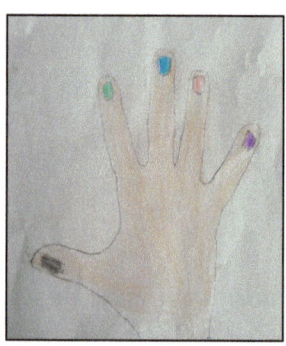
SARC41

Figure 8. Examples of children's drawings.

When Table 8 is examined, it is seen that children in Central Anatolia produced 38 different metaphors for their digital game. These metaphors are grouped under seven categories based on their similar characteristics. Among these, metaphors were produced the most in the Technology Images category (f = 28 (41.79%)). In comparison, the Value Images category (f = 1 (1.49%)) is where the most minor metaphors were produced. The children's expressions for the relevant categories and examples of pictures (Figure 9) are given below.

Table 8. Metaphoric perception categories of children attending preschool education institutions in Central Anatolia Region regarding "Digital Games".

Categories	Metaphors	No. of Metaphors	f	%
1. Affective/Motivational Images	Fun, Heart, Jumping Game	3	3	4.47
2. Fantasy/Supernatural Images	Mario, Roblox, Galaxy, Monster	4	4	5.97
3. Nature (Veh./Anim./Obj.) Images	Car (4), Sun (2), Bus, Rainbow, Slide, Money, Cat, Ladybug, Refrigerator, Pen	10	14	20.89
4. Technology Images	Tablet (9), Computer (6), Phone (5), Television (5), Internet, Technology, Cell Phone	7	28	41.79
5. Education Images	Game (3), Sauce Game (2), Shape (2), Wrestling, Tennis, Toy, Coloring, Chess, Intelligence Game	9	13	19.40
6. Struggle Images	Race Car, Football Racing, Car Game, Catch	4	4	5.97
7. Value Images	King	1	1	1.49
	Total	38	67	100

The numbers in brackets indicate how many times the metaphor is repeated.

CARC34   CARC11

Figure 9. Examples of children's drawings.

Category 1. "Affective/Motivational Images"
CARC11—"Digital game is like a heart. Because I love heart shape so much, I love the games too. Who does not love to have fun."

Category 2. "Fantasy/Supernatural Images"
CARC39—"Digital game is like a galaxy. Because you are going to galaxies, they are very mysterious places. You do not know what will happen. I think they send me nice things."

Category 3. "Nature (Veh./Anim./Obj.) Images"
CARC23—"Digital game is like a rainbow. Because the rainbow is colourful. A variety of games. All are very beautiful. Moreover, I always want to play with them, but my mother will not let me."

Category 4. "Technology Images"
CARC64—"Digital game is like a computer. Because there are many games on computers, I find everything I am looking for."

Category 5. "Education Images"
CARC51—"Digital game is like intelligence game. Because games develop our brains, I always play games like this. My mom will not let me play any other games."

Category 6. "Struggle Images"
CARC34—"Digital game is like a football match. Because they always compete in matches. I also compete with my friends. Sometimes I win; sometimes they win."

Category 7. "Value Images"
CARC7—"Digital gaming is like a king. Because kings punish people, when we make mistakes in games, we cannot progress. The bad guys are catching us."

4. Discussion

The results obtained from the children's pictures and discourses are handled and discussed in the framework of each category.

Among these categories, "Nature (Vehicle/Animal/Object) Images" is seen as the category in which the most metaphors were produced from all regions (f = 121, 28.34%). The Marmara Region (6.65%) ranks first in this category, followed by Eastern Anatolia (5.22%), Southeastern Anatolia (4.03%), Aegean (3.80%), Central Anatolia (3.32%), Mediterranean (3.08%), and Black Sea (2.61%) regions. In this context, it is remarkable that the children expressed digital games primarily using analogies with nature images. In support of these results, Hazar et al. [37] stated in their study that children try to explain digital games by using nature images. As the related literature states, nature is an unknown and abstract image for children. Many objects found in nature are intertwined with the game. Nature is children's most crucial educational channel [38]. Children's direct experiences with nature are compelling in their development and learning. At this point, qualified early childhood education programs should aim at their holistic development by ensuring children have these opportunities. The importance of children's experiences with nature constitutes a solid basis for establishing sustainable relations between humans and nature [39]. Since children interact with natural elements in their daily lives, it is an expected developmental feature that children include metaphors for nature images in their analogies about digital games.

Secondly, the metaphors produced by children regarding digital games in all regions were collected in the category of "Technology Images" (f = 70, 16.62%). The Central Anatolia Region (6.65%) ranks first in this category. The Southeastern Anatolia (3.56%), Black Sea (2.37%), Mediterranean (1.90%), Marmara (1.18%), Aegean, and Eastern Anatolia regions (0.47%) follow, respectively. In support of this category, Hazar, Tekkurşun Demir and Dalkıran [37] stated in their study that children try to explain digital games by using technology images. In the related literature, a digital game is generally defined as a game that uses technology [40]. Bers [41] sees technological resources as playgrounds. It is stated that the number and variety of digital games that children aged 3–6 years can play using many technological resources (such as tablets, electronic toys, and learning systems) in these playgrounds are increasing gradually [11]. Australian parents report that a third of preschool children have access to a tablet or smartphone [4]. According to a Common Sense Media [25] report, while 61% of children aged 0–8 years use a computer, almost all (91%) children aged 5–8 years use a computer (Common Sense Media, 2013, as cited in [42]). Since the game tools used in digital games are computer-based (tablet, phone, game console, etc.) game tools [37], it can be considered a natural result that children associate digital games with technological images.

Thirdly, the metaphors produced by children regarding digital games in all regions were collected in the category of "Fantasy/Supernatural Images" (f = 65, 15.43%). The Black Sea Region (3.08%) ranks first in this category, followed by the Aegean Region (2.85%), the Mediterranean and Eastern Anatolia regions (2.61%), the Southeast Anatolia and Marmara regions (1.66%), and finally, the Central Anatolia Region (0.95%). The related literature states that digital games, which develop rapidly with the development of technology, attract children's attention with their attractive, fantastic, and rich content [43]. Children have difficulty grasping the boundaries between fiction and reality. Nevertheless, game scenarios need to be more realistic regarding their content. Some game heroes have superpowers, while others contain magic and supernatural creatures. Often children take on the role of superheroes or invoke this mythical mood by settling in a fantasy space whose action is characterized by supernatural figures [44,45]. Children forget that they are small and weak for a while when they are playing games. By magically thinking of people and objects, they can recreate the world as they please [46]. In this respect, children's use of expressions related to fantasy/supernatural images in their metaphors for digital games is consistent with the relevant literature.

Fourthly, the metaphors produced by children regarding digital games in all regions were collected in the category of "Education Images" (f = 61, 14.48%). In this category, Central Anatolia (3.08%) ranks first, followed by Mediterranean and Southeastern Anatolia (2.61%), Marmara (2.13%), Aegean (1.90%), and Eastern Anatolia (0.95%). In support of these results, Hazar, Tekkurşun Demir and Dalkıran [37] stated in their study that children try to explain digital games by using educational images. In addition, it is noteworthy that in the relevant literature, digital games are seen as the main activity of children's imaginary and cognitive development, and digital games are not a low-level game mode [3]. A study conducted with children aged 8–10 years argued that digital games could be considered qualified educational tools that develop creative thinking skills [9]. Digital games offer children many learning opportunities and also enable children to participate actively in the learning process [47]. It supports the development of memory, attention, imagination, and manual skills in children and ensures that they are disciplined [48]. Interaction, control over action, feedback, external rewards or punishments, and identification with character(s) during play support children's interpersonal and internal learning and development [9]. Matching games with age, abilities and skills means they are used educationally [49]. Longitudinal research shows that using games as an educational tool increases learning and productivity [50,51]. From this point of view, it is a natural result that interactive and visually stimulating digital games are likened to educational images by children.

Fifth, the metaphors produced by children regarding digital games in all regions were collected in the category of "Affective/Motivational Images" (f = 48, 11.40%). The Black Sea Region (4.27%) ranks first in this category. It is followed by the Marmara Region (1.66%), the Aegean and Eastern Anatolia regions (1.42%), the Southeast Region (1.18%), and finally, the Central Anatolia and Mediterranean regions (0.71%). In support of these results, Hazar, Tekkurşun Demir and Dalkıran [37] stated in their study that children evaluate digital games through motivational images. When the relevant literature is examined, touch screen technologies have come to the forefront strongly as a way to engage and entertain children simultaneously [40]. In a broader framework, digital games combine education with entertainment and stimulate learning motivation [52]. Many emotional reactions, such as happiness, joy, fear, anxiety, pain, hatred, and independence, are gained through play. In a well-designed game, children learn because they need to play it, and motivation is built into it [20]. This provides interactive experiences that support children's skill development and learning [9]. These experiences include sensory stimuli that trigger children's motivation to focus on the goal and increase their familiarity with thematic concepts. In other words, digital games offer alternative ways of motivating children to understand complex topics better [53]. In this context, the relevant literature and research results are consistent with the current research findings. Children can evaluate digital games through affective/motivational images when these processes are reviewed.

Sixth, the metaphors produced by children regarding digital games in all regions were collected in the "Struggle Images" category (f = 42, 9.97%). The Southeastern Anatolia Region (2.85%) ranks first in this category. The Aegean (2.37%), Marmara, and Eastern Anatolia regions (1.42%), Black Sea and Central Anatolia regions (0.95%), and finally, the Mediterranean Region (0.7%) follow, respectively. In support of these results, Hazar, Tekkurşun Demir and Dalkıran [37] stated that children evaluate digital games through images of struggle. In the related literature, the struggle feature of digital games is emphasized. This emphasis is explained by expressing digital games as rule-based systems that involve the player's struggle to reach a goal [9,11]. In addition, games are based on challenge and curiosity [54]. Additionally, it is stated that one of the seven intrinsic motivation features to guide the processes in game design is a struggle [20]. According to Prensky (2001), it was stated that games have 12 elements that occupy the players, and these include winning situations and challenges (competition, conflict, etc.) (Prensky, 2001, as cited in [55]). As stated in the literature, the fact that there are elements such as struggle or competition among the features of digital games may have caused children to notice these features and to make similarities in this direction.

Seventh, the metaphors produced by children regarding digital games in all regions were collected in the "Value Images" category (f = 14, 3.32%). In this category, the Marmara Region (0.95%) ranks first. Next comes the Aegean Region (0.71%), then the Mediterranean and Eastern Anatolia regions (0.47%), and finally, the Central Anatolia, Southeastern Anatolia, and the Black Sea regions (0.23%). In support of these results, Hazar, Tekkurşun Demir and Dalkıran [37] stated that children evaluate digital games through value images in their study. In the literature, value is expressed as the thoughts, feelings, behaviours, and rules adopted or accepted in society. Children begin to assimilate values early in their lives. Values manifest themselves in playtime, an essential activity for the child [56]. Developing values and teaching behaviour is better achieved through digital games. Games can be used as an opportunity for children to gain and feel some values earlier [57]. Therefore, game-based learning models can be used and developed to guide the design and development of value-based digital games [56]. In light of this information, the analogies made by the children regarding digital games in the present study may indicate that they have a perception of value. In addition, the children's association of value images with daily life may show that they perceive play as a part of real life.

5. Conclusions

Today's children have a more comprehensive range of modern digital technologies, such as tablets and smartphones, than previous generations [1]. Digital technology has changed the platforms children can access to play and interact with materials. Digital technology must guide children to engage with information, navigate ideas, and represent their thoughts [4]. The advent of the digital age is changing the available game resources and improving different game genres. Among these game types, the digital game is accepted as a new game category [3]. Digital games are tools that have meaning and interest for children and are designed to fit their talents and desire to play and explore. In this way, children can develop critical thinking and creativity, develop representations, identify connections through actions, solve problems, gain cognitive skills, and, ultimately, construct knowledge [9]. In this respect, digital games have become an essential part of contemporary culture, and at the same time, children's games with technologies have turned into a controversial activity. Therefore, digital games have become a new topic of educational research [3].

Contemporary children benefit from various digital advances, including a new form of play called "digital play". Following the digital culture of children worldwide, numerous reports were conducted in different countries focusing on children's use of digital media. Most of these studies focus on children's use, participation, attitude, understanding, and interaction with digital technology [2]. International studies show that children use digital technology daily. Australian parents report that a third of preschool children have access to a tablet or smartphone and spend up to 26 h using that device per week [4].

For this reason, digital gaming is accepted as a form of gaming. Unlike traditional play activities, digital play was defined as a context in which children use digital technologies [8]. Recently, a growing interest has been in investigating the link between digital games and children. Therefore, more research is needed in this direction. However, gaining young children's perspectives is difficult, so more research should focus on digital games and young children. From this point of view, this research aims to use metaphors to examine the views of preschool children studying in different regions of Turkey on digital play.

Based on the study's purpose, 421 metaphors were obtained in this study. These metaphors produced by children were collected in the following seven categories for all regions: "Nature (Vehicle/Animal/Object) Images, Technology Images, Fantasy/Supernatural Images, Education Images, Affective/Motivational Images, Struggle Images, and Value Images".

It was determined that children developed metaphors in the category of "Nature Images" the most and "Value Images" the least. When evaluated based on regions, the Black Sea Region ranked first in the "Fantasy/Supernatural Images and Affective/Motivational Images" categories. In contrast, the Central Anatolia Region ranked first in the "Technology Images and Education Images" categories, and the Marmara Region ranked first in the "Nature Images and Value Images" categories. In addition, it was determined that the Southeast Anatolia Region ranked first in the "Struggle Images" category.

In line with the research results, it was seen that preschool children convey their feelings and thoughts about digital games through different metaphors. Thanks to the metaphors produced, meaningful information about how the children conceptualized the learned information was obtained. In short, it was seen that children used metaphors to express digital games from different perspectives. In light of all this information, asking children to directly produce metaphors about digital play and questioning the reason for this is considered the best way to evaluate them. The answers given by the children in this process show that they could produce detailed descriptions of digital games. Therefore, this study contributes to a more holistic understanding of how children make meanings out of digital games. As a result, it was determined that the students expressed digital games from different perspectives through metaphors.

6. Recommendations

In similar studies, using perceptions, interviews, and observation techniques related to digital games can be essential in obtaining more in-depth and comprehensive data about the subject. From another point of view, the data collected were obtained from children in the same age group. Comparisons can be made by collecting data from different grade levels. Findings containing multi-dimensional explanations for digital games can be reached by working with the parents of children attending preschool education institutions in different regions. Scientific research can be conducted to reveal the content of the games played by children and their contribution to education.

7. Limitations

Due to the nature of qualitative research, the inability to make generalizations is a limitation. In addition, this research is limited to the data collected from classroom teachers and the use of digital tools to send these data.

Author Contributions: Conceptualization, methodology, visualization, and writing, E.Y.A.; validation, review and editing, and supervision, S.P.; review and editing, and supervision, M.K. All authors have read and agreed to the published version of the manuscript.

Funding: This research received no external funding.

Institutional Review Board Statement: The study was conducted following the Declaration of Helsinki and approved by the Düzce University Scientific Research and Publication Ethics Committee (E-78187535-050.06-290724).

Data Availability Statement: The datasets used and/or analyzed during the current study are available from the corresponding author upon reasonable request.

Conflicts of Interest: The authors declare no conflict of interest.

References

1. Marsh, J.; Plowman, L.; Yamada-Rice, D.; Bishop, J.; Scott, F. Digital play: A new classification. *Early Years* **2016**, *36*, 242–253. [CrossRef]
2. Dong, P.I. Exploring Korean parents' meanings of digital play for young children. *Glob. Stud. Child.* **2018**, *8*, 238–251. [CrossRef]
3. Wang, X.; Rahman, M.N.A. An exploration of Chinese parental attitudes toward the digital play practices of children aged between 3 and 6 years old. *J. Posit. Sch. Psychol.* **2022**, *6*, 1012–1031.
4. Mantilla, A.; Edwards, S. Digital technology use by and with young children: A systematic review for the statement on young children and digital technologies. *Australas. J. Early Child.* **2019**, *44*, 182–195. [CrossRef]
5. Taghizadeh, M.; Vaezi, S.; Ravan, M. Digital games, songs and flashcards and their effects on vocabulary knowledge of Iranian preschoolers. *Int. J. Engl. Lang. Transl. Stud.* **2017**, *5*, 156–171.
6. Nevski, E.; Siibak, A. The role of parents and parental mediation on 0–3 year olds' digital play with smart devices: Estonian parents' attitudes and practices. *Early Years* **2016**, *36*, 227–241. [CrossRef]
7. Kinzie, M.B.; Joseph, D.R. Gender differences in game activity preferences of middle school children: Implications for educational game design. *Educ. Technol. Res. Dev.* **2008**, *56*, 643–663. [CrossRef]
8. Disney, L.; Geng, G. Investigating young children's social interactions during digital play. *Early Child. Educ. J.* **2022**, *50*, 1449–1459. [CrossRef]
9. Nikiforidou, Z. Digital games in the early childhood classroom: Theoretical and practical considerations. *Digit. Child. Technol. Child. Everyday Lives* **2018**, *22*, 253–265. [CrossRef]
10. Wu, C.S.T.; Fowler, C.; Lam, W.Y.Y.; Wong, H.T.; Wong, C.H.M.; Yuen Loke, A. Parenting approaches and digital technology use of preschool age children in a Chinese community. *Ital. J. Pediatr.* **2014**, *40*, 44. [CrossRef]
11. Lieberman, D.A.; Fisk, M.C.; Biely, E. Digital games for young children ages three to six: From research to design. *Comput. Sch.* **2009**, *26*, 299–313. [CrossRef]
12. Sivrikova, N.V.; Ptashko, T.G.; Perebeynos, A.E.; Chernikova, E.G.; Gilyazeva, N.V.; Vasilyeva, V.S. Parental reports on digital devices use in infancy and early childhood. *Educ. Inf. Technol.* **2020**, *25*, 3957–3973. [CrossRef]
13. Biricik, Z.; Abdulkadir, A. Geleneksleden dijitale değişen oyun kavramı ve çocuklrda oluşan dijital oyun kültürü. *Gümüşhane Üniversitesi İletişim Fakültesi Elektron. Derg.* **2021**, *9*, 445–469. [CrossRef]
14. Veresov, N.; Veraksa, N. Digital games and digital play in early childhood: A cultural-historical approach. *Early Years* **2022**, 1–13. [CrossRef]

15. Wooldridge, M.B.; Shapka, J. Playing with technology: Mother–toddler interaction scores lower during play with electronic toys. *J. Appl. Dev. Psychol.* **2012**, *33*, 211–218. [CrossRef]
16. Mustola, M.; Koivula, M.; Turja, L.; Laakso, M.-L. Reconsidering passivity and activity in children's digital play. *New Media Soc.* **2018**, *20*, 237–254. [CrossRef]
17. Plowman, L.; Stephen, C. Children, play, and computers in preschool education. *Br. J. Educ. Technol.* **2005**, *36*, 145–157. [CrossRef]
18. Li, M.-C.; Tsai, C.-C. Game-based learning in science education: A review of relevant research. *J. Sci. Educ. Technol.* **2013**, *22*, 877–898. [CrossRef]
19. Verenikina, I.; Kervin, L.; Rivera, M.C.; Lidbetter, A. Digital play: Exploring young children's perspectives on applications designed for preschoolers. *Glob. Stud. Child.* **2016**, *6*, 388–399. [CrossRef]
20. Lee, M.J.; Eustace, K.; Fellows, G.; Bytheway, A.; Irving, L. Rochester Castle MMORPG: Instructional gaming and collaborative learning at a Western Australian school. *Australas. J. Educ. Technol.* **2005**, *21*, 446–469. [CrossRef]
21. Stieler-Hunt, C.; Jones, C.M. Educators who believe: Understanding the enthusiasm of teachers who use digital games in the classroom. *Res. Learn. Technol.* **2015**, *23*, 26155. [CrossRef]
22. Slutsky, R.; DeShetler, L.M. How technology is transforming the ways in which children play. *Early Child Dev. Care* **2017**, *187*, 1138–1146. [CrossRef]
23. Edwards, S. Digital play in the early years: A contextual response to the problem of integrating technologies and play-based pedagogies in the early childhood curriculum. *Eur. Early Child. Educ. Res. J.* **2013**, *21*, 199–212. [CrossRef]
24. Mertala, P.; Meriläinen, M. The best game in the world: Exploring young children's digital game–related meaning-making via design activity. *Glob. Stud. Child.* **2019**, *9*, 275–289. [CrossRef]
25. Common Sense Media. *The Common Sense Census: Media Use by Kids Age Zero to Eight*; Common Sense Media: San Francisco, CA, USA, 2013.
26. Granic, I.; Lobel, A.; Engels, R. The benefits of playing video games. *Am. Psychol.* **2014**, *69*, 66–78. [CrossRef]
27. Johnson, J.E.; Christie, J.F. Play and digital media. *Comput. Sch.* **2009**, *26*, 284–289. [CrossRef]
28. Elson, M.; Ferguson, C.J. Twenty-five years of research on violence in digital games and aggression. *Eur. Psychol.* **2014**, *19*, 33–46. [CrossRef]
29. Zhu, R.; Gopnik, A. Preschoolers and Adults Learn From Novel Metaphors. *Psychol. Sci.* **2023**, *34*, 696–704. [CrossRef]
30. Thibodeau, P.H.; Hendricks, R.K.; Boroditsky, L. How linguistic metaphor scaffolds reasoning. *Trends Cogn. Sci.* **2017**, *21*, 852–863. [CrossRef]
31. Fraenkel, J.R.; Wallen, N.E.; Hyun, H.H. *How to Design and Evaluate Research in Education*; McGraw-Hill: New York, NY, USA, 2012; Volume 7.
32. Obilor, E.I. Convenience and purposive sampling techniques: Are they the same. *Int. J. Innov. Soc. Sci. Educ. Res.* **2023**, *11*, 1–7.
33. Barraza, L. Children's drawings about the environment. *Environ. Educ. Res.* **1999**, *5*, 49–66. [CrossRef]
34. Elo, S.; Kääriäinen, M.; Kanste, O.; Pölkki, T.; Utriainen, K.; Kyngäs, H. Qualitative content analysis: A focus on trustworthiness. *SAGE Open* **2014**, *4*, 2158244014522633. [CrossRef]
35. Saban, A. Okula ilişkin metaforlar. Kuram ve uygulamada eğitim yönetimi [Metaphors related to school. Educational management in theory and practice]. *Kuram Ve Uygulamada Eğitim Yönetimi* **2008**, *55*, 459–496.
36. Miles, M.B.; Huberman, A.M. *Qualitative Data Analysis: An Expanded Sourcebook*; Sage: Thousand Oaks, CA, USA, 1994.
37. Hazar, Z.; Tekkurşun Demir, G.; Dalkıran, H. Ortaokul öğrencilerinin geleneksel oyun ve dijital oyun algılarının incelenmesi: Karşılaştırmalı metafor çalışması. Sportmetre Beden Eğitimi ve Spor Bilimleri Dergisi [Examination of secondary school students' perceptions of traditional games and digital games: A comparative metaphor study]. *Spormetre J. Phys. Educ. Sport Sci.* **2017**, *15*, 179–190. [CrossRef]
38. Paslı, A.M. Doğal Çevre, Kent ve Çocuk İlişkisini Yeniden Kurmak "İskandinavya'da Doğa Temelli Eğitim ve İsveç Orman Okulu Örneği". Yüksek Lisans, İstanbul Şehir Üniversitesi [Re-Establishing the Relationship between the Natural Environment, the City and the Child "Nature-Based Education in Scandinavia and the Swedish Forest School Example". Master's Thesis, İstanbul Şehir University, İstanbul, Turkey, 2019.
39. Davis, J. Young children, environmental education and the future. In *Education And The Environment*; World Education Fellowship: London, UK, 1998; pp. 141–155.
40. Johnston, K. Engagement and ımmersion in digital play: Supporting young children's digital wellbeing. *Int. J. Environ. Res. Public Health* **2021**, *18*, 10179. [CrossRef]
41. Bers, M.U. *Designing Digital Experiences for Positive Youth Development: From Playpen to Playground*; Oxford University Press: Oxford, UK, 2012.
42. Flynn, R.M.; Richert, R.A. Parents support preschoolers' use of a novel interactive device. *Infant Child Dev.* **2015**, *24*, 624–642. [CrossRef]
43. Bülbül, H. Dijital oyunlar üzerine. *TRT Akad.* **2022**, *7*, 1173–1178. [CrossRef]
44. Darıcı, S. Dijital oyunlarda kullanılan subliminal mesajların gerçeklik algısı üzerindeki etkilerine yönelik bir çalışma: Gerçeklik eşiği kavramı [A study on the effects of subliminal messages used in digital games on perception of reality: The concept of reality threshold]. *Electron. Turk. Stud.* **2015**, *10*, 181–202.
45. Murray, J.H. Toward a cultural theory of gaming: Digital games and the co-evolution of media, mind, and culture. *Pop. Commun.* **2006**, *4*, 185–202. [CrossRef]

46. Subbotsky, E. Games with the supernatural: Magical reality in the everyday life. *Sententia Eur. J. Humanit. Soc. Sci.* **2017**, *2*, 38–55. [CrossRef]
47. Papadakis, S.; Kalogiannakis, M. Evaluating the effectiveness of a game-based learning approach in modifying students' behavioural outcomes and competence, in an introductory programming course. A case study in Greece. *Int. J. Teach. Case Stud.* **2019**, *10*, 235–250. [CrossRef]
48. Fang, M.; Tapalova, O.; Zhiyenbayeva, N.; Kozlovskaya, S. Impact of digital game-based learning on the social competence and behaviour of preschoolers. *Educ. Inf. Technol.* **2022**, *27*, 3065–3078. [CrossRef]
49. Thai, A.M.; Lowenstein, D.; Ching, D.; Rejeski, D. *Game Changer: Investing in Digital Play to Advance Childrens Learning and Health*; The Joan Ganz Cooney Center at Sesame Workshop: New York, NY, USA, 2009.
50. Kalogiannakis, M.; Papadakis, S.; Zourmpakis, A.-I. Gamification in science education. A systematic review of the literature. *Educ. Sci.* **2021**, *11*, 22. [CrossRef]
51. Tahir, R.; Wang, A.I. State of the art in game based learning: Dimensions for evaluating educational games. In Proceedings of the European Conference on Games Based Learning, Graz, Austria, 5–6 October 2017; Academic Conferences International Limited: Cambridge, MA, USA, 2017; pp. 641–650.
52. Wu, Q.; Zhang, J.; Wang, C. The effect of English vocabulary learning with digital games and its influencing factors based on the meta-analysis of 2160 test samples. *Int. J. Emerg. Technol. Learn.* **2020**, *15*, 85–100. [CrossRef]
53. Papanastasiou, G.P.; Drigas, A.S.; Skianis, C. Serious games in preschool and primary education: Benefits and impacts on curriculum course syllabus. *Int. J. Emerg. Technol. Learn.* **2017**, *12*, 44–56. [CrossRef]
54. Hussain, H.; Embi, Z.C.; Hashim, S. *A Conceptualized Framework for Edutainment*; InSite-Where Parallels Intersect; Informing Science: Santa Rosa, CA, USA, 2003; pp. 1077–1083.
55. Butler, Y.G. Motivational elements of digital instructional games: A study of young L2 learners' game designs. *Lang. Teach. Res.* **2017**, *21*, 735–750. [CrossRef]
56. Tilvawala, K.; Sundaram, D.; Myers, M.D. Values-Based Digital Games: Designing a digital game platform to foster sustainability in early childhood. In Proceedings of the 2016 Pacific Asia Conference on Information Systems (PACIS), Chiayi, Taiwan, 27 June–1 July 2016; Volume 368, p. 368.
57. Gündüz, M.; Aktepe, V.; Uzunoğlu, H.; Gündüz, D.D. Okul öncesi dönemdeki çocuklara eğitsel oyunlar yoluyla kazandırılan değerler. Muğla Sıtkı Koçman Üniversitesi Eğitim Fakültesi Dergisi [Values gained by preschool children through educational games]. *J. Muğla Sıtkı Koçman Univ. Fac. Educ.* **2017**, *4*, 62–70. [CrossRef]

Disclaimer/Publisher's Note: The statements, opinions and data contained in all publications are solely those of the individual author(s) and contributor(s) and not of MDPI and/or the editor(s). MDPI and/or the editor(s) disclaim responsibility for any injury to people or property resulting from any ideas, methods, instructions or products referred to in the content.

Article

Evaluating Video Games as Tools for Education on Fake News and Misinformation

Ruth S. Contreras-Espinosa [1,*] and Jose Luis Eguia-Gomez [2]

[1] Data and Signal Processing Group, University of Vic—Central University of Catalonia, c/de la Laura 13, 08500 Vic, Spain
[2] Departamento de Ingeniería de Proyectos y de la Construcción, Universitat Politècnica de Catalunya, 08029 Barcelona, Spain; eguia@ege.upc.edu
* Correspondence: ruth.contreras@uvic.cat

Abstract: Despite access to reliable information being essential for equal opportunities in our society, current school curricula only include some notions about media literacy in a limited context. Thus, it is necessary to create scenarios for reflection on and a well-founded analysis of misinformation. Video games may be an effective approach to foster these skills and can seamlessly integrate learning content into their design, enabling achieving multiple learning outcomes and building competencies that can transfer to real-life situations. We analyzed 24 video games about media literacy by studying their content, design, and characteristics that may affect their implementation in learning settings. Even though not all learning outcomes considered were equally addressed, the results show that media literacy video games currently on the market could be used as effective tools to achieve critical learning goals and may allow users to understand, practice, and implement skills to fight misinformation, regardless of their complexity in terms of game mechanics. However, we detected that certain characteristics of video games may affect their implementation in learning environments, such as their availability, estimated playing time, approach, or whether they include real or fictional worlds, variables that should be further considered by both developers and educators.

Keywords: fake news; media literacy; video games; media literacy skills

Citation: Contreras-Espinosa, R.S.; Eguia-Gomez, J.L. Evaluating Video Games as Tools for Education on Fake News and Misinformation. *Computers* **2023**, *12*, 188. https://doi.org/10.3390/computers12090188

Academic Editors: Carlos Vaz de Carvalho, Hariklia Tsalapatas and Ricardo Baptista

Received: 27 August 2023
Revised: 17 September 2023
Accepted: 18 September 2023
Published: 21 September 2023

Copyright: © 2023 by the authors. Licensee MDPI, Basel, Switzerland. This article is an open access article distributed under the terms and conditions of the Creative Commons Attribution (CC BY) license (https://creativecommons.org/licenses/by/4.0/).

1. Introduction

Access to credible information is crucial for achieving equal opportunities in our society. The proliferation of fake news in media and on social media has been the subject of concern in recent years [1,2]. Such news represents false information presented as truth, disseminated through the media, particularly online, with the intention to deceive and manipulate the public opinion [3,4]. Sunstein [5] states that the proliferation of information on social media has impacted our ability to access a wide range of political perspectives and argues that some political actors take advantage of this dynamic to influence our behaviors. As the volume of information available online continues to grow, and with our ease of access to information through the Internet [6], our ability to distinguish between truthful and false information becomes increasingly important [7,8].

The origin of fake news can be traced back to antiquity, with the Romans already disseminating false information to manipulate the population [9]. Today, its impact is even more drastic. Fake news has become a political tool used to win elections, influence public opinion, and generate disinformation [10,11]. Moreover, the impact of such news on society can be highly negative when dealing with health, politics, or security issues. According to Bin Naeem and Bhatti [12], the spread of fake news about COVID-19 has led to a decline in the adoption of preventive measures and has increased confusion and fear among the population, possibly resulting in a greater spread of the virus and an increase in cases. Similarly, fake news may influence political decision-making [8,13], with significant implications for social stability [14]. Addressing these issues requires promoting media literacy and critical thinking among the population [15,16] to help identify fake news or other types of misinformation [17].

Teachers have traditionally instructed students on journalism concepts, news-writing, and the differences between news genres [18]. School curriculums typically include some notions about news, but within a limited context, and most documented classroom interventions are the consequence of a personal commitment by teachers [1]. However, it is insufficient to seek formulas for incorporating media literacy into a curriculum already laden with academic knowledge; it is necessary to create new spaces for reflection on and well-founded analyses of misinformation [16]. This is paramount in the current landscape characterized by a lack of scenarios that promote reflection and initiatives that address the need for media literacy [19]. Given the need to cultivate the youth's journalistic literacy skills in our age of misinformation [20], without spaces for reflection [21] and with current initiatives not fully addressing these needs [22,23], it is necessary to re-think the strategies for delivering media literacy skills. Using video games can be a powerful approach to fostering these [24], as specific learning content (such as fake news, digital privacy, personal media habits, and practical media skills) may be seamlessly incorporated into game design to achieve multiple learning outcomes and the cultivation of competencies transferable to real life [25].

However, Glas et al. [26] point out that despite there being many games aimed at teaching about fake news or privacy, among others, it is not clear how they actually serve as educational tools for media literacy, which competencies or content they focus on, and how these are delivered through game design. Most studies are limited to the scope of one game and its effects or centered around one particular aspect of media literacy. Moreover, there is a lack of studies focusing on the practical aspects of game implementation in educational settings.

The purpose of this research is to contribute to filling this research gap by conducting a quantitative analysis of a pool of media literacy video games while adopting a broad and multifaceted understanding of media literacy. This will (1) provide an overview of which media literacy competencies and learning outcomes are covered by existing media literacy video games in relation to a curriculum of reference, (2) generate insights into the strategies used to promote these competencies through game design, and (3) understand how certain characteristics of these video games may influence their implementation as pedagogical tools in educational settings such as schools. Additionally, we provide a set of detailed practical examples of how content and design in video games may contribute to delivering learning experiences. For this study, we utilize the definition of "media literacy games" as games whose purpose extends beyond entertainment, focusing on media literacy, and which, through their design, are explicitly oriented towards one or more of the key themes, skills, or competencies associated with media literacy, thus connecting with a broader field of research centered on the use of digital or tabletop games for education or behavioral change [27].

This work is part of the initial phase of the YO-MEDIA project (Youngsters' Media Literacy in Times of Crisis) funded by the Calouste Gulbenkian Foundation (269094), designed to examine how games may be used to enhance media literacy among young people. This research seeks to shed light on the effectiveness and potential of media literacy video games as educational tools, addressing a critical need in our information-driven society. It is a vital component of the YO-MEDIA project.

2. Theoretical Framework

2.1. Media Literacy

Media literacy focuses on understanding news through the combination of journalism, citizenship, and technological concepts [28] and helps individuals become adept readers and producers [29]. Despite media literacy historically being understood as the ability to read, watch, listen, and comprehend media, the evolution of the traditional media landscape linked to new digital technologies has brought about a change [26]. Today, the concept of media literacy cannot be confined to an instrumental or functional literacy driven by the mere retrieval of information, but also needs to be construed as the capacity for the critical understanding of and active participation in the media [30]. As Buckingham [30] highlights, media literacy is much more than simply "accessing" these media or utilizing them as tools for learning. Instead, it entails cultivating a broader critical understanding that delves into the attributes of these media while also considering their social, economic, and cultural implications.

We understand media literacy from the perspective of "game literacy", as the manner in which players develop and apply different skills; not only as cognitive competencies that build informal learning skills by enabling players to think, converse, and read [31], but as competencies through which players also acquire critical thinking abilities, make decisions, and take action within a dynamically evolving environment [32] and have abundant opportunities to carry out informal learning activities, and, in some cases, achieve formal learning [33]. Ultimately, the capacity to engage with a game often encompasses more than mere familiarity with its rules, objectives, and interface; it also involves the aptitude to partake in social and communicational practices [34], regardless of whether the game is played on a computer, mobile device, or console.

2.2. Games and Media Literacy

In the growing field of gaming research, scholars have been addressing literacy practices and learning during gameplay in a range of games and game genres [31,35,36]. Gee [35] discussed the literacy possibilities offered by games, the kind of experiential learning they provide, and how players can engage with topics and concepts not easily accessible through conventional learning approaches.

Games can potentially serve as an effective tool for improving media literacy and the ability to distinguish between true and false information [37]. They have been shown to enhance people's ability to process and comprehend information and help develop critical thinking skills [38]. They also provide a safe environment for experimentation, serving as an effective tool for media literacy [39,40]. Video games have not only proven to be useful in acquiring these skills, but in general, they are tools that increase motivation [41], including for self-learning [42], thanks to game mechanics that can affect engagement [43]. They attract the player with the narrative, influencing their interest and fulfillment, key factors towards increasing commitment and satisfaction [44].

Some video games have been shown to help improve performance in critical thinking tests, like Minecraft, which requires players to think creatively and strategically to solve construction problems [45], or Portal 2, a puzzle video game in which players must find solutions to riddles [46]. Other benefits include improvements in problem-solving and decision-making [47]. Moreover, some have been specifically designed to foster media literacy and help distinguish between genuine and fake news, such as Bad News or Factitious—Pandemic Edition [48]. But their main advantage lies in their ability to demonstrate how things work in a practical manner by engaging users in a vivid experience [24]. However, it should not be surprising that their effectiveness relies on the inherent possibilities of each game [38].

Games are also media, and this becomes more pertinent when considering their extensive utilization by the youth [46]. This implies that the analysis of games requires new and distinctive methods that cannot simply be transferred from other media, though this is equally the case when we compare television and books, for example. While some elements are shared across media, others are distinctive to a specific medium [30].

2.3. Media Literacy Frameworks for Education

To enhance media literacy, it is necessary to tackle the underlying problems and promote a culture of fact-checking and critical thinking. Some authors have suggested that school education could include programs about responsible social media use and the importance of verifying information before sharing it [49]. Tools and technologies such as FactCheck.org [50] could be used to verify the authenticity of news, and pedagogical strategies could be implemented, like media analysis, evaluation of information quality, and understanding of media rhetoric and biases [51].

Other authors have proposed their own theoretical frameworks for media literacy in educational programs [16,52–54], and there have also been institutional efforts toward this end, such as the Digital Citizenship Education Handbook by the Council of Europe [55], written from a theoretical perspective. These initiatives and proposals are of a general nature and do not tackle media literacy in education from a practical and implementable point of view. The lack of clearly defined learning objectives hinders, to a certain extent, their implementation in formal educational settings, which are typically curriculum-driven and need to use learning objectives as evaluation measures.

The European Association for Viewers Interests (EAVI) [56] produced the report Get Your Facts Straight!: Toolkit for Educators and Training Providers, which includes a goal-oriented methodology and curriculum that breaks down media literacy education on misinformation and fake news into three main learning areas, each corresponding to a module that includes several learning outcomes, as shown in Table 1. Despite not being specifically designed to be implemented through video games, for this study, we will use its objective-driven educational methodology on media literacy as a reference to evaluate which learning outcomes the analyzed sample of video games contribute to delivering.

Table 1. Structure of Get Your Facts Straight!: Toolkit for Educators and Training Providers.

Module	Learning Outcomes
Module 1 What is disinformation	1.0 I can explain the difference between information and disinformation. 1.1 I can identify the types of misleading news.
Module 2 How social media make money and why disinformation and propaganda are vastly present on social media	2.1 I understand the consequences of believing and sharing false information for the society and for myself. 2.2 I understand the reasons why disinformation is published with the intention to mislead me. 2.3 I know there are some political or commercial interests that try to affect my behavior online. 2.4 I have a general idea about how algorithms affect what we see online.
Module 3 How to recognize and react to disinformation	3.1 I understand what are some examples of credible sources of information. 3.2 I know how to check information and I know the changes in the media landscape. 3.3 I know how to defend myself from threats and risks on social media. 3.4 I know what I can do to be a positive and responsible social media user.

3. Materials and Methods

This work applies the deductive method to provide, after an exploratory phase, a quantitative–descriptive view of the existing video games about media literacy in order to understand their usefulness for teaching purposes and their ease of implementation in learning environments. Figure 1 illustrates a summary of the methodology followed during the present study.

We selected a sample of video games after reviewing the JournalismGames.org database from the Interactive Media Department of the School of Communication at the University of Miami and after carrying out searches on the STEAM platform and on Google with the keywords "fake news", "media literacy", and "media". We discarded any video game that did not focus on addressing media literacy, the use of fake news, or tangentially fought misinformation. We prioritized video games that provided a deep and meaningful approach that would be of use for teaching purposes, offering an immersive and educational experience.

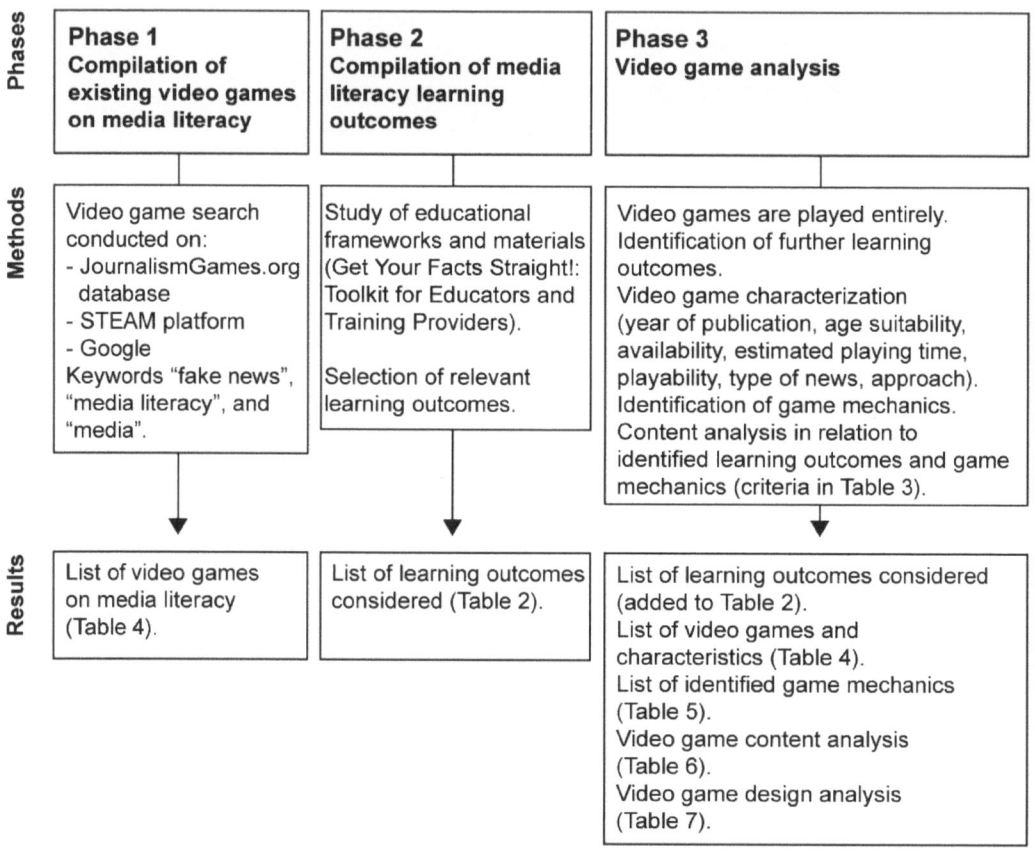

Figure 1. Summary of the methodology.

To understand the practicalities of how video games may be incorporated into learning environments, we analyzed the main characteristics of the selected sample regarding their playability, the type of news they depict (real vs. fictional worlds), their approach (constructivist/behaviorist), availability (whether they were free or had a fee, and their availability on the STEAM platform), age suitability, and estimated playing time. We also studied the games' content in relation to the curriculum set out in Get Your Facts Straight!: Toolkit for Educators and Training Providers [43], specifically in reference to the 8 learning outcomes suggested in Modules 2 and 3 of the curriculum (see Table 1), by playing each game entirely. We did not consider the learning outcomes included in Module 1, as these are focused on more theoretical content that cannot be delivered through video games alone, while Modules 2 and 3 revolve around the student and clearly describe educational objectives that may be delivered by video games in a measurable way. While playing and studying the games, we identified an additional possible learning outcome in the pool of video games (9. "Know how to create fake news"). The complete list of learning outcomes considered for the analysis is shown in Table 2. Last, we compiled a list of the game mechanics present in the sample.

Video game complexity and usefulness for media literacy teaching purposes were evaluated by scoring the content (learning outcomes considered) and design (game mechanics identified) on a scale from 0 to 3, following the criteria set out in Table 3.

Table 2. List of learning outcomes considered.

No.	Learning Outcomes Considered
1	Understand the consequences of believing and sharing false information for society and the individual.
2	Understand the reasons why disinformation is published with the intention to mislead.
3	Know that some political or commercial interests try to affect online behavior.
4	Have a general idea about how algorithms affect what we see online.
5	Understand what some examples of credible sources of information are.
6	Know how to check information and know the changes in the media landscape.
7	Know how to defend oneself from threats and risks on social media.
8	Know how to be a positive and responsible player on social media.
9	Know how to create fake news.

Table 3. Evaluation criteria for learning outcomes and game mechanics.

No.	Content: Learning Outcomes	Design: Game Mechanics
0	Non-consideration of learning outcome	Not present
1	Learning outcome addressed indirectly	Low impact on the game
2	Learning outcome addressed metaphorically	Relative impact on the game
3	Learning outcome addressed explicitly	High impact on the game

4. Results

4.1. Media Literacy Video Games and Characteristics

We identified 24 video games on media literacy and conducted a comprehensive evaluation of their general features. Table 4 summarizes the main characteristics of each video game, including its name, year of publication, age suitability, availability, estimated playing time, whether it is re-playable, the inclusion of real news in the content, and the learning approach.

We identified six re-playable games within the analyzed sample. Most video games (n = 14) used real news, while the rest focused on fictional worlds. We also analyzed the games' type of approach to determine whether the video games provided the player with the necessary tools to build their own procedures to solve a problematic situation, through a constructivist approach, or whether the games proposed activities with content that had to be learned by the player, promoted by appropriate stimuli at each moment, in a behaviorist approach. We found that the majority of the games (n = 18) presented a behaviorist approach. A large number of the games studied were intended for audiences above 12 years old (n = 11), while 12 were intended for audiences over 14 (n = 6) and 16 years old (n = 6). Only one of the games was aimed at players above 18 (G21). Most video games were free of charge (n = 17), but all the games found on the STEAM platform had a fee (n = 7). The majority of the games reported an estimated playing time of 30+ min or below (n = 15), while two games had estimated playing times of 45+ min, and seven games had estimated playing times of 60+ min.

Table 4. Media literacy video games and characteristics.

No.	Video Game	Year	Age	Availability	Playing Time (min)	Characteristics
G1	The Republica Times	2012	12+		15+	+ º
G2	Interland	2017	12+		30+	- º
G3	Bad News	2017	12+		30+	+ º
G4	Fake It to Make It	2017	16+		30+	* ^ -
G5	Go Viral!	2018	12+		30+	+ º
G6	Fakey	2018	12+		15+	+ º
G7	Post Facto	2018	12+		15+	+ º
G8	Cranky Uncle	2020	12+		15+	+ º
G9	Harmony Square	2020	12+		20+	+ º
G10	Choose your own fake news	2020	12+		20+	+ º
G11	Adventures of Literatus	2020	14+		30+	- º
G12	Stop the troll	-	14+		20+	+ º
G13	BBC iReporter	2020	14+		30+	^ -
G14	Cat Park	2022	14+		60+	^ -
G15	EU vs. Disinfo Quiz	2022	14+		15+	+ º
G16	NewsFeed Defenders	2023	16+		60+	+ º
G17	Julia: A Science Journey	2023	12+		20+	+ º
G18	Political Animals	2016	16+	ab	60+	* - ^
G19	Headliner	2017	14+	ab	45+	- º
G20	No Place for the Dissident	2020	16+	ab	60+	*- º
G21	Floor 13: Deep State	2020	18+	ab	60+	*- º
G22	Influence Inc.	2022	16+	ab	60+	*- ^
G23	Power & Revolution 2022 Edition	2022	16+	ab	60+	* +^
G24	Forge of Destiny	2023	12+	ab	45+	+ º

a Has a fee, b Available on STEAM, * Re-playable, + Uses real news, - Uses fictional news, ^ Constructivist approach, º Behaviorist approach.

4.2. Identification of Game Mechanics

After playing each video game entirely, we compiled a list of the six game mechanics identified in the analyzed pool (Table 5).

Table 5. Identified game mechanics.

No.	Game Mechanics
1	Allows players to investigate the news and find out which are real and which are fake.
2	Allows players to create news and includes consequences for the player if they spread fake news.
3	Challenges the player to spread fake news created by others.
4	Shows the player different news items (real and fake) and asks them to select the real ones.
5	Allows players to create fake news and spread them within the game.
6	Quizzes the player by challenging their knowledge of political or current topics.

4.3. Most Complete Video Games for Media Literacy

Table 6 displays the list of video games analyzed and the scores assigned for each of the learning outcomes (see Table 2) according to the scale presented in Table 3, as well as the total number of learning outcomes within each game and the total score for the whole amount of learning outcomes in each video game (total game score). It also shows in how many video games each learning outcome is included (outcome presence) as well as the total score of each learning outcome in the whole pool of video games (total score per outcome). We also included the number of times each learning outcome was assigned each scale value in the pool of games studied.

Table 6. Video games and learning outcomes.

Game/Mechanic	1	2	3	4	5	6	7	8	9	No. of Outcomes	Total Game Score
G22	3	3	3	3	2	3	3	3	3	9	26
G4	2	3	3	3	3	3	2	3	2	9	24
G14	2	2	2	3	3	3	3	3	3	9	24
G3	2	2	2	3	3	2	3	3	3	9	23
G13	3	3	3	1	3	3	1	3	2	9	22
G5	2	2	1	3	3	2	3	2	3	9	21
G16	1	1	1	3	3	3	3	3	3	9	21
G21	3	3	3	2	1	2	2	3	2	9	21
G19	2	3	1	1	1	3	3	3	3	9	20
G24	2	2	2	3	3	2	2	2	2	9	20
G20	3	3	2	1	1	1	2	3	2	9	18
G23	3	3	2	1	1	2	1	2	1	9	16
G9	1	1	1	3	2	1	2	1	3	9	15
G11	1	1	1	1	2	2	2	1	2	9	13
G10	3	3	2		1	1	1	2	1	8	14
G18	2	2	2	2		2	2		1	7	13
G8				1	3	3		2	3	5	12
G7					3		3	2	3	4	11
G6	1						3	3	2	4	9
G17					2		2	2	2	4	8
G15			3	3					2	3	8
G2	1				3		3			3	7
G1			1		1	1				3	3
G12					3	3				2	6
Outcome presence	18	16	17	18	21	20	20	19	20		
Total score per outcome	37	37	32	40	47	45	45	47	45		
Max. score (3)	6	8	4	10	11	9	8	11	8		
Med. score (2)	7	5	7	2	4	7	9	6	9		
Min. score (1)	5	3	6	6	6	4	3	2	3		

As shown in Table 6, more than half of the video games analyzed (n = 13) include the nine learning outcomes (see Table 2). When also considering the total score assigned to each video game according to the learning outcomes present, we can determine which items among these have the best total score and should, therefore, be suitable for a more complete and robust educational experience. Up to 11 video games include all the learning outcomes in quite a direct manner in general (total game score of 18 or above): these are the most comprehensive games identified in the sample. Moreover, G22 approaches the maximum total score with 26 points, making it the most robust game analyzed in terms of media literacy learning outcomes by explicitly meeting and addressing nearly all the outcomes except for one, which is addressed metaphorically. This video game would thus constitute the most complete media literacy educational tool available in the pool of games analyzed.

4.4. Most Common and Robust Learning Outcomes

According to Table 6, the outcome with the greatest presence in the video game pool is number 5 ("Understand what some examples of credible sources of information are"), which is included in 21 of the games analyzed. On the other hand, the least represented is outcome 2 ("Understand the reasons why disinformation is published with the intention to mislead"), present in 16 games.

The total score per outcome indicates which learning outcomes are generally better incorporated into the sample of video games analyzed. There are two outcomes with a score of 47 (maximum score is 72), which is the highest score achieved: outcomes 5 and 8 ("Understand what some examples of credible sources of information are" and "Know how to be a positive and responsible player on social media", respectively); these are the learning outcomes most explicitly addressed in the sample. This analysis also shows that outcome 9 ("Know how to create fake news") is present in a large number of the games (n = 20) and with a relatively high score (45 points), justifying its inclusion in the list of outcomes considered. The learning outcome with the lowest total score is number 3 ("Know that some political or commercial interests try to affect online behavior") with 32 points, also on the list of outcomes that appear in fewer games (n = 17).

Most outcomes are more often addressed explicitly than indirectly, except for number 3 ("Know that some political or commercial interests try to affect online behavior"). Five of the outcomes are predominantly addressed directly in the pool: the number of times an outcome is explicitly addressed surpasses the occasions when it is addressed metaphorically and greatly exceeds indirect treatment (learning outcomes 2, 4, 5, 6, and 8). Outcomes 8 ("Know how to be a positive and responsible player on social media"), 4 ("Have a general idea about how algorithms affect what we see online"), and 5 ("Understand what some examples of credible sources of information are") stand out, as their explicit treatment surpasses both metaphorical and indirect approaches combined.

4.5. Learning Outcomes and Game Mechanics

Table 7 displays the list of video games analyzed and the scores assigned for each of the mechanics identified (see Table 5) according to the scale presented in Table 3, as well as the total amount of mechanics within a game and the total score for the whole number of mechanics in each video game (total game score). It also shows in how many video games each mechanic is included (mechanic presence) as well as the total score of each mechanic in the whole pool of video games (total score per mechanic). We also included the number of times each mechanic was assigned each scale value in the pool of games studied.

As shown in Table 7, we found that six video games in the analyzed sample include the six mechanics identified (see Table 5). These games (G19, G20, G21, G22, G23, G24) also present the nine learning outcomes set out in Table 2 (see Table 6). It should be noted that, although a game may include all the mechanics detected, this does not guarantee that it will meet all the learning outcomes. For instance, G3 addresses the nine outcomes with only one mechanic (mechanic 1: "Allows players to investigate the news and find out which are real and which are fake"). Additionally, G4, G5, and G9 address the nine outcomes with only two mechanics (mechanic 2: "Allows players to create news and includes consequences for the player if they spread fake news" and mechanic 4: "Shows the player different news items (real and fake) and asks them to select the real ones"). The opposite is also true; there are games with a high variety of mechanics (n = 5) and, therefore, presumably have more complexity, but they address fewer than half of the objectives (such as G6).

Table 7. Video games and mechanics.

Game/Mechanic	1	2	3	4	5	6	No. of Mechanics	Total Game Score
G22	3	3	3	3	3	3	6	18
G21	3	2	2	3	3	2	6	15
G19	3	3	2	3	3	2	6	16
G24	3	3	2	3	3	2	6	16
G20	2	3	2	3	3	2	6	15
G23	2	3	2	2	3	2	6	14
G13	3	3	3	3	3		5	15
G16	3	3	3	3	3		5	15
G6	3		3	3	3	3	5	15
G8	1	2	1	1		3	5	8
G17	3		1	1	1	1	5	7
G14	3	3		3	3		4	12
G10	3		3	1	1		4	8
G11	3		3		1		3	7
G7	3	1		3			3	7
G18	3	1				2	3	6
G15	3			1	1	3	3	8
G9		3		3			2	6
G4		3		3			2	6
G5		3		3			2	6
G12	3		3				2	6
G2	3				2		2	5
G3	3						1	3
G1		1					1	1
Mechanic presence	20	16	15	19	13	11		
Total score per mechanic	56	40	34	47	33	25		
Max. score (3)	17	11	7	13	10	4		
Med. score (2)	2	2	2	2	0	6		
Med. score (1)	1	2	3	4	3	1		

The most frequently used mechanics are 1 ("Allows players to investigate the news and find out which are real and which are fake") and 4 ("Shows the player different news items (real and fake) and asks them to select the real ones"), which appear on 20 and 19 occasions, respectively. The least used mechanic is number 6 ("Quizzes the player by challenging their knowledge of political or current topics"), which appears in only 11 games. The most impactful mechanics in the pool of games are also 1 and 4, with total scores of 56 (maximum score is 72). The mechanic that seems to have a lower overall impact is also number 6, with a score of 25 points. In general, nearly all mechanics, when used, are employed with high impact in the video game in which they appear.

5. Discussion

In recent years, the number of games that address fake news and media literacy has grown, which aligns with the observations made by Adams et al. [1] and Shehata [2], who highlight a global concern. Game developers respond to these preoccupations by creating experiences that promote critical awareness. In the following subsections, we provide a comprehensive summary of the main characteristics of the video games focused on media literacy currently available on the market, including content (learning outcomes) and design (game mechanics), and discuss their suitability and usefulness for learning purposes. Last, we provide a series of examples of how learning outcomes are incorporated into video game content.

5.1. Characterization of Media Literacy Video Games

Certain game characteristics may influence the suitability and usefulness of video games for media literacy teaching and their ease of implementation in the classroom (see Table 4). For instance, despite re-playability being beneficial for educational purposes [57], we found that only six out of the twenty-four video games analyzed were re-playable. This variable is also closely linked to the estimated playing time, as most games with a longer duration (45 min or more) were also re-playable.

The estimated playing time also affects the extent to which video games can be implemented as a tool in the classroom. Shorter games (30 min or less) may be played during a typical one-hour class, while longer video games could be suitable, for example, as homework. We found that most of the games in the sample, fifteen in total, had a low estimated playing time and could easily be used in a classroom environment, while two of them had a longer duration (45 min or more) that could still enable their use in this setting with certain constraints. These video games could be incorporated into the resources and materials of Modules 2 and 3 from Get Your Facts Straight!: Toolkit for Educators and Training Providers [56], which suggest other resources such as videos, mobile games, websites, or slides to help achieve specific learning outcomes with an estimated time for group work and discussion with students between 10 and 40 min. Games with playing times above 60 min (n = 7) would be difficult to use in typical school settings.

Target audiences determine whether a video game can be used as a teaching tool within the educational system or not. Nearly all the video games were suitable for audiences within school and high school ages (n = 23). Games aimed at players over 18 years old could only be used in countries where the schooling age is extended (for instance, some Nordic countries) or in educational settings other than schools and high schools, such as at university level. Only one of the video games analyzed (G21) fell into the latter category; it is a complex simulator that showcases a dystopian thriller with game mechanics that force players to challenge authority to achieve their objectives and in which they must control people's opinions and will.

Other variables related to game availability (pricing and platform) may influence accessibility and, therefore, affect or hinder their implementation in the classroom. Free items are more accessible and more likely to be implemented. They can also easily be used for homework, as there is no additional cost for the student or parents and guardians. We found that the majority of the video games analyzed (n = 17) were free of charge. While the integration of media literacy topics into platforms such as STEAM increases their reach and promotes public awareness of the problem of misinformation in society, all the games that were available on the STEAM platform (n = 7) required a fee. This may represent an obstacle for their implementation: first, due to a fee that may affect their incorporation into school settings but also prevent their usage at home for those who cannot afford it; and second, because STEAM is a platform that requires individual accounts and subscriptions, which would complicate its use in a collective setting such a classroom.

Within the analyzed sample, most free video games were suitable for younger audiences (over 12 years old) and had shorter estimated playing times (approx. 30 min on average). These games are designed to deliver quick play rounds and are more suitable for broader audiences, which makes them great candidates for incorporation into the classroom. In contrast, paid video games seem to be aimed at older audiences (over 14 years old) and have longer playing times of more than 45 min, often exceeding 60 min in re-playable games. However, they may offer a more in-depth learning experience.

Another key characteristic is the learning approach within these games, which determines how these tools may be implemented and the role of teachers and students. Behaviorism focuses on repetition and constant practice, while constructivism encourages critical thinking, active participation, and problem-solving. In behaviorism, the student assumes a passive role, whereas in constructivism their participation is active. In this regard, constructivist approaches, like games and projects which involve student participation, are more suitable for learning purposes, as they allow the student to be an active element that generates knowledge while the teacher only plays the role of a guide. In the case of using video games in the classroom, the teacher's role would thus vary: in a behaviorist approach, the teacher would be an information transmitter, while in a constructivist approach, they would act as facilitators and guides of the learning process [58,59]. Within the sample, the majority of video games (n = 18) presented a behaviorist approach.

The effectiveness of video games is achieved by integrating accurate and credible information, persuasively conveyed [60] through the context and documentation provided [61]. However, the type of world a video game depicts may also affect its implementation in a learning setting. Games that include real news may contribute to a greater awareness of the world at a given time and promote critical and grounded discussions. In this regard, it is important to mention G23, which employs news about the war in Ukraine. Other games (G6 and G15) also feature real current issues, including general mentions of the war, while G17 specifically addressed the COVID-19 pandemic. However, the content of these games ages as reality and the knowledge of a given event change and evolve, which may render certain video games obsolete or unusable in the long term for teaching purposes. Most of the games analyzed used real news (n = 14) and, although these might be a desirable tool in current times, video games using fictional worlds (n = 10) may be more suitable for inclusion in a curriculum in the long run. The latter mainly focus on detecting fake news and combating misinformation, and they do not include current political issues so as not to distract from the learning goals with the tensions and divisions that merely talking about politics, war, and the like, can create.

5.2. Learning Outcomes and Game Mechanics

More than half of the video games analyzed (n = 13) included all the learning outcomes considered (see Tables 2 and 6) and up to 11 video games addressed all the learning outcomes in a metaphorical or explicit manner, which is more suitable for media literacy teaching purposes than the inclusion of learning outcomes in an indirect fashion. These constitute the most comprehensive media literacy video games within the sample. It is noteworthy to mention G22, the most robust game detected, which addresses nearly all the selected learning outcomes (8 out of 9) explicitly, making it the most complete media literacy educational asset in the sample. This game also has a constructivist approach and is re-playable, ideal characteristics for a video game to become a learning tool. Nevertheless, the fact that there are so many items that address all the learning outcomes provides teachers and educators with a wide array of options to choose from for media literacy education, and most of those are free of charge (n = 8), are not restricted to any platform (n = 8), or are short enough for use in the classroom (n = 6). Unfortunately, the most complete and robust game (G22) carries several constraints to its classroom implementation: it is a long game (more than 60 min of estimated playing time), it has a fee, and is only available on STEAM. These characteristics may hinder its use for educational purposes, but these disadvantages can certainly be overcome.

The outcome with the greatest presence in the video game pool is number 5 ("Understand what some examples of credible sources of information are"), present in 21 items. The least represented is outcome 2 ("Understand the reasons why disinformation is published with the intention to mislead"), available in 16 games. Two learning outcomes stand out as the most explicitly addressed in general in the whole sample: outcomes 5 and 8 ("Understand what some examples of credible sources of information are" and "Know how to be a positive and responsible player on social media", respectively). The least robustly treated topic within the video game sample seems to be outcome number 3 ("Know that some political or commercial interests try to affect online behavior"), which also appears in few games as compared to others (n = 17). Nevertheless, most learning outcomes are more often addressed explicitly than indirectly, except for number 3, which is more often addressed indirectly in the analyzed games. Learning outcomes 8 ("Know how to be a positive and responsible player on social media"), 4 ("Have a general idea about how algorithms affect what we see online"), and 5 ("Understand what some examples of credible sources of information are") are the most explicitly addressed in general. However, it seems that not all learning outcomes are treated equally, and the crucial topic of the manipulation of information and why it exists (outcomes 2 and 3) seems to be more difficult to deliver through video games as a tool.

Regarding game design, during the analysis, we detected six different game mechanics (Table 5). The presence and impact of these mechanics were evaluated in each game (Table 7). We determined that a total amount of six games included all the mechanics. It is worth noting that the appearance of all the mechanics does not imply that the game will meet all the learning outcomes. Some games address all the learning topics with very few mechanics (G3, G4, G5, and G9), while other games with a high number of mechanics cover fewer than half of the outcomes (G6). The most frequent and impactful game mechanics detected were number 1 ("Allows players to investigate the news and find out which are real and which are fake") and number 4 ("Shows the player different news items (real and fake) and asks them to select the real ones"). On the other hand, the most underused and least relevant mechanic was number 6 ("Quizzes the player by challenging their knowledge of political or current topics"). In general, nearly all the mechanics, when used, were employed with a high impact in the video game in which they appear. Nevertheless, it is paramount for educators to remember, when choosing a video game for media literacy teaching, that there does not seem to be a direct relationship between the complexity of the game design in terms of mechanics and their robustness as a teaching tool in terms of learning outcomes.

5.3. Examples of Learning Outcomes in Media Literacy Games

Longer re-playable games offer more prolonged experiences in time and, in general, include a more complex design in terms of game mechanics, which requires a greater commitment from the players [43]. They may stimulate immersion to raise awareness in the players, making them the protagonists of the actions and putting them in the shoes of the person who manipulates information. For example, in "No Place for the Dissident" (G20), the player must expand an ideology to dominate the world. The game mechanics allow the user to adopt ideological policies, manage a country, and compete with other players who also want to impose their ideologies. This way, some video games manage to put the users in the role of each involved party to help them understand the consequences of spreading misinformation (learning outcome 1). But this immersive strategy is not exclusive to very complex games in terms of design. For instance, in "Choose your own fake news" (G10), the player embarks on a journey where they choose their adventure by exploring news and data about job opportunities and vaccines and constantly challenges the player to learn to discern between truth and manipulation and develop the ability to make informed decisions in an environment plagued by deception.

To foster discernment skills in relation to media [51] so users can acquire tools to question the information they receive, identify biases, and make informed decisions, other games explore the motives behind disinformation with the intention to mislead (learning outcome 2). They may offer a deep understanding of why some political or commercial parties seek to influence our behavior (learning outcome 3) through information manipulation [5] and focus on showing how algorithms can affect what we see online (learning outcome 4). For example, in "Fake It to Make It" (G4) the plot is focused on the field of advertising. Players seek strategies to influence people's perceptions and behaviors to achieve their commercial objectives. The in-game experience immerses the player in a world where ethics and responsibility are questioned, providing deep insight into the power and consequences of manipulating real-life information and discussing how algorithms work.

To promote media literacy, it is crucial to encourage critical thinking and a culture of verification (learning outcomes 5 and 6). Games with quizzes and narratives that include real news emphasize this. Such an approach helps users identify fake news and other types of misinformation [17]. Video games that present methods to verify information and understand changes in the media landscape provide players with tools to discern between reliable input and misinformation. For example, in "BBC's iReporter" (G13, Figure 2), the player is a reporter who must contrast their information sources under pressure to produce impactful news as quickly as possible. Players need to select predefined responses and interact with characters to resolve situations based on a narrative. These games highlight the importance of verifying information before sharing it, even when in a rush [49].

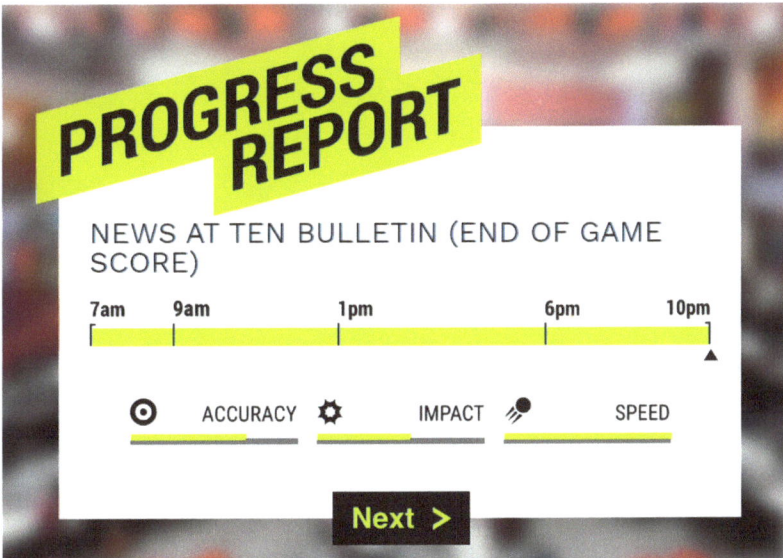

Figure 2. BBC's iReporter (G13). Screen capture.

Regarding the defense against threats and risks on social media (learning outcomes 7 and 8), certain games provide relevant information on how fake news is constructed (learning outcome 9). In "Cat Park" (G14), with typical RPG mechanics, the player travels through a city where they will encounter an eclectic group of characters sharing a common goal: to bring down a cat park and spread it on social media. However, as the players create fake news, they will also learn counter-misinformation techniques and understand the benefits and limitations of different approaches. The goal is to solve the problems in the plot in a creative and strategic way that fosters creative solutions [46]. This experience provides a perfect blend of strategy and learning, allowing the player to explore the intricacies of information and misinformation.

In respect of video games that include current issues in the real world, it is worth highlighting "Power and Revolution 2022 Edition" (G23, Figure 3). It offers a simulation where players take on the role of political leaders and explore the Ukraine war, conspiracy theories, fake news, the interference of secret services, animal welfare, and global warming. Players can thus understand the complexities and implications of their actions in the media and the social environment. The game management panel displays, through icons, how the country's economy functions, how waste recycling is carried out, the level of democracy, resources, education, and voting, as well as changes in political party and religion.

Figure 3. Power and Revolution 2022 Edition (G23). Screen capture.

Last, in "Influence Inc. 2022" (G22, Figure 4), the most robust game in terms of learning outcomes, the player takes on the role of the manager of a communication agency that resorts to all kinds of tactics to meet their goals. The user manipulates social media and news to promote celebrities and even influence elections and must think strategically about how to manage a digital influence agency using propaganda and advertising. Thus, while the users are amusing themselves with the video game, they are also carrying out actions within the game that produce valuable results for them [45]. This experience allows players to immerse themselves in a wide range of educational content, and although their malicious actions are carried out in a fictitious world, it provides a deep understanding of information control in the real world. By exploring the complexities of communication management, the player acquires critical awareness of how to handle and manipulate information, providing a valuable perspective for the world outside the game. The knowledge gained and the consequences observed in the game could be applied ethically in the classroom to ensure effective and contextualized learning.

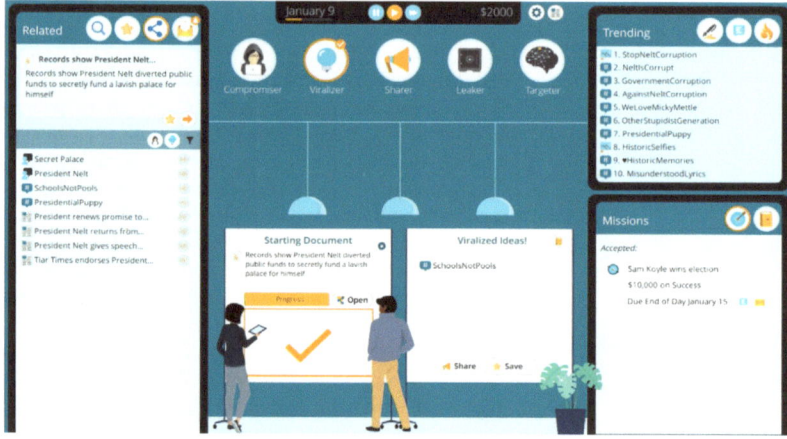

Figure 4. Influence Inc. 2022 (G22). Screen capture.

6. Conclusions

In the present study, we analyzed 24 video games focused on media literacy by scrutinizing their content (learning outcomes according to the curriculum of reference), the way in which the content is delivered through their design (game mechanics), and the features that might impact their implementation within educational environments.

The growth in the number of games focused on fake news reflects the increasing importance of media literacy and the fight against misinformation in our current times. The content, mechanics, and dynamics of video games could offer a high degree of effectiveness in achieving critical learning outcomes for media literacy [27], allowing users to understand, practice, and implement these skills to combat misinformation. Likewise, these can foster sharper critical thinking and greater responsibility when consuming and sharing information, which is essential in an increasingly connected world. By integrating elements of news creation and distribution, video games offer a more immersive and challenging experience [45] than other tools, and players directly experience the consequences of their actions in a controlled environment [31,35]. Although in some cases the objectives within a video game may be considered negative, such as obtaining powers through violence (as in G21, G22, and G23 in the analyzed sample), the knowledge acquired can be applied ethically in the real world. Ultimately, the insights gained through video games can promote the development of critical thinking skills [26] and making ethical decisions in various contexts outside the game.

However, we found that some game characteristics may influence the suitability and usefulness of video games for media literacy teaching and their ease of implementation in the classroom. Estimated playing time should be considered when choosing a media literacy video game in a formal education setting, as re-playable games, though extremely beneficial for learning purposes, tend to be longer (60 min or more) and are, therefore, more difficult to incorporate into typical one-hour subject sessions or toolkits for educators, such as Get Your Facts Straight!: Toolkit for Educators and Training Providers [56]. Fortunately, most of the media literacy games found on the market today have estimated playing times that would fit these frameworks.

Target audiences also determine whether a video game can be used as a tool within the educational system or not, as only games aimed at players under 18 could typically be employed. It seems, though, that nearly all media literacy video games available on the market as of today (except for one item) are directed toward younger audiences and, therefore, are suitable for media literacy teaching purposes in schools.

Video game availability, particularly pricing and distribution platform, is a variable that educators should also evaluate before choosing a video game, as it greatly affects accessibility. While popular platforms such as STEAM increase the reach of video games and topics, these typically require a fee and individual subscriptions, which may hinder their implementation in collective settings such as educational environments. However, most media literacy video games are free and accessible.

The learning approach of a video game should be carefully considered before choosing it as a tool. Those based on behaviorism focus on repetition and constant practice, and the student plays a somewhat passive role. On the other hand, constructivist video games encourage critical thinking, active participation, and problem-solving. With the latter, educators should mainly act as guides. Even though the constructivist approach is more desirable for learning purposes, the majority of the video games on media literacy on the market incorporate a behaviorist approach.

Last, educators should ponder whether they prefer using a tool that employs real information, or video games depicting fictional worlds. Those that include real news may contribute to more awareness of the world at a given time and promote critical and grounded discussions, but they may rapidly become obsolete in our ever-changing world and constitute a distraction from learning objectives due to controversial issues being brought up. Most of the available video games on media literacy focus on real news.

Regarding crucial learning outcomes for media literacy, we discovered that more than half of the video games on the market were very complete and addressed all the learning objectives as set out in the curriculum of reference (Get Your Facts Straight!: Toolkit for Educators and Trainers); most of them did so in a metaphorical or explicit manner. These comprehensive tools were mostly free of charge, not restricted to a platform, and short enough to facilitate their implementation in the classroom, offering educators a broad range to choose from.

Most of the video games addressed the issue of credible sources of information (learning outcome 5). Other robust learning outcomes identified within the analyzed sample that tended to be addressed explicitly include knowing how to distinguish real and fake news (learning outcome 4) and knowing how to be a responsible player on social media (learning outcome 8). In contrast, the topics that seemed represented least and in a poorer manner were why disinformation is published with the intention to mislead (learning outcome 2) and the issue of political and commercial manipulation of information (learning outcome 3), which are crucial to navigating the Internet and social media nowadays. While not all considered learning outcomes were equally addressed, the findings demonstrate that certain media literacy video games currently available in the market could be harnessed as effective tools to attain critical learning objectives.

By playing the video games in their entirety, we were able to identify the most common game mechanics present in media literacy games. Most video games opt to focus on requesting the player to differentiate between true and false news (game mechanics 1 and 4), but those that better engage the player incorporate a variety of mechanics [25,41], including news creation and distribution on social media and facing the consequences of these actions (game mechanics 2, 3, and 5). This allows players to be involved in a more active way, encouraging decision-making and reflection on the impact of misinformation in society. Challenging the player about their knowledge about political or current topics was the least used method (game mechanic 6), which is somewhat positive, as focusing too much on controversial issues may distract students from the learning process. We also observed that the complexity of game design in terms of mechanics did not have a direct relationship with how comprehensive a video game was regarding the inclusion of media literacy learning outcomes. Thus, video games may empower users to grasp, exercise, and deploy skills to counter misinformation, regardless of the intricacies of gameplay mechanics.

As a practical example of the implementation of video games as an educational tool, educators may use the most robust game in terms of learning outcomes identified in this study (G22) to present students with a fictional scenario and situations involving the evaluation and verification of information, challenging them to discern between real and fake news. Through questions, decisions, and searches for reliable sources (outside the game), students can develop skills to identify and question misleading news in the real world.

After analyzing the main characteristics of media literacy video games, we can conclude that, despite the great effort of developers to incorporate this issue as a response to an increasing preoccupation, developing teams would benefit from a closer collaboration with educators to ensure that the video games produced can become effective and useful tools that can be smoothly integrated into formal education. Understanding how critical skills for media literacy are best addressed through video games and what mechanics best deliver these is a paramount area for future research, to ensure that video games can be effectively incorporated into curricula in this age of mis- and disinformation characterized by a lack of scenarios that promote critical thinking and initiatives that address media literacy. The development of research in the area of video games that employ fictional worlds and constructivist approaches is also of great importance for the determination of their educational potential as pedagogical tools, as the current media landscape is likely to be filled with new fictive worlds that have a great prospect for pedagogical purposes besides mere entertainment. It would also be crucial to develop more studies on video games that focus on crises, such as wars and pandemics, very relevant issues that motivated some of the objectives of the YO-MEDIA project. However, research in this area faces certain limitations, such as the reduced amount of video games available on media literacy as compared to other topics and the lack of databases or repositories that compile such initiatives.

Finally, in the future, the use of other data analysis or processing techniques could be considered, once we have a larger dataset and a greater number of analyzed video games.

Author Contributions: Conceptualization, R.S.C.-E. and J.L.E.-G.; methodology, J.L.E.-G.; software, J.L.E.-G.; validation, R.S.C.-E.; formal analysis, R.S.C.-E.; investigation, J.L.E.-G.; resources, J.L.E.-G.; data curation, J.L.E.-G.; writing—original draft preparation, R.S.C.-E.; writing—review and editing, R.S.C.-E.; visualization, J.L.E.-G.; supervision, J.L.E.-G.; project administration, R.S.C.-E.; funding acquisition, R.S.C.-E. All authors have read and agreed to the published version of the manuscript.

Funding: This research was funded by the Calouste Gulbenkian Foundation, grant number 269094.

Data Availability Statement: The data presented is available upon request.

Conflicts of Interest: The authors declare no conflict of interest.

References

1. Adams, Z.; Osman, M.; Bechlivanidis, C.; Meder, B. (Why) Is Misinformation a Problem? *Perspect. Psychol. Sci.* **2023**. [CrossRef] [PubMed]
2. Shehata, A.M. The Problem of Misinformation and Fake News. In *Mass Communications and the Influence of Information During Times of Crises*; Al-Suqri, M., Alsalmi, J., Al-Shaqsi, O., Eds.; IGI Global: Hershey, PA, USA, 2022; pp. 99–122. [CrossRef]
3. Terian, S. What Is Fake News: A New Definition. *Transilvania* **2021**, *11–12*, 112–120. [CrossRef]
4. Jaster, R.; Lanius, D. Speaking of Fake News: Definitions and Dimensions 2021. In *The Epistemology of Fake News*; Oxford University Press: Oxford, UK, 2021. [CrossRef]
5. Sunstein, C.R. *Republic: Divided Democracy in the Age of Social Media*; Princeton University Press: Princeton, NJ, USA, 2017.
6. Majerczak, P.; Strzelecki, A. Trust, Media Credibility, Social Ties, and the Intention to Share towards Information Verification in an Age of Fake News. *Behav. Sci.* **2022**, *12*, 51. [CrossRef] [PubMed]
7. Johnson, N.F.; Velasquez, N.; Restrepo, N.F.; Leahy, R.; Gabriel, N.; El Oud, S.; Zheng, M.; Manrique, P.; Wuchty, S.; Muñoz, R.; et al. The online competition between pro-and anti-vaccination views. *Nature* **2020**, *582*, 230–233. [CrossRef]
8. Pennycook, G.; Rand, D.G. Who falls for fake news? The roles of bullshit receptivity, overclaiming, familiarity, and analytic thinking. *J. Pers.* **2019**, *88*, 185–200. [CrossRef]
9. Posetti, J.; Matthews, A. A Short Guide to the History of 'Fake News' and Disinformation. *Int. Cent. J.* **2018**. Available online: https://www.icfj.org/news/short-guide-history-fake-news-and-disinformation-new-icfj-learning-module (accessed on 24 March 2023).
10. Miró-Llinares, F.; Aguerri, J.C. Misinformation about fake news: A systematic critical review of empirical studies on the phenomenon and its status as a 'threat'. *Eur. J. Criminol.* **2023**, *20*, 356–374. [CrossRef]
11. Van der Linden, S.; Panagopoulos, C.; Roozenbeek, J. You are fake news: Political bias in perceptions of fake news. *Media Cult. Soc.* **2020**, *42*, 460–470. [CrossRef]
12. Bin Naeem, S.; Bhatti, R. The COVID-19 'infodemic': A new front for information professionals. *Health Inf. Libr. J.* **2020**, *37*, 233–239. [CrossRef]
13. Cover, R.; Haw, A.; Thompson, J.D. *Fake News in Digital Cultures: Technology, Populism and Digital Misinformation*; Emerald Group Publishing: Bingley, UK, 2022; p. 20.
14. Kubin, E.; von Sikorski, C. The role of (social) media in political polarization: A systematic review. *Ann. Int. Commun. Assoc.* **2021**, *45*, 188–206. [CrossRef]
15. Jones-Jang, S.M.; Mortensen, T.; Liu, J. Does media literacy help identification of fake news? Information literacy helps, but other literacies don't. *Am. Behav. Sci.* **2021**, *65*, 371–388. [CrossRef]
16. Kellner, D.; Share, J. *The Critical Media Literacy Guide: Engaging Media and Transforming Education*; Brill: Leiden, The Netherlands, 2019; p. 34.
17. Gaozhao, D. Flagging fake news on social media: An experimental study of media consumers' identification of fake news. *Gov. Inf. Q.* **2021**, *38*, 101591. [CrossRef]
18. Frey, N.; Fisher, D. Junior Journalists: Reading and Writing News in the Primary Grades. In *Teaching New Literacies in Grades K-3. Resources for 21st-Century Classrooms*; Moss, B., Lapp, D., Roser, N., Fuhrken, C., Dybdahl, C., Eds.; Guilford Press: New York, NY, USA, 2010; pp. 71–83.

19. Herrero Curiel, E.; La Rosa, L. Guía Docente para el Profesorado de Educación Secundaria. Alfabetización Mediática e Informacional 2021. Universidad Carlos III de Madrid-Fundación BBVA. Available online: https://www.uc3m.es/uc3m/media/uc3m/doc/archivo/doc_guia-docente-/2guia-docente-para-el-profesorado-de-educacion-secundaria.pdf (accessed on 30 January 2023).
20. Roozenbeek, J.; Van der Linden, S. Fake news game confers psychological resistance against online misinformation. *Palgrave Commun.* **2019**, *5*, 65. Available online: https://www.nature.com/articles/s41599-019-0279-9 (accessed on 20 January 2023). [CrossRef]
21. Stringer, K. Push for Media Literacy Takes on Urgency Amid Rise of 'Fake News' 2018. Education Writers Association. Available online: https://www.ewa.org/blog-educated-reporter/push-media-literacy-takes-urgency-amid-rise-fake-news?utm_source=the74&utm_medium=linkback&utm_campaign=74-hosted (accessed on 24 January 2023).
22. Bulger, M.; Davison, P. *The Promises, Challenges, and Futures of Media Literacy*; Data & Society Institute: New York, NY, USA, 2018; Available online: https://datasociety.net/output/the-promises-challenges-and-futures-of-media-literacy/ (accessed on 20 January 2023).
23. Culver, S.H.; Redmond, T. Media Literacy Snapshot 2019. National Association for Media Literacy Education. Available online: https://namle.net/wp-content/uploads/2019/06/SOML_FINAL.pdf (accessed on 30 January 2023).
24. Basol, M.; Roozenbeek, J.; Van der Linden, S. Good news about bad news: Gamified inoculation boosts confidence and cognitive immunity against fake news. *J. Cogn.* **2020**, *3*, 1–9. [CrossRef]
25. Molnar, A.; Kostkova, P. On effective integration of educational content in serious games: Text vs. game mechanics. In Proceedings of the 2013 IEEE 13th International Conference on Advanced Learning Technologies, Beijing, China, 18–19 July 2013. [CrossRef]
26. Glas, R.; van Vught, J.; Fluitsma, T.; De La Hera, T.; Gómez-García, S. Literacy at play: An analysis of media literacy games used to foster media literacy competencies. *Front. Commun.* **2023**, *8*, 1155840. [CrossRef]
27. Egenfeldt-Nielsen, S.; Smith, J.H.; Tosca, S.P. *Understanding Video Games: The Essential Introduction*; Routledge: New York, NY, USA, 2020; p. 50.
28. Mihailidis, P. Introduction: News Literacy in the Dawn of a Hypermedia Age. In *News Literacy: Global Perspectives for the Newsroom and the Classroom*; Mihailidis, P., Ed.; Peter Lang: New York, NY, USA, 2012; pp. 1–20.
29. Guerrero, M.A.; Luengas, M. Media Literate "Prodiences": Binding the knot of news content and production for an open society. In *News Literacy: Global Perspectives for the Newsroom and the Classroom*; Mihailidis, P., Ed.; Peter Lang: New York, NY, USA, 2012; pp. 41–63.
30. Buckingham, D. Defining Digital Literacy: What Do Young People Need to Know about Digital Media. *Nord. J. Digit. Lit.* **2006**, *4*, 263–276. Available online: https://www.idunn.no/doi/10.18261/ISSN1891-943X-2006-04-03 (accessed on 24 January 2023). [CrossRef]
31. Gee, J.P. *Good Video Games and Good Learning. Collected Essays on Video Games, Learning, and Literacy*; Peter Lang: New York, NY, USA, 2013; p. 35.
32. Pérez, O.; Contreras Espinosa, R. Performative skills. In *Teens, Media and Collaborative Cultures*; Scolari, C.A., Ed.; CeGe: Barcelona, Spain, 2018; pp. 44–51.
33. Lammers, J.C.; Kurwood, J.S.; Magnifico, A.M. Toward an affinity spacemethodology: Considerations for literacy research. *J. Educ. Pract.* **2012**, *11*, 44–58.
34. Scolari, C.; Contreras-Espinosa, R. How do teens learn to play video games? Informal learning strategies and video game literacy. *J. Inf. Lit.* **2019**, *13*, 45–61. [CrossRef]
35. Gee, J.P. *What Video Games Have to Teach Us about Learning and Literacy*; Palgrave Macmillan: New York, NY, USA, 2003; p. 34.
36. Squire, K. Cultural framing of computer/video games. *Game Stud.* **2002**, *2*, 90.
37. Scheibenzuber, C.; Nistor, N. Media Literacy Training Against Fake News in Online Media. In Proceedings of the 14th European Conference on Technology Enhanced Learning EC-TEL 2019, Delft, The Netherlands, 16–19 September 2019. [CrossRef]
38. Mao, W.; Cui, Y.; Chiu, M.; Lei, H. Effects of Game-Based Learning on Students' Critical Thinking: A Meta-Analysis. *J. Educ. Comput. Res.* **2021**, *59*, 073563312110070. [CrossRef]
39. Jenson, J.; Burrell-Kim, D. Digital literacies & multimodal learning. In *The International Encyclopedia of Education*, 4th ed.; Jenson, J., Burrell-Kim, D., Eds.; Elsevier: Oxford, UK, 2023; pp. 538–589. [CrossRef]
40. Hobbs, R. *Media Literacy in Action: Questioning the Media*; Rowman & Littlefield Publishers: London, UK, 2021; p. 13.
41. Ustaoğlu, A.; Çelik, H. High school students' video game involvement and their English language learning motivation: A correlation study. *J. Educ. Online* **2023**, *20*, 17. [CrossRef]
42. Karpova, S.I.; Chirich, I.V.; Avtsinova, G.I.; Shtukareva, E.B.; Ukhina, T.V.; Gordeeva, T.A. Information and communication technologies in education: Video games as an effective environment for the development of self-directed learning of students. *Webology* **2021**, *18*, 116–128. [CrossRef]
43. Salen, K.; Zimmerman, E. *Rules of Play: Game Design Fundamentals*; The MIT Press: London, UK, 2004; p. 30.
44. Nacke, L.E.; Bateman, C.; Mandryk, R.L. BrainHex: A Neurobiological Gamer Typology Survey. *Entertain Comput.* **2014**, *5*, 55–62. [CrossRef]
45. Petry, A. Playing in Minecraft: An exploratory study. *Rev. FAMECOS* **2018**, *25*, 1–18. [CrossRef]

46. Bunt, B.; Grosser, M. Puzzle video games and the benefits for critical thinking: Developing skills and dispositions towards self-directed learning. In *Self-Directed Learning Research and Its Impact on Educational Practice*; Mentz, E., Bailey, R., Eds.; Aosis Academy: Fallon, NV, USA, 2020; pp. 155–195. [CrossRef]
47. Dobrowolski, P.; Skorko, M.; Myśliwiec, M.; Kowalczyk-Grębska, N.; Michalak, J.; Brzezicka, A. Perceptual, Attentional, and Executive Functioning after Real-Time Strategy Video Game Training: Efficacy and Relation to In-Game Behavior. *J. Cogn. Enhanc.* **2021**, *5*, 397–410. Available online: https://link.springer.com/article/10.1007/s41465-021-00211-w (accessed on 20 January 2023).
48. Chang, Y.K.; Literat, I.; Price, C.; Eisman, J.I.; Gardner, J.; Chapman, A.; Truss, A. News literacy education in a polarized political climate: How games can teach youth to spot misinformation. *Harv. Kennedy Sch. HKS Misinformation Rev.* **2020**, *4*, 1–9. [CrossRef]
49. Rasi, P.; Vuojärvi, H.; Ruokamo, H. Media Literacy Education for All Ages. *J. Media Lit. Educ.* **2019**, *11*, 1–19. [CrossRef]
50. Nakov, P.; Corney, D.P.; Hasanain, M.; Alam, F.; Elsayed, T.; Barron-Cedeno, A.; Papotti, P.; Shaar, S.; Da San Martino, G. Automated Fact-Checking for Assisting Human Fact-Checkers. In Proceedings of the International Joint Conference on Artificial Intelligence, Montreal, QC, Canada, 19–27 August 2021.
51. Mcdougall, J. Media literacy versus fake news: Critical thinking, resilience and civic engagement. *Media Stud.* **2019**, *10*, 29–45. [CrossRef]
52. Martin, A. A European framework for digital literacy. *Nord. J. Digit.* **2006**, *1*, 151–161. [CrossRef]
53. Lin, T.; Li, J.; Deng, F.; Lee, L. Understanding New Media Literacy: An Explorative Theoretical Framework. *Educ. Technol. Soc.* **2013**, *16*, 160–170. Available online: http://www.jstor.org/stable/jeductechsoci.16.4.160 (accessed on 30 January 2023).
54. von Gillern, S.; Gleason, B.; Hutchison, A. Digital Citizenship, Media Literacy, and the ACTS Framework. *Read Teac.* **2022**, *76*, 145–158. [CrossRef]
55. Richardson, J.; Milovidov, E. *Digital Citizenship Education Handbook. Being Online. Well-Being Online. Rights Online*; Document Council of Europe: Strasbourg, France, 2019; Available online: https://rm.coe.int/16809382f9 (accessed on 30 January 2023).
56. EAVI European Association for Viewers Interests. Get Your Facts Straight!: Toolkit for Educators and Training Providers; Deliverable n.3. 2020. Available online: https://www.alldigitalweek.eu/get-facts/ (accessed on 30 January 2023).
57. Kelle, S.; Klemke, R.; Specht, M. Design patterns for learning games. *Int. J. Technol. Enhanc. Learn.* **2011**, *3*, 555–569. [CrossRef]
58. Piaget, J. *El Desarrollo del Pensamiento: Equilibración de las Estructuras Cognitivas*; Fondo de Cultura Económica: Mexico City, Mexico, 1972; p. 43.
59. Vygotsky, L.S. *Mind in Society: The Development of Higher Psychological Processes*; Harvard University Press: Cambridge, UK, 1978; p. 67.
60. García-Avilés, J.; Ferrer-Conill, R.; García-Ortega, A. Gamification and Newsgames as Narrative Innovations in Journalism. In *Total Journalism Models, Techniques and Challenges*; Vázquez Herrero, J., Silva-Rodríguez, A., Negreira-Rey, M., Toural-Bran, C., López-García, X., Eds.; Springer: Cham, Switzerland, 2022; pp. 53–67. [CrossRef]
61. Gomez-García, S.; Cabeza, J. El discurso informativo de los newsgames: El caso Bárcenas en los juegos para dispositivos móviles. *Cuadernos.info* **2016**, *38*, 137–148. [CrossRef]

Disclaimer/Publisher's Note: The statements, opinions and data contained in all publications are solely those of the individual author(s) and contributor(s) and not of MDPI and/or the editor(s). MDPI and/or the editor(s) disclaim responsibility for any injury to people or property resulting from any ideas, methods, instructions or products referred to in the content.

Article

Prospective ICT Teachers' Perceptions on the Didactic Utility and Player Experience of a Serious Game for Safe Internet Use and Digital Intelligence Competencies

Aikaterini Georgiadou and Stelios Xinogalos *

Department of Applied Informatics, University of Macedonia, GR-54636 Thessaloniki, Greece; dai18138@uom.edu.gr
* Correspondence: stelios@uom.edu.gr

Abstract: Nowadays, young students spend a lot of time playing video games and browsing on the Internet. Using the Internet has become even more widespread for young students due to the COVID-19 pandemic lockdown, which resulted in transferring several educational activities online. The Internet and generally the digital world that we live in offers many possibilities in our everyday lives, but it also entails dangers such as cyber threats and unethical use of personal data. It is widely accepted that everyone, especially young students, should be educated on safe Internet use and should be supported on acquiring other Digital Intelligence (DI) competencies as well. Towards this goal, we present the design and evaluation of the game "Follow the Paws" that aims to educate primary school students on safe Internet use and support them in acquiring relevant DI competencies. The game was designed taking into account relevant literature and was evaluated by 213 prospective Information and Communication Technology (ICT) teachers. The participants playtested the game and evaluated it through an online questionnaire that was based on validated instruments proposed in the literature. The participants evaluated positively to the didactic utility of the game and the anticipated player experience, while they highlighted several improvements to be taken into consideration in a future revision of the game. Based on the results, proposals for further research are presented, including DI competencies detection through the game and evaluating its actual effectiveness in the classroom.

Keywords: serious games; digital intelligence; DQ Institute; safe internet use; SGDA framework; evaluation

Citation: Georgiadou, A.; Xinogalos, S. Prospective ICT Teachers' Perceptions on the Didactic Utility and Player Experience of a Serious Game for Safe Internet Use and Digital Intelligence Competencies. *Computers* **2023**, *12*, 193. https://doi.org/10.3390/computers12100193

Academic Editors: Paolo Bellavista, Ricardo Baptista, Hariklia Tsalapatas and Carlos Vaz de Carvalho

Received: 31 July 2023
Revised: 8 September 2023
Accepted: 21 September 2023
Published: 26 September 2023

Copyright: © 2023 by the authors. Licensee MDPI, Basel, Switzerland. This article is an open access article distributed under the terms and conditions of the Creative Commons Attribution (CC BY) license (https:// creativecommons.org/licenses/by/ 4.0/).

1. Introduction

The current period of time is distinguished by rapid technological evolution, unequal access to modern Information and Communication Technologies (ICT) around the world and a constantly declining time for educating people to use technology properly. Consequently, a gap is being created, between the digital innovations and the knowledge that is required to handle them. This gap, combined with the need for quick transition of primary and secondary school students to the digital world due to the COVID-19 pandemic, resulted in increased cyber risks. The COVID-19 pandemic has caused rushed digitalization of primary and secondary student education, bringing even more in the forefront the younger ages and cyber-risks—such as bullying, technology addiction, and misinformation—that must be addressed [1].

It is important for children to cultivate digital skills, or else to develop their Digital Intelligence (DI), in order to adapt to the new requirements of the digital world and protect themselves from underlying threats. DI, as defined by the DQ (Digital Intelligence Quotient) Institute [2], stands for a comprehensive set of technical, cognitive, meta-cognitive, and socio-emotional competencies that are grounded in universal moral values, which enable individuals to face the challenges and harness the opportunities of digital life. Thus, individuals equipped with DQ become wise, competent, and future-ready digital citizens who successfully use, control, and create technology to enhance humanity.

On the other hand, serious games (SGs) are referred to as entertaining tools with a purpose of education, where players cultivate their knowledge and practice their skills through overcoming numerous hindrances during gaming [3]. Therefore, SGs provide the chance to create a simulated environment for players to experience situations that are impossible to achieve in the real world for reasons such as safety and cost [4].

SGs have proved to be useful in encouraging attitude change, in supporting the development of critical thinking, and in problem solving and developing decision-making skills [5]. Given the above, a SG within the prism of the digital world will not only educate and empower children's digital skills, but will also entertain and give them the ability to recognize and curb real life risks.

In this paper, we present the design and evaluation of the game "Follow the Paws", which aims to educate primary school students on safe Internet use and support them in acquiring relevant DI competencies. The game was designed by taking into account the Serious Game Design Assessment Framework proposed by Mitgutsch and Alvarado [6] and was evaluated by 213 undergraduate Informatics students attending a "Didactics of Informatics" course for prospective ICT teachers. The participants playtested the game at their homes and evaluated it through an online questionnaire based on the relevant literature [7–9]. The aim of this pilot evaluation was to investigate the potential of this game prior to using it in the classroom. The following research questions were investigated:

RQ1: How do prospective ICT teachers evaluate the didactic utility of "Follow the Paws" as a tool for educating primary school students on safe internet use and relevant digital intelligence competencies?

RQ2: What are prospective ICT teachers' expectations of primary school students' experience when playing "Follow the Paws"?

The rest of the paper is structured as follows. In Section 2 a brief literature review is presented. In Section 3 the design of "Follow the Paws" is presented, while in Section 4 materials and methods are briefly described. This is followed by the results of the study and a discussion in Sections 5 and 6, respectively. The limitations of the study and plans for future work are presented in Section 7.

2. Literature Review

2.1. Digital Intelligence

DI, as defined by the DQ Institute, refers to a person's ability to adapt to the modern digital world and to act consciously and responsibly within it, using digital media as tools to achieve their goals and improve their life quality. Digitally intelligent people can be defined as individuals able to effectively and flexibly manage advanced technology, as well as successfully collaborate with others in the digital world. Social, emotional and cognitive adaptation to digital reality are the main axes of DI.

To structure DI and embed digital competencies in a universal standard that can be used across organizations, incorporating the digital literacy, digital skills, readiness and flexibility, a global framework was required. The DQ Institute met this challenge by integrating over 25 leading digital proficiency frameworks worldwide, creating the DQ Framework.

In Figure 1, the 24 digital competencies that are held within the framework are presented in tabular form, where in the vertical axis, the three levels of DQ understanding and mastery are defined, and on the horizontal, the eight DQ areas.

In a closer view, this framework is structured around two categories: the eight areas of digital intelligence it examines and the three levels that govern them. These eight areas and three levels form the matrix of 24 digital intelligence competencies (Figure 1).

These eight domains and three levels form the matrix of 24 digital intelligence competencies (Figure 1).

	Digital Identity	Digital Use	Digital Safety	Digital Security	Digital Emotional Intelligence	Digital Communication	Digital Literacy	Digital Rights
Digital Citizenship	1 Digital Citizen Identity	2 Balanced Use of Technology	3 Behavioral Cyber-Risk Management	4 Personal Cyber Security Management	5 Digital Empathy	6 Digital Footprint Management	7 Media and Information Literacy	8 Privacy Management
Digital Creativity	9 Digital Co-Creator Identity	10 Healthy Use of Technology	11 Content Cyber-Risk Management	12 Network Security Management	13 Self-Awareness and Management	14 Online Communication and Collaboration	15 Content Creation and Computational Literacy	16 Intellectual Property Rights Management
Digital Competitiveness	17 Digital Changemaker Identity	18 Civic Use of Technology	19 Commercial and Community Cyber-Risk Management	20 Organizational Cyber Security Management	21 Relationship Management	22 Public and Mass Communication	23 Data and AI Literacy	24 Participatory Rights Management

Figure 1. 24 DQ Competencies as defined in [2].

2.2. Advantages of Serious Games

Although the range of applications and usability of SGs may suggest a range of benefits, in reality there is limited evidence to support the benefits of using SGs. Despite this, one of the advantages of SGs, according to Corti [4], is their ability to simulate environments, systems, situations, etc., allowing players to experience situations that might have been impossible to experience in real-world conditions. In this way, risks involved in various activities and behaviors can be avoided, while at the same time allowing players to make their own decisions, having as much time as they need, without the pressure of real life. Also, as pointed out, simulation games have been found to include use of metacognition and mental models, improved strategic thinking and insight, better psychomotor skills, and development of analytical and spatial skills, iconic skills, visual selective attention, computer skills, etc. [5]. Moreover, the work of Blumberg and Ismailer [10] attests, through an extensive body of research, the efficacy of digital games for promoting players' learning. Finally, through the use of SGs, the participation of even the most timid students is being encouraged and, as empirically supported by Robertson and Miller [11], the less able children tended to be even more availed.

2.3. Relevant Educational Games

SGs development is attracting increased attention, leading to the creation of many educational games in the field of internet safety. Such a game is Google's Interland (Be Internet Awesome), a 3D game that aims to help students to be safe and successful citizens in our networked world. The concepts taught are anti-bullying, strong passwords, being careful what you post online, and phishing detection [12]. Another SG around cyber safety and digital citizenship is "Pledge Planets", a game within Messenger Kids launched by Meta (formerly Facebook) that helps kids learn and practice how to make healthy decisions online, stay safe and build resilience [13], aiming to develop their digital citizenship identity. In addition, Minecraft's "CyberSafe: Home Sweet Hmm", introduces fundamental cyber safety principles and demonstrates ways to stay safer online [14]. In DI, however, there is limited research in means of measure and detection, although the DQ Institute created a questionnaire that evaluates the digital citizenship level of individuals, which is called the Digital Citizenship Test (DCT). The DCT is an online/mobile test introduced to establish new ethical standards for individuals, particularly children and teens, by modeling for them the characteristics of a responsible digital citizen, in order to better protect them from threats posed by the Internet [15].

The aforementioned SGs and questionnaire aim to pursue internet safety and digital citizenship (digital intelligence life skills) separately and each addressing specific purposes. The SG that is presented in this paper is an attempt to combine, through a storyline, both technology usage and cyber-risk exposure within a realistic environment, proposing a means of addressing unitedly aspects of digital intelligent living.

3. Analysis and Design of Follow the Paws

3.1. Analysis of Follow the Paws

"Follow the Paws" is an SG or, more specifically, an educational game, aiming to consider whether the player can transfer the knowledge from the real world to the digital one and handle cyber risks, along with analyzing the player's DI skills. The game starts with the main character of the game finding a little dog in his/her neighborhood park, while he/she is out for a walk. The player has to successfully find and return the dog back to its owner. Throughout the game, the main character opposes cyber risks such as phishing, exposure of private details and communication with face profiles on-line, meeting a secondary goal of the game, collecting information about fundamental principles of safe internet browsing. Meanwhile, another contribution of the game is to educate the user by providing informative messages during playing based on his/her actions, concerning safe browsing on the internet and how it can be improved. The main purpose is to either reward and justify correct actions of the user, or offer guidance and hints regarding incorrect practices of online behavior and use. Regarding the social aspect of the game, the core message through the plot is raising awareness on stray animals, while safe browsing and the protection of players from malicious users are at the epicenter of the concept.

"Follow the Paws" is targeted to primary school students as an educational tool on safe internet use, a simulator of real life cyber risks to better prepare students' response on these types of matters, while, at the same time, can indicate some signs of DI.

The game can be installed on a standalone PC and was created using low-poly 3D animations and Third-to-First Person Cameras, in order to change the perspective along play and place even greater emphasis on the correlation of behaviors and actions of the real world with the digital one. The play time depends on the player and its perception of the game's purpose. During the game, if needed, the player has at his/her disposal aids giving hints to continue to the next steps and complete the game. The only limitation in the provision of aids is the time required between the use of one aid to the next one, in order to give the player time to proceed with different solutions. An estimation of the time needed to complete the game is approximately 30 min.

In Figure 2, the main areas of the game board are presented: (A) the game stage progress; and (B) the message box where the dialog that takes place with every interaction is displayed, where the player can interact with other characters or objects he/she encounters by guiding the character near the interactive object and pressing the space button; (C) informative messages upon user's actions; (D) the help area that displays the remaining time until the next available hint; and (E) the information collected.

Figure 2. The environment of the game: (A) Achieved game stage name: Information Collection, (B) Message Box: the text in Greek refers to the information the player receives, (C) Informative messages, (D) Help area: Hints, (E) Information collected.

The game, and the way that DI is applied in it, is presented in sections based on each one of the game's stages. The game is divided into six stages, and each stage consists of different tasks that the player must complete. These stages also correspond to some digital areas, goals and skills, through the achievement of which, information can be collected for the detection of the player's DI. At the same time, the player is asked to apply basic principles of safe internet use in order to protect himself/herself from cyber threats. In this way, it is also possible to collect certain information that can reflect the player's knowledge and readiness in the area of safe internet use.

3.1.1. Information Collection

In the first stage, the player must try to collect information about the dog from his/her surroundings. This should be performed by checking if the dog is wearing a collar with contact details of the dog's owner, and by trying to find someone around the park to ask if they have seen anyone looking for a dog.

At this stage of the game there is a combination of digital capabilities being explored, those of *Digital Co-Creator Identity* and *Online Communication and Collaboration*. The competencies considered derive from the actions of the player, by trying to simulate a logical act of the real world to the digital one. Specifically, the player alone has to think about approaching the dog and searching its collar as a result of investigating and identifying real problems in the digital world, whereas with the action of asking passers-by if they have seen this dog before or if anyone is looking for it, the ability to interact and cooperate in the digital world is examined.

3.1.2. Internet Search

As a continuation of the first stage, after the collection of information is completed the gameplay is transferred to the house so that the search process continues on the internet, on websites for lost dogs (Figure 3). During the player's use of the search engine there will be suggestions regarding keywords that the player should select in order to perform the search. The player is asked to choose the most suitable ones, and then select a navigation site among the resulting links. The purpose is to select the most suitable sites based on the titles displayed.

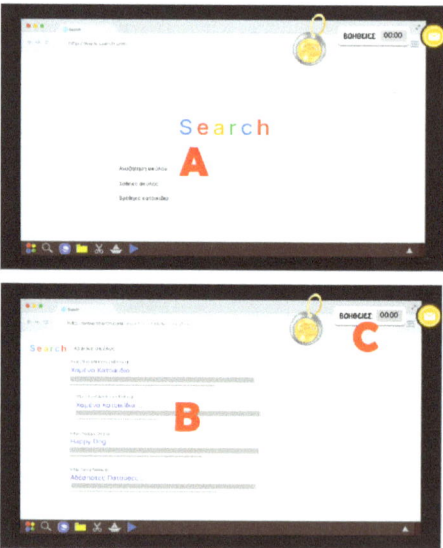

Figure 3. The use of the search engine and the resulting websites: (A) Search keywords options from top to bottom: Dog search, Lost dog, Found pets, (B) Resulted websites after search: the text in Greek represents sites names. (C) Help area: Hints.

After navigating to several sites, one of them displays a pop-up window in which the player is asked to accept certain terms in order to continue browsing the site (Figure 4). However, the conditions he/she is asked to accept violate player's privacy, basically it is a form of attack. In order not to fall into the "trap" the player must not accept the terms, or notice at the edge of the pop-up window the exit button to close it.

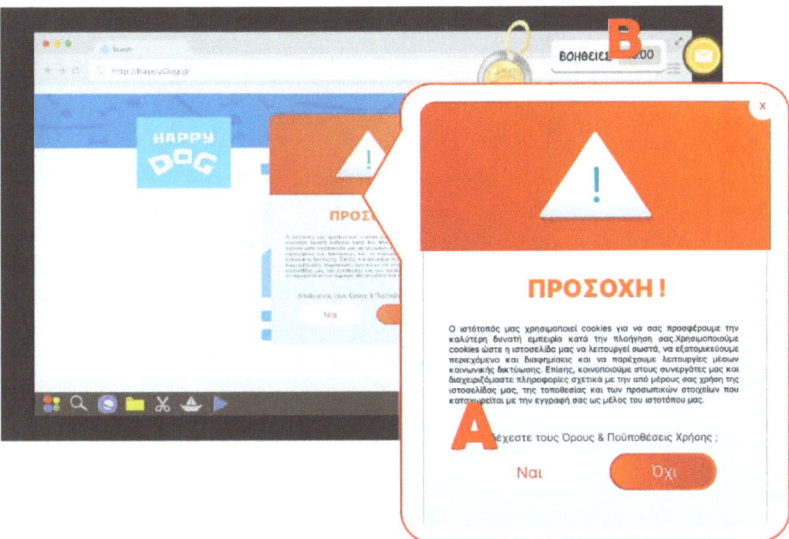

Figure 4. The malicious pop-up window: (A) Pop-up window with terms to be accepted or not: the text in Greek refers to the message body of the terms the player is asked to accept or not, (B) Help area: Hints.

The digital areas explored at this stage are *Digital Literacy* and specifically *Media and Information Literacy*. It examines the player's ability to search and evaluate the information found based on the chosen keywords, the links to navigate, and use critical thinking. The field of *Digital Security* is also being investigated, with the aim of managing personal protection in cyberspace (*Personal Cyber Security Management*), as it is pointed out whether the player can perceive and avoid a potential threat in order to protect his/her personal data.

3.1.3. Ad Posting

In the 3rd stage, since nobody has appeared looking for the dog on the websites visited, the next step is for the main character to create and post an ad about the dog. At this point, the player is asked to create an ad on his/her own containing information and a photo of the dog, as well as some contact details. The whole process is completed as the player chooses from the given options, the most suitable profile for the ad to be posted (Figure 5).

Digital Literacy and *Digital Rights* are, at this stage, the areas of DI tested. Specifically, *Content Creation and Computational Literacy* are examined through the player's ability to combine and create material and information in an effective and creative way online. *Privacy Management* refers to the user's ability to discreetly manage personal information online.

Figure 5. The post options: (A) Post option for the player to choose from: the text in Greek refers to the context of each ad, (B) Help area: Hints.

3.1.4. Owner Authentication

The 4th stage focuses on how the player will manage the responses received after posting the ad (Figure 6). The player is asked to find out which of the people who answered to the ad is the real owner of the dog and which malicious users that are trying to mislead and take advantage of him/her. The game at the end of this stage leads the player to the real owner, with the use of the informative messages, even if the player follows incorrect actions. The aim of the stage is to put the player through the process of authenticating the owner, so even if he/she does not manage to avoid the cyber risks, the information of the real owner is being given to him/her to be able to continue to the next stage.

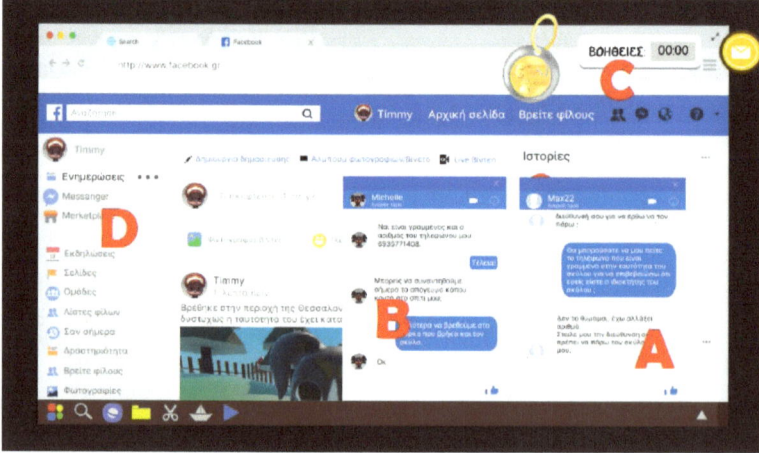

Figure 6. The communication with other internet users: (A) Message box for internet user Max22: the text in Greek refers to the communication being held between the player and the internet user, (B) Message box for internet user Michelle: the text in Greek refers to the communication being held between the player and the internet user, (C) Help area: Hints, (D) Social media web page: the text in Greek refers to a translation of the web page in Greek.

Through the process of verifying the real owner, with the help of the information collected in the first stage, the selection of a meeting point and the arrangement of how to recognize the owner there, the domain of *Digital Rights* is approached. The focus is on *Privacy Management*, which refers to the player's ability to discreetly manage the personal information of others online. Also, it is linked to the goal of *Digital Safety*, with *Behavioral Cyber-Risk Management* being analyzed through the player's ability to perceive and manage cybersecurity risks related to interpersonal online behaviors. The third and final area examined at this stage of the game is *Digital Communication*, based on *Online Communications and Collaboration*, which derives data through the player's ability to communicate and collaborate efficiently with others using the internet.

3.1.5. Returning the Dog to Its Owner

In this stage, which follows the authentication of the owner, the selection of the meeting point for delivering the dog takes place (Figure 7). The player must ensure that getting to the meeting point does not require using any means of transport in which the dog cannot be carried. Upon the arrival of the player and the dog to the meeting point, an animation takes place displaying the dog running to its owner, indicating to the player the dog's real owner.

Figure 7. Returning the dog to its owner: (A) Help area: Hints.

The digital skill that we examine with this action concerns the *Systems analysis and evaluation*, in the composition of which the digital skill of *Digital Co-Creator Identity* is included in a secondary stage. The goal is for the player to connect real-world information, such as not being able to use certain public transportation with the dog, which will help him/her reach areas far from the neighborhood. For this reason, a suitable meeting point must be chosen based on the available means of transportation.

3.1.6. Deletion of the Post & Completion of the Game

Finally, to complete the game, after returning the dog to its owner, the player must consider deleting the post that was uploaded on the previous step (Figure 8).

Digital Communication, in terms of *Digital Footprint Management*, is the digital area under consideration. In detail, it is being examined whether the player considers deleting the post from the internet, realizing the parameters of keeping a no longer valid post on the internet, as well as some personal details.

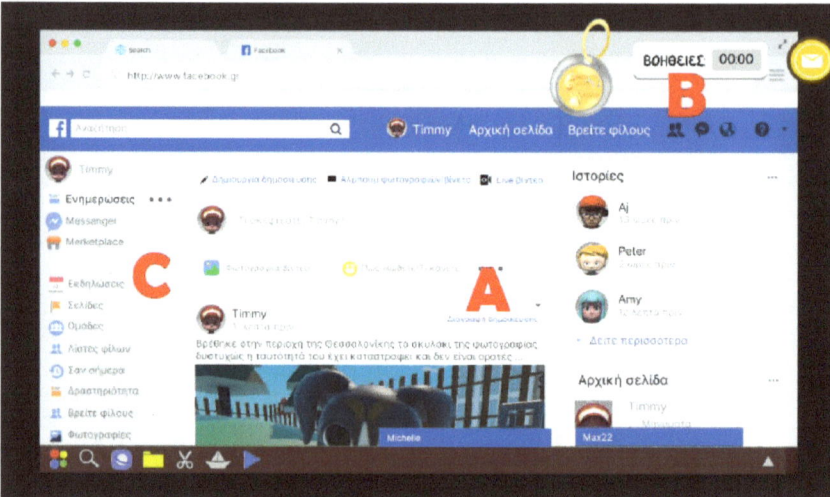

Figure 8. The deletion of the internet post: (A) Post deletion button: Delete Post, (B) Help area: Hints, (C) Social media web page: the text in Greek refers to a translation of the web page in Greek.

3.1.7. Summary of DQ Competencies Covered in the Game

In Table 1, a summary of the DQ competencies covered in each one of the game's stages and player's actions is presented, along with the corresponding competency number and the digital maturity level.

Table 1. "Follow the Paws" stages (column 1) and the corresponding DQ domains (row 1), DQ Competencies (row 2) & Player's actions (competency number—DQ level).

	Digital Identity	Digital Safety	Digital Security	Digital Communication		Digital Literacy		Digital Rights
	Digital Co-Creator Identity	Behavioral Cyber-Risk Management	Personal Cyber Security Management	Online Communications and Collaboration	Digital Footprint Management	Content Creation and Computational Literacy	Media and Information Literacy	Privacy Management
Information Collection	Dog's collar check for contact details. (9—Digital Creativity)			Collect information from surroundings. (14—Digital Creativity)				
Internet Search			Avoidance of maliculus pop-up window. (4—Digital Citizenship)				Correct use of keywords and site selection, while browsing the internet. (7—Digital Citizenship)	
Ad Posting						Efficiently creating the ad by combining the information gathered with the given options (15—Digital Creativity)		Discreet handling of personal information in the ad creation (8—Digital Citizenship)
Owner Authentication		Handling online behaviors and verification of the real owner. (3—Digital Citizenship)		Information collection through online communication and collaboration with internet users. (14—Digital Creativity)				Discreet handling of third-party personal information online. (8—Digital Citizenship)
Returning the Dog to its Owner	Proper selection of a meeting place, for returning the dog to its owner, based on the transport options available. (9—Digital Creativity)							
Deletion of the Post & Completion of the game					Deletion of the post from the internet when it is no longer valid. (5—Digital Citizenship)			

3.2. Design of Follow the Paws

Mitgutsch et al. [6] developed a holistic assessment framework, Serious Game Design Assessment Framework (SGDA), that focuses on the cohesiveness among the essential design elements and the coherence in relation to the games' purpose. Games that are designed with a specific purpose and with the intention to impact the players (purpose-based games) need to conceptualize this purpose in their design process. Following a two-step approach, firstly, the elements that the SGDA consists of, presented in Figure 9, are defined. Secondly, the coherence and cohesiveness is considered based on the overall consistency of the aforementioned design elements with the purpose of the game and the general formation of a confluent whole.

Figure 9. Serious Game Design Assessment Framework [6].

On this basis and in addition to the main aim of "Follow the Paws", to provide an environment similar to the one of the real world, SGDA was selected. In the following paragraphs, we present how the game applied the framework's axes.

- The *Purpose* of this game is to offer a life-like environment that represents realistic cyber threats and contains game missions corresponding to specific DQ competencies. As a secondary impact "Follow the Paws" sensitizes the player to the matter of stray and lost animals.
- The *Content & Information* of the game, refers to the data visible to the player, provided by the game or perceived by him/her during gameplay. Such data include supplementary information about the characters of the game, the back story, the time left to use the aid button, the educational messages that the player receives upon his actions during the game, the dialogues between characters, the messages at the completion of each game stage, the data that the player gathers during the gameplay in order to complete each step, etc.
- The *Game's Framing*, given the young ages of the game's audience, is highlighted to consist of a basic play literacy when no specific knowledge is required from the player to complete the game. It is of high importance to have in mind that we are looking for the spontaneous and unforced behavior (own will) of the player, during the simulation of a physical condition taken from the real world.
- The *Mechanics*, which govern the game, involve all the choices the player makes in order to achieve the main goal of the game, which is to return the lost dog to its owner.

It includes all the actions that the player applies within the operations scope of the game, as well as the time it takes to move from one step to another. What is more is that all these actions contribute to the formation of a profile on the DQ axis, while also weaknesses in player's behavior can be identified, in terms of safe internet use.

- Regarding *Fiction & Narrative*, the storyline that uses a child and a dog as protagonists has been chosen in order to appeal to children of primary school ages regardless of their gender. Representing the main character in an age close to the age of the game's target group makes the game even more relatable to the player, bringing the realism in the forefront. In parallel, the ending goal of the game familiarizes the player with the process required to return a lost pet to its owner, in an environment as close to reality as possible.
- The *Aesthetics & Graphics* of "Follow the Paws" consist of animated cartoon figures, graphics with bright colors and music, corresponding to the age group in which the game is addressed and in favor of constructing an image and content interesting for a child.

Analyzing the coherence and cohesiveness between the SGDA elements (Figure 10), the purpose of offering an entertaining and stress free environment, simulating realistic cyber risks and underscoring DI skills is coherent with the core elements of the game. The infrastructure of the game guides the player, without delivering simplistic answers for the player's next move, reflecting on the ability given to apply his/her own understanding. Moreover, the educational content that is given during the gameplay, based on the player's activity on the internet, allows him/her to first perform even incorrect actions that, in reality, might place him/her at risk, emphasizing on achieving a realistic but yet educational environment.

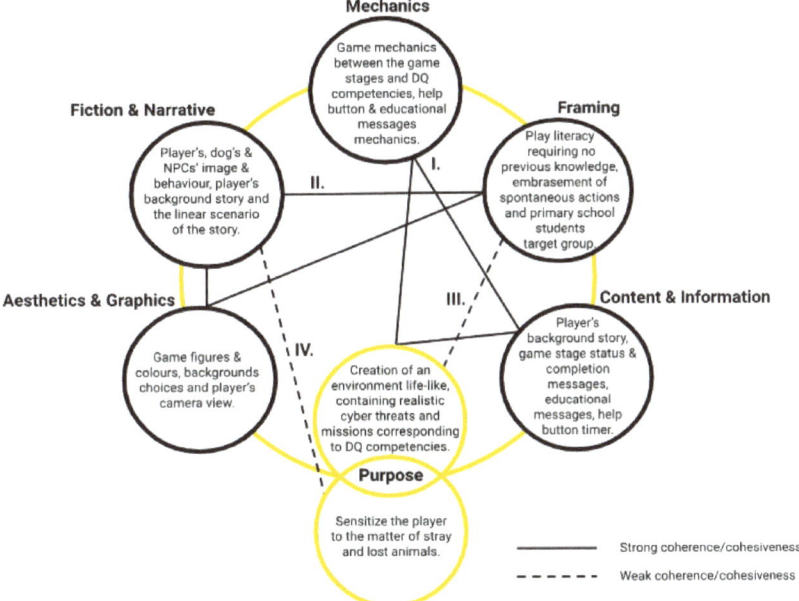

Figure 10. Coherence & Cohesion between SGDA elements for "Follow the Paws".

On a closer look, a strong connection is formed between the game *Mechanics*, the *Content & Information* and the *Purpose* of the game (I in Figure 10), as the game mechanics encourage the player to act freely, while the content provides guidance and feedback. As a result, the correspondence with the DQ competencies remains unbiased and the education on safe internet use takes place. In addition, another triangle of coherence is held between the *Fiction & Narrative*, the *Framing* and the *Aesthetics & Graphics* (II in Figure 10). As a SG designed for primary school ages, it was important to be applicable to all children without being predetermined by the sex or the former knowledge of the player. This framing is represented through the selection of the *Aesthetics & Graphics* and the *Fiction & Narrative*, using bright colors, cartoon-like characters, background and character roles familiar to a child and a linear story that is applicable to real life.

Less distinct connections that are observed among the elements of the game are between the *Framing* and the *Purpose* (III in Figure 10) and between the *Fiction & Narrative* and *Purpose* (IV in Figure 10). Separately, both cases embrace the secondary purpose of the game by providing a storyline that engages the player in taking action to a situation of a lost dog, and make it possible even for a child to contribute to a real life event like this.

4. Materials and Methods

4.1. Context of the Study

The study took place as an activity in the context of an elective undergraduate course on the "Didactics of Informatics" at the Department of Applied Informatics, University of Macedonia, Greece. This course is part of a "Pedagogical and Teaching Proficiency Program" offered by the Department with the aim of preparing students as prospective Information and Communications Technologies (ICT) teachers in public and private schools. One of the modules in this course refers to game-based learning and serious/educational games. Specifically, the benefits of educational games as tools for supporting the teaching and learning of various fields of Informatics, ways of incorporating educational games in the learning process, educational game design and evaluation frameworks are analyzed and critically discussed with students. Among other issues, the MEEGA+ model for evaluating the quality of educational games [7] and the key criteria for game design proposed by Sanchez [9] were presented to students.

During the spring semester of the academic year 2022–2023, the students attending the course were asked to play-test and evaluate the game "Follow the Paws" after the module on educational games and game-based learning had been completed. An announcement was made and delivered to the institutional email addresses of students through the Learning Management System (LMS) used by the Institution and more specifically the course. Students were provided with a link for downloading the game and a link for the online questionnaire (Google form) and had a whole month for play-testing and evaluating the game.

4.2. Research Questions

The study aimed to investigate prospective ICT teachers' acceptance of "Follow the Paws" as a tool for educating primary school students on safe internet use and relevant DI competencies, as well as the anticipated player experience of the game's target group. Consequently, the study aimed at investigating the following research questions:

RQ1: How do prospective ICT teachers evaluate the didactic utility of "Follow the Paws" as a tool for educating primary school students on safe internet use and relevant digital intelligence competencies?

RQ2: What are prospective ICT teachers' expectations of primary school students' experience when playing "Follow the Paws"?

4.3. Participants

"Follow the Paws" was play-tested and evaluated by 213 undergraduate students. The majority of the students were in the 4th year of studies ($N = 178$, 83.5%), which is actually the year of studies that the course is offered (the program of studies lasts for 4 years). The age of the students was in the range of 19 to 46 years old, while the mean age was 22.2 (sd = 2.34). Finally, 70% of the students were male and 30% female.

4.4. Data Collection and Analysis

Data were collected through an online questionnaire prepared as a google form. The questionnaire was based on the MEEGA+ model for evaluating the quality of educational games [7,8] and the key criteria for game design proposed by Sanchez [9], since both instruments have been validated and heavily used for evaluating educational games. The MEEGA+ model and the accompanying questionnaire investigate two quality factors, namely player experience and perceived short term-learning. In our study, we utilized the questions that refer to player experience appropriately adjusted so as to record the perceptions of prospective ICT teachers on the anticipated experience of primary school students when playing "Follow the Paws" (RQ2). From the various key criteria for serious game design and the accompanying questionnaire proposed by Sanchez [9], we utilized questions that refer to game acceptance, usability and didactic utility that as a whole would help us draw conclusions on the overall didactic utility of the game. All the aforementioned questions were actually statements and the participants expressed their level of agreement in a 5-point Likert scale, where 1 = totally disagree and 5 = totally agree. Finally, the questionnaire included demographics questions, questions on the participants' game play habits and open source questions on the games' positive and negative aspects, as well as problems encountered during game play.

5. Results of the Evaluation

In this section the results of the questionnaire on game acceptance and didactic utility (Section 5.1) and the anticipated player experience (Section 5.2) are presented. Before presenting the main results of the questionnaire, we would like to present some results on the participants' game playing habits and the play-testing process of "Follow the Paws".

The majority of the participants are game players. Specifically, only 7% of the participants stated that they never play games, 36% of them rarely play games and 57% of them play games at least once a month (with 20% playing daily). Consequently, the vast majority of the participants were still active game players. Moreover, the majority of the participants (55%) finished the game "Follow the Paws" in a time range of 10 to 45 min. The time devoted by the majority of the participants (61%) on play testing (no matter if they finished the game or not) was approximately 15 min, while 30% of them played for approximately 30 min and 7% for 45 min.

5.1. Game Acceptance and Didactic Utility (RQ1)

In Table 2, the results on game acceptance and didactic utility of "Follow the Paws" are presented. The participants consider that the educational content of the game (median = 4) is relevant to its purpose and without errors; fits the characteristics of its target group; and the main principles of safe Internet use. What is more important is that the participants agree that there is a balance between game elements and the achievement of the educational objective of the game (median = 4), which is widely accepted to be a great challenge for educational games [16,17]. The tasks within the game are considered to be appropriate to the educational background of primary school students (median = 4), while the game provides scaffolding features. The majority of the participants agree that the game provides both guidance and adequate help (median = 4) and clear and relevant feedback to students about their choices during game playing (median = 4). As a consequence, prospective ICT teachers believe that the game can improve students' knowledge on safe Internet use (median = 4) and, as such, can be used as a supplement in the educational process (median = 4).

Table 2. Game acceptance and didactic utility of the game.

Dimension	Statement	Mean	Std.Dev.	Median
Game acceptance	The content is relevant (no errors)	3.6	0.86	4
	The content fits the characteristics of the students (age, prior knowledge, etc.)	4.1	0.73	4
	The content fits the main principles of safe Internet browsing	4.3	0.7	4
Usability	The game provides guidance and adequate help to the students	3.6	1.05	4
	The game provides clear and relevant feedback to students about their choices during game playing	3.8	0.96	4
Didactic utility	The game can be used as a supplement during the educational process	4.1	0.83	4
	The game is suited to the pedagogical objectives of the teacher	4	0.81	4
	The tasks of the students within the game are relevant with their educational background	3.9	0.83	4
	There is a balance between game elements and the achievement of the educational objective	3.9	0.73	4
	Through this game students improve their knowledge on safe Internet browsing	4.1	0.81	4

5.2. Anticipated Player Experience (RQ2)

In Table 3, the results on the anticipated player experience are summarized. The results are quite encouraging, since 26 out of the 29 questions had a median value of 4 and 3 questions a median value of 3. Next, we briefly analyze the results for each dimension of the MEEGA+ model that refers to player experience.

Aesthetics. For the first dimension, the evaluation indicated that the game design is quite attractive for young students (median = 4) and that the text font and colors are well blended and consistent (median = 4).

Learnability. The participants are divided as to whether the students will have to learn a few things before they can play the game (median = 3). However, they agree that learning to play the game will be easy for the students (median = 4) and they will learn to play it very quickly (median = 4).

Operability. Regarding the operability of the game, both questions managed to reach a median of 4 (agree), meaning that the game is easy to play and the rules are clear and easy to understand. However, as will be explained in the results of the open-ended questions, the participants proposed to provide more guidance to the player about the tasks that have to be carried out as the story evolves.

Accessibility. For the accessibility dimension, the participants agree (median = 4) that the fonts used in the game are easy to read and the colors are meaningful.

Error prevention and recovery. The participants agree (median = 4) that the game prevents students from making mistakes, and in case of a mistake, it is easy to recover from it quickly. Nevertheless, the mean value of the answers received ranges from 3.5 to 3.7 and this might be another alert for providing further guidance to the players in the case of unsuccessful actions.

Confidence. The confidence dimension has been evaluated through questions regarding the impression of the players on how easy the game will be and how confident players are that they will learn with this game. For both questions, the median was 4 (agree).

Challenge. Regarding the challenge dimension, the median is 4 as well (agree) for all its items, while the mean ranges from 3.5 to 3.9 and the standard deviation ranges from 0.88 to 1.02. The participants consider that the game is appropriately challenging for young students and provides new challenges at an appropriate pace, but when it comes on to how monotonous the game can become as it progresses, the mean is 3.5 and the responses were not so close to the median (Std.Dev.: 1.02).

Satisfaction. The participants agree (median = 4) that players will have the feeling of accomplishment by completing the tasks based on their own efforts; they will learn from the game and they would recommend it to other students.

Table 3. Anticipated player experience.

Dimension	Statement	Mean	Std.Dev.	Median
Usability: Aesthetics	The game design is attractive for young students (interface. graphics., etc.).	4.1	0.64	4
	The text font and colors are well blended and consistent.	4	0.92	4
Usability: Learnability	Students will have to learn a few things before they can play the game.	3.2	1.02	3
	Learning to play this game will be easy for students.	3.9	0.87	4
	I think that most students will learn to play this game very quickly.	3.8	0.90	4
Usability: Operability	I think that the game is easy to play for young students.	3.8	0.90	4
	The game rules are clear and easy to understand.	3.8	0.96	4
Usability: Accessibility	The fonts (size and style) used in the game are easy to read.	4.2	0.78	4
	The colors used in the game are meaningful.	4	0.84	4
Usability: Error prevention and recovery	The game prevents students from making mistakes.	3.5	0.99	4
	When the student makes a mistake it is easy to recover from it quickly.	3.7	0.92	4
Confidence	When the student first looks at the game, s/he will have the impression that it will be easy for him/her.	3.9	0.89	4
	The contents and structure will help the student to become confident that he/she will learn with this game.	3.8	0.78	4
Challenge	This game is appropriately challenging for a young student.	3.9	0.91	4
	The game provides new challenges (offers new obstacles. situations or variations) at an appropriate pace.	3.7	0.88	4
	The game does not become monotonous as it progresses (repetitive or boring tasks).	3.5	1.02	4
Satisfaction	Completing the game tasks will give students a satisfying feeling of accomplishment.	4.1	0.84	4
	Managing to advance in the game will be due to the student's personal effort.	4	0.79	4
	Students will feel satisfied with the things that they learned from the game.	3.8	0.8	4
	I would recommend this game for young students.	4.1	0.85	4
Fun	Students will have fun with the game.	4	0.86	4
	During the game I believe that there are features that would make a young student smile (game elements. competition., etc.).	3.9	0.82	4
Focused Attention	There is something interesting at the beginning of the game that will capture the attention of a young student.	3.8	0.92	4
	A young student could be so involved in the gaming task that would lose track of time.	2.8	1.13	3
	A student could forget about his/her immediate surroundings while playing this game.	2.7	1.08	3
Relevance	The game contents are relevant to the interests of a young student.	3.7	0.84	4
	It is clear to me how the contents of the game are related to the investigation of digital competencies.	4	0.8	4
	This game is an adequate method for investigating digital intelligence.	3.6	0.86	4
	I suggest investigating and teaching students about the correct digital behavior with this game in comparison to learning through other ways (e.g., other teaching methods).	4.1	0.79	4

Fun. The participants believe that young students will have fun while playing the game, while there are also some features that will make them smile (median = 4).

Focused attention. In this dimension, the answers ranged from indifferent (median = 3) to agree (median = 4). Although the participants believe that the game will capture the attention of the players at the beginning (median = 4), they are divided as to whether the players will lose track of time (median = 3), or forget their immediate surroundings while playing the game (median = 3). This is a clear indication that effort should be made in order to achieve a higher immersion of players.

Relevance. The participants agree (median = 4) that the game is relevant to the interests of young students, it is related to educating them on digital competencies and DI and it is recommended as a teaching method for correct digital behavior.

5.3. Positive and Negative Aspects of the Game

The questionnaire included three open-ended questions on the positive and negative aspects of the game, as well as the problems encountered during game play.

The most positive aspects of the game according to the participants are the graphics, the colors and the overall environment of the game. The game is considered to be interesting, entertaining and interactive. Another strong aspect of the game is its educational nature and at the same time its simplicity.

The participants' responses regarding the negative aspects of the game and the problems encountered during game playing were at a high degree similar as expected. We must note that 48% of the participants did not report any problems during game playing. The most prominent negative aspect or else problem refers to the guidance offered to players regarding the next steps that they should follow. Several participants mentioned the interaction with the virtual computer in the game and the hitboxes (e.g., buttons that should be pressed) as problematic, which required restrictive precision from the player. Although the game provides help/hints for the next steps, this feature is not available at all time and some participants proposed guiding the player with other ways as well, such as *"highlighting with an arrow image the button that should be pressed in the virtual computer"* or *"a bright outline rectangle enclosing elements in the game that the player can interact with"*. The participants proposed several improvements of this type that will be taken into account for improving the game prior to its usage in the classroom. We have to note, however, that some of the interactions that were characterized as vague were conscious design choices and were designed as puzzles for cultivating the players' problem solving capabilities. Another problem mentioned was the movement of the protagonist of the game with the mouse instead of the arrow keys or WASD, as well as the fact that the mouse cursor was visible in the game environment. Some participants mentioned as a negative aspect the limited number of levels and the duration of the game and proposed adding more levels for covering other aspects of safe Internet use and DI as well. Finally, the participants mentioned some problems with the colliders of specific game objects, as well as problems that might be attributed to bugs. Taking into account the number of the participants (213), we consider the feedback invaluable and it will be utilized for implementing the final version of the game.

6. Discussion

Based on the results of the evaluation by prospective ICT teachers, "Follow the Paws" received, for all dimensions of the questionnaire, a median of 4 (agree), except for the learnability and the focused attention dimensions that received an average of 3.67 and 3.33, respectively. In general, the participants positively evaluated the didactic utility of the game based on close-typed questions adopted by key criteria for game design and the corresponding questions proposed by Sanchez [9], implying the acceptance of the game as a supplement in the educational process. Furthermore, the questions adopted from the MEEGA+ model on the anticipated player experience indicated that the game's user interface and experience is suitable for primary school students.

However, it should be noted that through the open-type questions commented on the game, there were some interesting points to be noted for further improvements and expansion of the game. These comments included:

- Uncertainties on the use of the game by students unfamiliar with the use of computers and social media.
- Improvements on the game controls, in order to simplify player's movement while playing.
- Providing the player with an introduction to the game's purpose, in order to familiarize him/her with the aim of not receiving direct guidance through the game, but only minor hints to encourage their own thinking and problem solving skills.

Of course, in order to draw safe conclusions, the game should be tested also by primary school students themselves, in order to obtain feedback from the game's main target group on their experience and also assess the impact of the game on acquiring safe Internet use and related DI competencies. Furthermore, the game might be good to be translated to English (currently it supports only Greek) to help expand the use of the game.

7. Limitations and Future Work

The study investigated the acceptance of "Follow the Paws" by prospective ICT teachers in terms of its didactic utility and the anticipated player experience. The participants evaluated the game positively and pointed out various improvements. Since the majority of the participants were active players and were also knowledgeable about game-based learning and the instruments utilized in the evaluation of the game, the results are considered trustworthy. A major limitation of the study lies in the fact that the game was not evaluated by the actual target group, which is primary school students. Consequently, the actual effectiveness of the game on raising awareness on safe internet use and acquiring DI competencies remains to be investigated in the classroom after revising the game based on the recommendations recorded in its pilot evaluation.

"Follow the Paws" could be further enriched with new missions and intermediate steps between stages in order to include even more areas of DI. Other design frameworks could be utilized in this re-design process, such as the Activity Theory-based Model of Serious Games (ATMSG) [18], for achieving even better results in terms of achieving the pedagogical goals of the game. Additionally, it would be of great interest to measure, display and save the results and the progress of the player in the game. Saving the player's progress and results in a database will give him/her the opportunity of playing the game at any time by continuing from the last saved point and saving his/her progress for further use.

In order to measure and display the progress of the player in the game, a digital skills "map", which captures the progress of players and their actions throughout the game, could be added. More specifically, the purpose of this map will be to display the in-game performance of the player, based on the digital goals that have been set in advance, during the stages and sub-missions of the game from the beginning to its end. In the form of a polygonal web-map, the proficiency level of the skills under consideration will be determined, based on the distance from the center of the web-map to its endings, as shown in Figure 11.

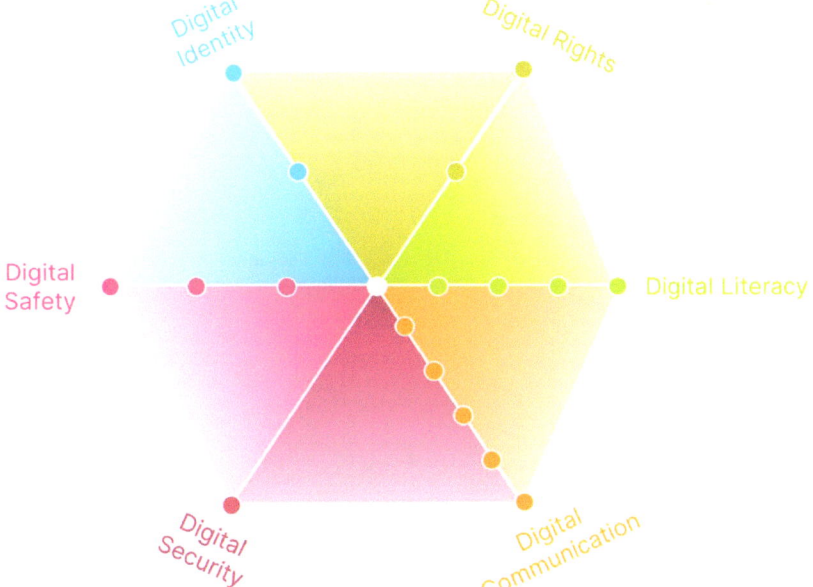

Figure 11. Results Map.

Author Contributions: Conceptualization, A.G.; methodology, A.G. and S.X.; software, A.G.; validation, A.G. and S.X.; investigation, A.G. and S.X.; resources, A.G. and S.X.; data curation, A.G. and S.X.; writing—original draft preparation, A.G. and S.X.; writing—review and editing, S.X.; visualization, A.G.; supervision, S.X. All authors have read and agreed to the published version of the manuscript.

Funding: This research received no external funding.

Data Availability Statement: The data presented in this study are available on request from the corresponding author.

Conflicts of Interest: The authors declare no conflict of interest.

References

1. Jackman, J.A.; Gentile, D.A.; Cho, N.J.; Park, Y. Addressing the digital skills gap for future education. *Nat. Hum. Behav.* **2021**, *5*, 542–545. [CrossRef] [PubMed]
2. Park, Y.; DQ Institute. DQ Global Standards Report 2019: Common Framework for Digital Literacy, Skills and Readiness. Available online: https://www.dqinstitute.org/wp-content/uploads/2019/03/DQGlobalStandardsReport2019.pdf (accessed on 5 May 2023).
3. Zhonggen, Y. A Meta-Analysis of Use of Serious Games in Education over a Decade. *Int. J. Comput. Games Technol.* **2019**, *2019*, 4797032. [CrossRef]
4. Corti, K. Games-based Learning; a serious business application. *PIXELearning* **2006**, *34*, 1–20.
5. Mitchell, A.; Savill-Smith, C. *The Use of Computer and Video Games for Learning*; Learning and Skills Development Agency: London, UK, 2004; Volume 88, pp. 1397–1399, ISBN 1-85338-904-8.
6. Mitgutsch, K.; Alvarado, N. Purposeful by design?: A serious game design assessment framework. In Proceedings of the International Conference on the Foundations of Digital Games (FDG '12), New York, NY, USA, 29 May 2012; pp. 121–128.
7. Petri, G.; von Wangenheim, C.; Borgatto, A. *MEEGA+: An Evolution of a Model for the Evaluation of Educational Games*; Brazilian Institute for Digital Convergence: Florianópolis, Brazil, 2016.
8. Petri, G.; von Wangenheim, C.G. MEEGA+: A Method for the Evaluation of the Quality of Games for Computing Education. In Proceedings of the SBGames, Rio de Janeiro, Brazil, 28–31 October 2019; pp. 28–31.
9. Sanchez, E. *Key Criteria for Game Design: A Framework*; IFE/Ecole Normale Supérieure: Lyon, France, 2011.
10. Blumberg, F.C.; Ismailer, S.S. What do children learn from playing digital games? In *Serious Games: Mechanisms and Effects*; Ritterfeld, U., Cody, M., Vorderer, P., Eds.; Routledge: Oxfordshire, UK, 2009; p. 135, ISBN 978-113-584-891-0.
11. Robertson, D.; Miller, D. Learning gains from using games consoles in primary classrooms: A randomized controlled study. In Proceedings of the World Conference on Educational Sciences: New Trends and Issues in Educational Sciences, Nicosia, North Cyprus, 4–7 February 2009; pp. 1641–1644.
12. Hill, W.A., Jr.; Fanuel, M.; Yuan, X.; Zhang, J.; Sajad, S. A survey of serious games for cybersecurity education and training. In Proceedings of the 2020 KSU Conference on Cybersecurity Education, Research and Practice, Kennesaw, GA, USA, 23–24 October 2020.
13. Weitzman, E.M. Teaching Good Digital Citizenship with Pledge Planets, an Intergalactic Journey from Messenger Kids. Available online: https://messengernews.fb.com/2022/01/20/teaching-good-digital-citizenship-with-pledge-planets-an-intergalactic-journey-from-messenger-kids/ (accessed on 5 May 2023).
14. McCarthy, D. Minecraft: Education Edition Launches a New World to Teach Students About Internet Safety in Honor of Safer Internet Day. Available online: https://news.xbox.com/en-us/2022/02/07/minecraft-education-edition-safer-internet-day-course/ (accessed on 5 May 2023).
15. DQ Institute. Digital Citizenship Test: Cyber-Risk and Digital Skills Assessment Launch. Available online: https://www.dqinstitute.org/news-post/digital-citizenship-test-cyber-risk-and-digital-skills-assessment-launch/ (accessed on 5 May 2023).
16. Silva, F.G.M. Practical Methodology for the Design of Educational Serious Games. *Information* **2020**, *11*, 14. [CrossRef]
17. Natucci, G.C.; Borges, M.A. Balancing Pedagogy, Emotions and Game Design in Serious Game Development. In Proceedings of the 20th Brazilian Symposium on Games and Digital Entertainment, Gramado, Brazil, 18–21 October 2021; pp. 1013–1016.
18. Carvalho, M.B.; Bellotti, F.; Berta, R.; De Gloria, A.; Islas Sedano, C.; Baalsrud Hauge, J.; Hu, J.; Rauterberg, M. An Activity Theory-based Model for Serious Games Analysis And-conceptual Design. *Comput. Educ.* **2015**, *87*, 166–181. [CrossRef]

Disclaimer/Publisher's Note: The statements, opinions and data contained in all publications are solely those of the individual author(s) and contributor(s) and not of MDPI and/or the editor(s). MDPI and/or the editor(s) disclaim responsibility for any injury to people or property resulting from any ideas, methods, instructions or products referred to in the content.

Article

An Educational Escape Room Game to Develop Cybersecurity Skills

Alessia Spatafora [1], Markus Wagemann [2], Charlotte Sandoval [3], Manfred Leisenberg [3] and Carlos Vaz de Carvalho [4,*]

1. Finance & Banking—Associazione per lo Sviluppo Organizzativo e delle Risorse Umane, Effebi Association, 00135 Roma, Italy; effebi@asseffebi.eu
2. ASW Norddeutschland e.V., 22547 Hamburg, Germany; markus.wagemann@aswnord.de
3. Research & Development, Fachhochschule des Mittelstands, 33602 Bielefeld, Germany; charlotte.sandoval@fh-mittelstand.de (C.S.); manfred.leisenberg@fh-mittelstand.de (M.L.)
4. GILT R&D, Instituto Superior de Engenharia do Porto, 4249-015 Porto, Portugal
* Correspondence: cmc@isep.ipp.pt

Abstract: The global rise in cybercrime is fueled by the pervasive digitization of work and personal life, compounded by the shift to online formats during the COVID-19 pandemic. As digital channels flourish, so too do the opportunities for cyberattacks, particularly those exposing small and medium-sized enterprises (SMEs) to potential economic devastation. These businesses often lack comprehensive defense strategies and/or the necessary resources to implement effective cybersecurity measures. The authors have addressed this issue by developing an Educational Escape Room (EER) that supports scenario-based learning to enhance cybersecurity awareness among SME employees, enabling them to handle cyber threats more effectively. By integrating hands-on scenarios based on real-life examples, the authors aimed to improve the knowledge retention and the operational performance of SME staff in terms of cybersafe practices. The results achieved during pilot testing with more than 200 participants suggest that the EER approach engaged the trainees and boosted their cybersecurity awareness, marking a step forward in cybersecurity education.

Keywords: cybersecurity; serious games; escape rooms; educational games; small and mid-sized enterprises

Citation: Spatafora, A.; Wagemann, M.; Sandoval, C.; Leisenberg, M.; Vaz de Carvalho, C. An Educational Escape Room Game to Develop Cybersecurity Skills. *Computers* **2024**, *13*, 205. https://doi.org/10.3390/computers13080205

Academic Editor: Paolo Bellavista

Received: 23 June 2024
Revised: 12 August 2024
Accepted: 14 August 2024
Published: 19 August 2024

Copyright: © 2024 by the authors. Licensee MDPI, Basel, Switzerland. This article is an open access article distributed under the terms and conditions of the Creative Commons Attribution (CC BY) license (https://creativecommons.org/licenses/by/4.0/).

1. Introduction

The frequency of cybercrime cases around the world has increased significantly in the recent years, together with the severity of their consequences. One of the reasons is the ongoing digitization of nearly all spheres of work and personal life, which has not been followed by the needed protective measures and has opened vulnerabilities to cyberattacks. For small and medium-sized enterprises (SMEs), attacks that tap into trade secrets or prevent the regular use of information, resources, or communication channels can mean economic ruin. As such, this rapid evolution of cyber threats requires SMEs to adopt robust cybersecurity measures to protect their online transactions and their sensitive data, to safeguard their operations and maintain customer trust [1]. SMEs must also navigate the complex landscape of legal and regulatory requirements, especially in the context of protecting not only their businesses, but also their customers' data and privacy [2]. But even when SMEs recognize the dangers of cyberattacks, they often lack comprehensive defense mechanisms, and/or they do not have the adequate resources or knowledge to implement them [3]. Yet, SMEs need proactive, informed, and strategic approaches to cybersecurity and must adopt scalable security measures that are adjusted to the rapidly changing cyber threat landscape [4–6]. Establishing tailored cybersecurity frameworks is crucial for SMEs to address their unique vulnerabilities and protect critical assets. Such frameworks can help SMEs identify their sensitive data, assess the risks, and implement appropriate safeguards based on organizational needs and asset sensitivity [7].

Effective cybersecurity for SMEs involves not only adopting technological solutions, but also fostering a culture of security within the organization. So, it is highly important to inform employees about the effects and consequences of a cyberattack, to raise their awareness so that they engage in proactive risk management practices. Employees should receive in-house training at regular intervals so that the probability of a successful attack is limited and, in the event of an attack, adequate procedures are immediately taken. This training should be tailored to the specific needs and contexts of the SMEs, considering their resource constraints [8–11].

The EyesOnCS (enhancing cybersecurity through the development of training using an escape room model) project aimed to develop innovative solutions and concrete educational products to prepare current and future SME employees for working safely in the digital world. The main project result was a digital Educational Escape Room (EER) focused on cybersecurity that fostered active and experiential learning through a scenario-based learning approach depicting realistic situations from everyday work and life. The reason for choosing this method was the acknowledgement that this type of tool is gradually gaining attention within education and training as it provides an engaging approach that allows learners to efficiently achieve the predefined learning objectives [12,13]. Also, training with these tools allows learners to have a close reference to the working reality, which translates into higher knowledge retention and the immediate transfer/application of this knowledge to the workplace [14,15]. This learning approach gives the learners a self-managed and self-determined role as it is an active process of making choices, assessing options, observing the consequences, and reflecting on them. By acting and reacting to those situations, learners gain experience and develop their skills.

A large-scale testing and validation process with more than 200 participants was carried out in Germany, Italy, and Portugal with the engagement of the main target groups. After playing the game, participants provided feedback on their satisfaction and acceptance of the proposed methodology and tools, as well as their perception of the relevance and effectiveness of the approach.

2. Materials and Methods

Video games, which emerged as consumer products around half a century ago, have evolved significantly from their original form as leisure activities to become integral elements of social and cultural landscapes. The influence and presence of games in society have been steadily escalating, marking them as crucial components in various aspects of life. According to Goertz et al., these interactive mediums have transcended mere entertainment to embed themselves deeply within the fabric of modern culture [16]. Games, inherently endogenous systems, are composed of structured problem-solving activities governed by specific mechanics and rules. These elements collectively foster engagement, with players participating out of internal motivation. Gameplay dynamics and mechanics, the core of these systems, facilitate varied interactions between players and game elements, leading to diverse behaviors and outcomes which engender a sense of immersion, as well as a state of deep mental involvement that contributes significantly to critical thinking and logical reasoning. Furthermore, games are instrumental in honing other cognitive, intrapersonal, and interpersonal skills, including perceptiveness, attention, memory, and various analytical abilities [17].

Videogames have, since then, been repurposed beyond entertainment. Recognizing the motivational and engaging aspects of gameplay, designers have leveraged these attributes for educational and training purposes. These games, also known as "serious games", are not primarily aimed at entertainment—they are committed to capitalizing on the inherent motivation and immersive experience of players, and use effective game mechanics to develop skills, knowledge, and/or awareness in an entertaining environment [18].

The education sector has achieved great success with the integration of these games (also referred to as educational games), which has led to the concept of "game-based learning" (GBL). For this purpose, games are designed with explicit learning objectives. As Prensky notes, GBL combines the engaging aspects of gaming with educational content, creating an interactive learning experience that motivates users to explore and deepen their knowledge [19]. Abt outlines the benefits of educational games and emphasizes the alignment of game objectives with educational goals; the promotion of the understanding of abstract concepts; the promotion of critical thinking; real-time feedback; and the provision of safe environments for the exploration of consequences and authentic assessment [20]. Educational games allow students to improve their social skills, teamwork, leadership, and collaboration. In addition, they are beneficial in training for hazardous processes (as is the case with cybersecurity applications and training) or situations where physical classrooms are too expensive. In short, the advantages of game-based learning can be summed up as follows:

- A playful addition to teaching methods, with a high motivation potential among learners through a sense of achievement in the game.
- Trackable active participation, widening the scope for action through interaction.
- Sustainable engagement with learning content.

Some of the disadvantages of game-based learning include the following:

- Learning content can be (partly) inferior.
- Requirements for technical equipment in some cases imply high costs.
- The games are very specific to a certain subject or domain.
- The teaching staff requires previous training [21].

Experiential learning, originally proposed by David Kolb as the process of learning by doing in a safe environment [22], provides the theoretical ground for game-based learning together with the active learning theory, proposed by Bonwell and Eison [23]. By engaging students in hands-on experiences and reflection, they become better able to connect theories and knowledge learned in the classroom to real-world situations. Educational Escape Room (EER) games are an experiential and active educational variety of the escape room game concept, targeted at the development of problem-solving, communication, collaboration, creativity, and critical thinking skills. Escape room games present an exciting narrative set in one or more fictional locations, such as prison cells, dungeons, laboratories, or even space stations, and the team of players is required to discover clues and solve puzzles aligned with the overall theme to achieve the final victory (normally escaping the room) within a limited time. The game normally begins with a brief introduction to the rules of the game, delivered by video, audio, or by a live gamemaster, and then players enter the room or area where they will be locked throughout the gameplay period. The game's challenges are generally more mental than physical, and different knowledge and skills are required for different types of puzzles. If players get stuck, there is normally a mechanism in place through which they can ask for hints that can be delivered in written, video, or audio form, or by the live gamemaster. Good endings are usually represented by escaping "alive" within the time limit (that is, completing the room's objective), while bad endings usually involve the players getting "killed" by the main driving force of the story or an antagonist coming to get the players once the timer has run out.

Virtual, digital, or online escape rooms are digital counterparts of physical escape rooms and take place through a computer and network. Like in physical escape rooms, the players solve riddles and complete puzzles within a limited amount of time. Escape room game software is used, which is either run by the players alone or by the gamemaster. This means that such games can be played by one player alone or by several players as a team. More complex digital escape rooms can use virtual reality to increase the sense of immersion of the players.

For some years now, the academic and vocational training sectors have recognized the benefits of escape rooms and have been using them for their own purposes. There are already various scientific studies worldwide on the effectiveness and use of Educational Escape Rooms (EERs), although this is far from systematic, as shown by Tercanli et al. [24]. These studies show that EERs can be used in different phases of the learning process, so while some EERs do not require any prior knowledge and enable the basics to be learnt, others require prior knowledge and are designed to deepen that knowledge [25]. Both physical and virtual EERs are effective learning methods and show a significant increase in students' knowledge after the experience and a higher retention rate [26]. For instance, the Cyber Defense Tower Game allows players to defend servers from various cyberattacks, promoting strategic thinking and problem-solving skills [27]. Identically, the CyberHero game, which incorporates adaptive learning techniques, offers personalized experiences that can significantly improve user engagement and learning outcomes [28]. Similarly, the Cyber Secured game is designed for cybersecurity novices and has been shown to enhance learning and retention of cybersecurity concepts [29]. The CyberNEXS game provides a platform for simulating cyber challenges that cater to a wide range of users, from casual computer users to advanced cybersecurity professionals, illustrating the versatility of games in cybersecurity education [30]. The educational impact of cybersecurity games is further enhanced by their ability to simulate adversarial thinking, a critical component in cybersecurity training, as they can improve strategic reasoning by simulating hacker strategies [31]. Games can also be designed to boost interest in cybersecurity careers, especially among younger audiences. The GenCyber program demonstrates how games can increase awareness and interest in cybersecurity, potentially addressing the skills gap in this critical field [32]. The virtual escape room CySecEscape 2.0, designed to raise cybersecurity awareness among small and medium-sized enterprises, demonstrates how physical escape rooms can be successfully adapted into virtual environments without losing player immersion [33]. The escape room designed by the Institut National des Sciences Appliquées de Toulouse offers hands-on experience in recognizing and mitigating cyber threats, emphasizing the importance of choosing strong passwords and identifying phishing emails [34].

On the other hand, scenario-based learning (SBL) allows the integration of realistic situations from everyday working life into learning scenarios, according to the learning objectives. Scenario-based learning in combination with the EER game model is very effective for learning, as it can provide a realistic context together with an emotional connection, which increases motivation and accelerates the acquisition and retention of knowledge. The reference to realistic situations has a very positive learning effect, as students can better transfer the learned lessons to the working environment in a company. By using this approach, it is possible to provide scientifically evaluated and content-tested scenarios for educational purposes.

The adoption of this combination (EER + SBL) within the EyesOnCS project was expected to provide a better understanding of cybersecurity while increasing the engagement and the learning experience through interactive and emotional learning, as it...:

- Provides real-life experiences: Players are confronted with real-life experiences that may occur daily while working with a computer connected to the Internet. In addition, there is a convergence between the use of the Internet during work in the office and daily life.
- Provides immediate feedback on the consequences of online activities: During the game experience, the player can make their moves or decisions in a safe space where it is even possible to observe the direct consequences of various decisions. This makes it possible to gain experience and further information on the subject. The possible outcomes, such as identity theft, credit card fraud, etc., can be experienced in a game without real consequences.
- Ensures improved knowledge retention: game immersion and the immediate observation of eventual consequences lead directly to increased knowledge retention as the player/student is able to experience each choice and consequence during their playtime.

2.1. EyesOnCS Educational Escape Room

The EyesOnCS game was planned to be mostly oriented towards self-diagnostics and self-assessment in relation to cybersecurity, as it was mostly thought to be used in an informal, autonomous learning process by SME staff. In the scope of vocational training, the game can be used by VET trainers as an educational tool. But then, a formal assessment should be conducted through an external procedure.

Following a requirement study conducted with a set of SMEs and desk research analyzing a large number of cyberattacks, the specific learning objectives for the EyesOnCS EER were defined as:

- Expanding the knowledge of cyberattacks and cybercrime methods.
- Strengthening security awareness while dealing with Internet applications like e-mails, instant messengers, etc., on computers, smartphones, or any other connected devices.
- Acquiring knowledge of basic relevant behavioral principles and how these are used by cyber criminals, e.g., in social engineering.
- Strengthening physical awareness like the importance of the clean desk policy, information security, etc.
- Recognizing the important functions of social media while acknowledging its potential for use for criminal purposes.

These learning objectives were broken down in more detail by taking a closer look at the different attack methods identified from real cases of cybercrime, like phishing, password attacks, vishing, smishing, and social engineering. Something that all these attacks have in common is the exploitation of the so-called human factor, i.e., people who might commit mistakes out of ignorance or gullibility.

Phishing: Phishing is the most common attack method used by cyber criminals to gain access to other computer systems, and phishing mails are by far the number one attack method. The learning objectives regarding phishing for the project were:

- Building knowledge of typical phishing characteristics in emails.
- Understanding basic psychological principles used by criminals to deceive or manipulate people into not following security protocols.
- Raising awareness of the risk of unknown data attachments and the disclosure of personal data.
- Understanding the importance of clean desk policies and information security.
- Know how to get first aid and how to protect themselves.

Password attacks: Easy-to-guess passwords are, even today, one of the main reasons why accounts are so easily hacked. Easy passwords can be cracked by brute force attacks within seconds or less. Yet, the characteristics of a strong password are barely known in the public or workplace. To make the whole situation even more difficult, there is often misinformation circulating around passwords. Even the German Federal Office for Information Security advised a few years ago that it is important to change passwords on a regular basis. This advice fails to take two aspects into account. The first is that passwords are not like vegetables, so they cannot go bad that quickly. The second point is the general behavior of people when they need to change their passwords—whether they like it or not—as they tend to just swap out a few numbers, and passwords become easier to guess over time. The learning objectives for this vector attack were:

- Acquisition of basic knowledge about secure passwords.
- Understanding the principles of different methods for guessing, cracking, or otherwise obtaining passwords.
- Knowledge of clean desk policies.
- Understanding the basic principles of social engineering on social media to obtain passwords.
- Recognizing the importance of data that can be shared via social media.

Vishing: Voice phishing, or vishing, is a variation on the classic phishing email. These are phone calls made by psychologically trained criminals to obtain secret information over the phone or to manipulate the victim into doing something that violates security protocols. Since AI voice generators became available, they have also been used to mimic the voices of CEOs to give orders over the phone. The learning objectives for vishing were:

- Understanding the basic principles of social engineering.
- Gaining awareness of suspicious calls and typical call patterns.
- Building knowledge of AI and voice generators.
- Obtaining general information about fraud.
- Gaining knowledge about measures for self-protection.

Smishing: Smishing is a variation of phishing and vishing. Attackers use instant messengers or text messages to involve the victim in a dialogue that is ultimately intended to lead to a demand for money via text, for example. This method is used in both professional and private environments and is one of the latest methods of attack in the world of cybercrime. The learning objectives for smishing are:

- Increasing the ability to recognize suspicious text messages.
- Acquiring the ability to check whether a text message is authentic or not.
- Developing the ability to protect oneself.

Social Engineering: Social engineering exploits psychological and behavioral aspects of human nature [35]. It involves deceiving or manipulating people into revealing sensitive information or performing actions that could jeopardize security. Almost every cybercrime attack method is accompanied or supported by social engineering capabilities. The learning objectives for social engineering are:

- Building knowledge about the topic in general.
- Raising awareness of dangerous messages, emails, and calls from unknown people.
- Acquisition of relevant basic psychological knowledge.

The chosen approach for the training process, together with the set of identified learning objectives, led to the decision to create a new game as none of the existing ones (namely those previously revised) covered all the goals and/or learning objectives, and their source code was not available; therefore, they could not be changed. The EyesOnCS game (Available at: https://www.eyesoncs.eu/results, accessed on 12 August 2024) follows a narrative where the player takes on the role of a new employee in the security department of a bank. During their work, they face various challenges related to the learning objectives previously described.

Episode 1—The Test: The first episode begins with the selection of an avatar and a name. To enter their new workplace, the player must enter an access code. They receive their first clue from the security guard at the door (Figure 1). The player must combine the letters used for the name of the bank, ECS (work place), with the alphabet to obtain the code with which they can enter the bank. This task gives some notions about the cryptographic process. After entering the bank, the player must set up their email account using their first password. Hints are provided around the desk and once the correct password has been used, the player can then select a new password. The player is asked whether the password should be saved on a computer and whether the anti-virus software should be switched off. The answer given then leads to further clues as to how the game could continue. The next task is to check certain emails to see whether they are possible phishing e-mails or not. The player can read and accept the emails as legitimate emails or reveal them as phishing e-mails. If the player identifies all e-mails correctly as legitimate or phishing e-mails, they will remain part of the game. If the player makes mistakes with the allocation of e-mails, they lose time, which might make it difficult to complete the EER. The episode ends with a summary of the topics learnt.

Figure 1. Starting the first episode.

Episode 2—The Job: The second episode begins with a relatively simple task for the player. The initial code for the building has been changed due to a hacker intervention. The player is given a hint about a specific date related to cybersecurity. After a quick Google search, the player can—with a little thought—enter the correct date as the new code. At their workplace, the player now receives several messages from colleagues that introduce them to the topics of phishing and vishing (Figure 2). The first e-mail leads to a suspicious fake website where the player has to find three typical markers for a fake website. Once the player has found these markers, they receive the next message. They are told to go to their colleague and leave their workplace. The player must now fulfil the clean desk policy (take their ID and smartphone with them, shut down their computer) before they can leave the workplace. The next scene in this episode is a phone conversation between the employee and their "supervisor" in which vishing markers are shown. It appears that the bank has been compromised. After the dialogue, the player receives a new phishing e-mail attempt (which they can also accept or reject). The next tasks focus on social media awareness, as the player must find three crucial pieces of information on their supervisor's social media account. Here, they will have to find clues to the password the supervisor uses for his e-mail account.

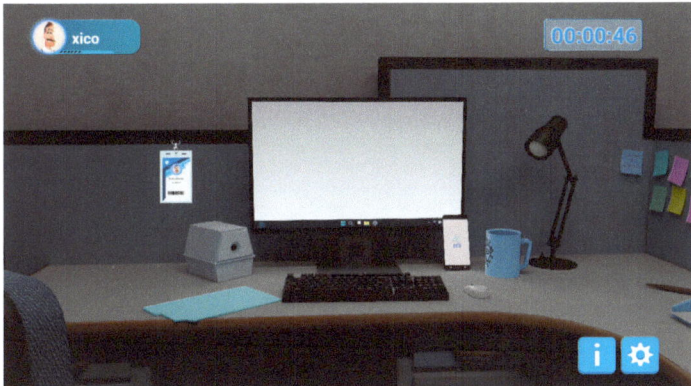

Figure 2. The player's work environment in the bank.

Episode 3—The Hacker: The third episode is called "The White Hacker". In this episode the player role-switches to a so-called "white hacker" story, so that the player can get to know the attacker's point of view. The episode starts with the player receiving a message from the CEO of the ECS bank. He tells us that he knows why the player was fired but he cannot do anything about it but he wants to use our expertise from the outside. As a white hacker he wants us to test the security of the bank systems. We accept the task and start investigating. The first job is to go to the Dark Web and find out what information is available about the company. We are able to retrieve the list of emails addresses of the company and phone numbers of the employees. With this information we prepare and start phishing and smishing campaigns for the employees. While the campaigns are running we start investigating the social media of the employees and realize that one of them posts a lot of information online. With that information we try to discover his access password. We contact again the CEO of the ECS bank and inform him that the information of the company is available on the Dark Web and that some accounts are hackable just by following the social media information of some employees. The CEO is then informed about the procedures he should adopt to make the bank systems secure. He is also informed about the importance of having a CSO (Chief Security Officer) that has full knowledge about cybersecurity.

Episode 4—The Expert: This is a quiz-based challenge that players can take at any moment. Therefore, it can be used for diagnostic purposes (if taken before playing a scenario) or for self-assessment if taken after playing a scenario. The questions for each round are randomly chosen from a large set of questions (more than 100 questions), which ensures that players have a different challenge every time they play it. Questions are associated with one or more of the three other episodes so that, if the player just finished one of those episodes and takes the quiz, the questions will be chosen from those associated with that episode.

2.2. Methodology

The validation of the game was done through a pilot testing process with primary objectives including assessing the educational value of the game, its playability, and its usability; identifying technical issues; and gathering user feedback to improve the game. The pilot testing lasted for three months (October to December 2023) and was conducted in three European countries: Germany, Italy, and Portugal. The participants were trainers and trainees from vocational schools and SME staff in those countries, as we had the intention of collecting data from both points of view. The selection of participants was done by project partners of organizations (schools and SMEs) that volunteered to participate and disseminated the information about the game internally. A hybrid implementation was adopted, combining face-to-face sessions (only in the schools) with the autonomous online playing of the game. In the face-to-face sessions, participants were informed about the project and game, they received a brief tutorial on the gameplay, and then they were able to play it themselves for about one hour. All participants were then allowed to play the game independently and autonomously after the session for about a week. In total, 384 participants were involved in the testing, as depicted in Figure 3.

On average, players spent 46 min playing the game (Figure 4), which shows the involvement of the players. Each scenario has a time limit of 30 min and the expected average play time for each scenario is about 20 min. This means that the players were involved, playing at least 3 scenarios on average.

Also, 84% of the players spent more than 1 min in the game, which shows that the number of "false testers" (that is, testers that just entered out of curiosity and immediately left) was quite reduced (Figure 5).

The total number of plays was 534, which means that each participant played the game between 1 and 3 times.

After playing the game, participants were asked to participate in an anonymous online survey. This meant that some players answered the questionnaire immediately after the face-to-face session, and others only answered it after one week. Unfortunately, due to the hybrid mode of the pilot testing, it was not possible to ensure that all the players did provide an answer so, in the end, only 77 participants provided feedback and comments.

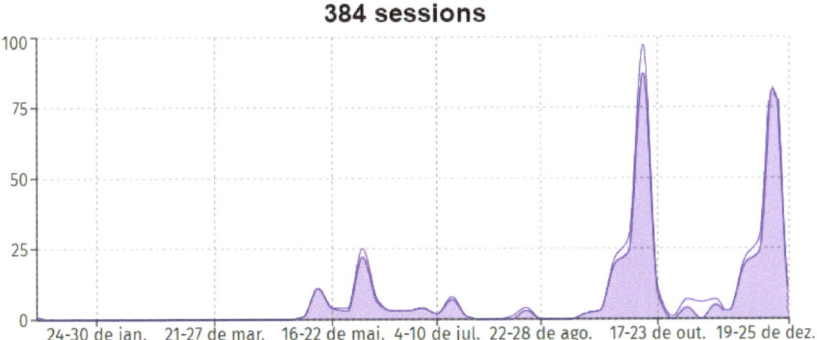

Figure 3. Game players throughout the entire testing period with special emphasis in pilot testing from October to December.

Figure 4. Average playing time during the same period.

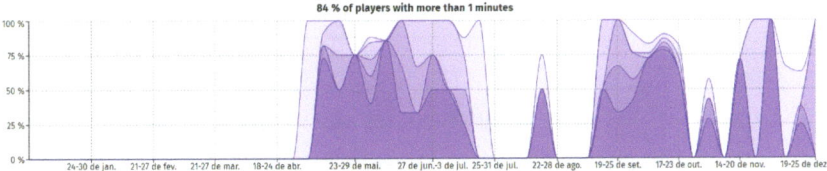

Figure 5. Players with more than 1 min of playing time during the same period.

3. Results

The questionnaire addressed some demographic aspects of the participants, as well as their perceptions of the usability, playability, and efficacy of the game.

Country of the respondents

- Portugal (5)
- Germany (23)
- Italy (12)
- Ukraine (7)—this group was part of the German pilot test.

The distribution of the players is slightly unbalanced between the different countries involved (more in Portugal and Germany, less in Italy) but, as there was no intention of doing a country-based analysis, this distribution did not introduce any issues in the analysis of the results.

Age
- 10–20: 38
- 20–30: 16
- 30+: 23

The age distribution also roughly shows the difference between trainees and trainers/SME staff, which can be established as 1/3 and 2/3 of the total participants. This distribution was planned at the beginning of the project. In Italy, all the participants were younger than 30 years old, while all the Ukrainian participants were older than 30 years.

Sex
- Male: 46
- Female: 27

There was some predominance of male participants among the trainees (34 vs. 17), while for the older participants, the distribution was more balanced (13 vs. 10).

Professional occupation
- Trainer, teacher, equivalent (19)
- Student, trainee (51)
- Educational manager (3)
- Other (4)

These numbers reflect the distribution of students/trainees and teachers/trainers/SME staff that was also observed across the age categories. In Italy there was only one trainer and all the Ukrainian participants were either trainers or SME staff.

Usability of the game

Feedback on the usability of the game was collected using a standard System Usability Scale (SUS) questionnaire where participants quantified their agreement with a set of 10 statements [36]. Their answers were then normalized to the interval 1–5, with 1 representing strong disagreement and 5 representing strong agreement (Table 1).

Table 1. Usability of the game.

Statement	Score (1–5)
I think that I would like to use this game frequently.	3.82
I found the game unnecessarily complex.	2.12
I thought the game was easy to use.	3.68
I think that I would need the support of a technical person to be able to use this game.	2.12
I found that the various functions in this game were well integrated.	3.83
I thought there was too much inconsistency in this game.	1.96
I would imagine that most people would learn to use this game very quickly.	4.81
I found the game very cumbersome to use.	2.12
I felt very confident using the game.	3.71
I needed to learn a lot of things before I could get going with this game.	2.21
Final SUS score	**76.7 (GOOD)**

The SUS score is close to 80—this means that the usability of the game is GOOD according to the standard analysis, but close to VERY GOOD (ratio: 80). For the trainees, the SUS score was higher than the one reported by the trainers/SME staff (77.8 vs. 74.4), although this difference is not significative and is expectable as they had more game-playing experience. Participants found the game easy to learn to use, with the functions well integrated, and they would use it frequently. They felt the need to learn a lot, which is connected to the learning nature of the game. Nevertheless, there are clearly some improvements to be made in relation to the usability and learning curve of the game.

Playability of the game

The playability of the game was analyzed using the standard Game Experience Questionnaire (GEQ) [37,38]. A set of 14 statements was used and players indicated their agreement with the statements through a Likert scale (5 levels), where 1 represented strong disagreement and 5 represented strong agreement (Table 2).

Table 2. Playability of the game.

Statement	Score (1–5)
I was interested in the game's story	4.05
I felt successful	3.77
I felt bored	2.12
I found it impressive	3.29
I forgot everything around me	2.66
I felt frustrated	2.16
I found it tiresome	2.01
I felt irritable	2.04
I felt skillful	3.61
I felt completely absorbed	2.88
I felt content	3.60
I felt challenged	3.87
I had to put a lot of effort into it	3.82
I felt good	3.83

On the positive side, players were interested in the game's story (4.05), they felt challenged (3.87), and felt good while playing the game (3.83). They did not feel bored, tired, or irritable. However, they were not totally absorbed by the game (2.85) to the point of forgetting everything around them (2.66). Older participants (trainers and staff) had a higher degree of interest in the game's story (4.25 vs. 3.96), felt better playing the game (3.96 vs. 3.77), and felt much more absorbed (3.42 vs. 2.64). Younger players (trainees) did not feel irritable (1.89 vs. 2.38) or bored (2.00 vs. 2.38) and felt more successful (3.85 vs. 3.58). But in general, the differences between the two groups were not big.

Completing the game

Considering the participants that provided feedback, 59.7% (46) of them completed the game (completing the game meant that players went through the three scenarios and answered the quiz (one or more times)), which shows that most of the players were able to receive all the information and training about cybersecurity. Naturally, the other 31 participants (40.3%) did not finish the game, which means that they did not get through all the challenges and, therefore, were not able to acquire all the knowledge and skills. This was to be expected as this is a game and should be challenging, but there was a consensus that the hint system embedded in the game should be improved to ensure a higher finishing ratio. On average, players took 25 min to complete a scenario. Players that finished the game used five hints, on average, which was the expected number of hints for a smooth progression.

Players that did not finish the game considered it either too difficult (18) or too hard to understand how to play (4). Just a few considered it boring (2) or not interesting (1), which is a good ratio.

Qualitative feedback

Participants were asked to provide some qualitative comments about the game and the pedagogical methodology, and also some suggestions to improve the game. In relation to the game, the comments provided by the participants were very positive (in parenthesis are the number of participants that made a certain statement):

- A great idea, a positive game, it works very well (14).
- A very interesting game, I liked the challenges that made me think (7).
- I like the graphics but they can still be improved (4).
- I had fun (2).
- It was easy to use (2).
- A nice idea, it still needs some improvement but I would definitely use it with my students (2).
- The game was easy to use and I learned a lot. The part with the password was very interesting.
- I really liked the game but some puzzles are too difficult (9).
- Give more freedom to the player so that we do not need to follow a predefined path (3).
- More hints are needed at the beginning so you know what to do.

Equally positive feedback was received in relation to the learning process:

- I like the idea, it makes sense with our students, it is a nice and interesting way to learn (15).
- Very positive (13).
- Adequate and effective (7).
- I learned a lot about cybersecurity (7).
- It is interesting to have the learning outcomes. It explains what we have learned (2).
- I liked the learning approach with the riddles (2).
- I could use it in my classes to discuss concepts with my students (2).
- It works very well, I enjoy learning with games (2).
- It gave me some interesting guidance on creating a secure password. I think I will use the game's method in the future.
- It is cool to see that some of the points we made are useful for everyday use, and you remember the game if you are in a similar position at work.

4. Discussion

The SUS tool provided an overall score of 78.7, categorizing the game as "GOOD" and bordering on "VERY GOOD" (score 80). This high score suggests that users generally found the game easy to understand and use (4.81), with well-integrated functionalities (3.83). This indicates that the game's design facilitates a smooth learning curve and cohesive user experience. However, some users felt the game was complex (2.12) and would occasionally require technical support (2.12), suggesting areas where simplification could improve accessibility. The confidence levels in using the game were reasonably high (3.71), although there is still room for improvement to make the interface more intuitive.

The GEQ tool results offer insights into the emotional and cognitive aspects of the game experience. The game effectively engaged users, with high scores for story interest story (4.05) and feeling challenged (3.87). This engagement is crucial for educational games where user interest directly impacts learning outcomes. Participants reported lower scores for feeling absorbed (2.88) and forgetting surroundings (2.66) which suggests that the game may struggle to maintain deep immersion, which could impact the sustained attention and learning. Positive emotional responses such as feeling successful (3.77) and good (3.83) outnumber negative reactions like frustration (2.16) and irritability (2.04), indicating a generally positive user experience.

About 59.7% of players completed the game, which is indicative of a well-balanced challenge level. However, the completion time was shorter than expected, and the hint system was noted as an area for potential improvement. Enhancements in these areas could lead to increased completion rates and deeper engagement.

User comments highlighted the educational value of the game, with many appreciating its challenge and the learning outcomes. Positive feedback emphasized the game's potential as a teaching tool in cybersecurity education. This follows the beliefs expressed by most of the literature sources, which agree that EERs can be valuable didactic learning methods and that they increase the students' interest in the topics covered, as they not only allow a deeper understanding of the course material already taught, but also help learners to understand connections between these topics [24]. Furthermore, EERs have been shown to positively influence confidence not only in academics, but also in the operational application of what has been learnt, as they can help students see a broader picture of course material by enhancing their understanding of additional interrelationships between topics [24–34].

5. Conclusions

Overall, the game was well received, with substantial strengths reported for usability, player engagement, and educational value. Improvements in game depth, immersion, and interface simplicity could further enhance its effectiveness. This feedback serves as a valuable guide for future development, ensuring the game not only entertains, but also educates effectively in the realm of cybersecurity.

As stated in the reviewed literature, the escape room approach promotes soft skills in general, increasing motivation and improving skills such as problem solving, as demonstrated by the studies of Veldkamp et al., and Fotaris and Mastoras [12,39]. Team building, out-of-the-box thinking, and critical questioning are also enhanced with EERs, as shown by Cain, Clarke and Peel, and Eukel [40–42]. In addition to this, learning with the escape room approach also creates an awareness of a specific topic or domain [43].

This is particularly desirable for the EyesOnCS project, which aims to support awareness of the responsible handling of cybersecurity issues. The escape room approach demonstrates significant pedagogical value, aligning with research that supports its use as an effective educational method. The game's ability to engage users with its high story interest score and the feeling of being challenged is crucial in educational settings, where sustained engagement directly impacts learning outcomes. By integrating these elements, the game encouraged active participation and problem-solving, which are essential for effective learning. Participants' feedback highlighted the educational benefits of the game, the game's challenges, and the depth of learning outcomes, reinforcing its potential as a teaching tool. This positive response is consistent with the aforementioned literature, which suggests that escape rooms can significantly increase students' interest in topics by allowing them to explore course material in a dynamic and interactive way. Such an approach not only deepens understanding, but also helps students identify connections between different topics, promoting a more holistic view of the subject matter.

Moreover, based on the qualitative comments received, the game positively influenced learners' confidence levels in the practical application of cybersecurity knowledge. By providing a platform for players to see the broader picture and understand the interrelationships between topics, the game fosters a deeper comprehension and higher retention of information. This aligns with studies indicating that Educational Escape Rooms enhance students' ability to apply learned concepts in real-world scenarios, thereby bridging the gap between theoretical knowledge and practical skills.

Overall, the EyesOnCS EER game represents a promising pedagogical approach, offering a unique blend of challenge, engagement, and educational value that can effectively enhance learning experiences.

Author Contributions: Conceptualization, A.S. and M.L.; methodology, M.L. and C.V.d.C.; software, C.V.d.C.; validation, A.S., M.L., M.W. and C.V.d.C.; formal analysis, C.V.d.C.; investigation, C.S.; data curation, C.S.; writing—original draft preparation, C.S.; writing—review and editing, C.V.d.C.; supervision, C.V.d.C.; project administration, M.L.; funding acquisition, M.L. All authors have read and agreed to the published version of the manuscript.

Funding: This research was co-funded by the European Union through the Erasmus+ Programme, grant number 2021-1-DE02-KA220-VET-000033003.

Data Availability Statement: The research data can be obtained by contacting the authors.

Conflicts of Interest: The authors declare no conflicts of interest.

References

1. Saleem, J.; Adebisi, B.; Ande, R.; Hammoudeh, M. A state of the art survey—Impact of cyber attacks on SME's. In Proceedings of the International Conference on Future Networks and Distributed Systems, Cambridge, UK, 19–20 July 2017. [CrossRef]
2. Kasl, F. Cybersecurity of Small and Medium Enterprises in the Era of Internet of Things. *Lawyer Q.* **2018**, *8*, 165–188.
3. Wallang, M.; Shariffuddin, M.; Mokhtar, M. Cyber Security in Small and Medium Enterprises (SMEs). *J. Gov. Dev. (JGD)* **2022**, *18*, 75–87. [CrossRef]
4. van Tooren, M.; Reti, D.; Schneider, D.; Bassem, C.; de la Cámara, R.S.; Schotten, H.-D. Research Questions in the Acceptance of Cybersecurity by SMEs in the EU. In Proceedings of the Computer Safety, Reliability, and Security—SAFECOMP 2022, Munich, Germany, 6–9 September 2022; pp. 247–255.
5. Manzoor, J.; Waleed, A.; Fareed Jamali, A.; Masood, A. Cybersecurity on a Budget: Evaluating Security and Performance of Open-Source SIEM Solutions for SMEs. *PLoS ONE* **2024**. Available online: https://journals.plos.org/plosone/article?id=10.1371/journal.pone.0301183 (accessed on 8 April 2024). [CrossRef]
6. Erdogan, G.; Halvorsrud, R.; Boletsis, C.; Tverdal, S.; Pickering, J. Cybersecurity Awareness and Capacities of SMEs. In Proceedings of the 9th International Conference on Information Systems Security and Privacy (ICISSP 2023), Lisbon, Portugal, 22–24 February 2023; pp. 296–304. [CrossRef]
7. Ajmi, L.H.; AlQahtani, N.; Rahman, A.; Mahmud, M. A Novel Cybersecurity Framework for Countermeasure of SME's in Saudi Arabia. 2019 2nd International Conference on Computer Applications & Information Security (ICCAIS), Riyadh, Saudi Arabia, 1–3 May 2019; pp. 1–9. [CrossRef]
8. Pieczywok, A. Training employees on risks in the area of cybersecurity. *Cybersecur. Law* **2022**, *7*, 261–271. [CrossRef]
9. Corradini, I. *Building a Cybersecurity Culture in Organizations*; Springer International Publishing: Berlin/Heidelberg, Germany, 2020; pp. 63–86. [CrossRef]
10. Tolossa, D. Importance of cybersecurity awareness training for employees in business. *Vidya* **2023**, *2*, 104–107. [CrossRef]
11. Trim, P.; Upton, D. *Cyber Security Culture: Counteracting Cyber Threats through Organizational Learning and Training*, 1st ed.; Routledge: London, UK, 2016. [CrossRef]
12. Veldkamp, A.; van de Grint, L.; Knippels, M.-C.; van Joolingen, W. Escape education: A systematic review on escape rooms in education. *Educ. Res. Rev.* **2020**, *31*, 100364. [CrossRef]
13. Pornsakulpaisal, R.; Ahmed, Z.; Bok, H.; Carvalho Filho, M.A.; Goka, S.; Li, L.; Patki, A.; Salari, S.; Sooknarine, V.; Woon Yap, S.; et al. Building Digital Escape Rooms for Learning: From Theory to Practice. The Clinical Teacher. Available online: https://asmepublications.onlinelibrary.wiley.com/doi/full/10.1111/tct.13559 (accessed on 8 April 2024).
14. Acharya, S.; Maxim, B.; Yackley, J. Applied Knowledge Retention—Are Active Learning Tools the Solution? In Proceedings of the 2019 ASEE Annual Conference & Exposition, Tampa, FL, USA, 16–19 June 2019. [CrossRef]
15. Chen, M. Research on the Relationship between Training and Knowledge Worker Retention. *DEStech Trans. Soc. Sci. Educ. Hum. Sci.* **2016**. [CrossRef] [PubMed]
16. Goertz, L.; Fehling, C.; Hagenhofer, T. Didaktische Konzepte Identifizieren—Community of Practice zum Lernen mit AR und VR. In Proceedings of the Social Virtual Learning; 2020; p. 3. Available online: https://www.social-augmented-learning.de/wp-content/downloads/210225-Coplar-Leitfaden_final.pdf (accessed on 8 April 2024).
17. Vaz de Carvalho, C.; Coelho, A. Game-Based Learning, Gamification in Education and Serious Games. *Computers* **2022**, *11*, 36. [CrossRef]
18. Baptista, R.; Coelho, A.; Vaz de Carvalho, C. Relationship between game categories and skills development: Contributions for serious game design. In Proceedings of the European Conference on Game Based Learning, Steinkjer, Norway, 8–9 October 2015; Volume 1, pp. 34–42.
19. Prensky, M. *Digital Game-Based Learning*; McGraw-Hill: New York, NY, USA, 2001; Volume 1, p. 1. [CrossRef]
20. Abt, C. *Serious Games*; University Press of America: Lanham, MD, USA, 1987.
21. Pohl, M.; Rester, M.; Judmaier, P. Interactive Game Based Learning: Advantages and Disadvantages. In *Universal Access in Human-Computer Interaction. Applications and Services. UAHCI 2009. Lecture Notes in Computer Science*; Stephanidis, C., Ed.; Springer: Berlin/Heidelberg, Germany, 2009; Volume 5616. [CrossRef]
22. Kolb, D. *Experiential Learning: Experience as the Source of Learning and Development*; Prentice Hall: Englewood Cliffs, NJ, USA, 1984.

23. Bonwell, C.C.; Eison, J.A. *Active Learning: Creating Excitement in the Classroom. ASH#-ERIC Higher Education Report No. 1*; The George Washington University, School of Education and Human Development: Washington, DC, USA, 1991.
24. Tercanli, H.; Martina, R.; Ferreira Dias, M.; Reuter, J.; Amorim, M.; Madaleno, M.; Magueta, D.; Vieira, E.; Veloso, C.; Figueiredo, C.; et al. *Educational Escape Rooms in Practice: Research, Experiences and Recommendations*; UA Editoria: Tucson, AZ, USA, 2021. [CrossRef]
25. Guckian, J.; Sridhar, A.; Meggitt, S.J. Exploring the perspectives of dermatology undergraduades with an escape room game. *Clin. Exp. Dermatol.* **2020**, *45*, 153–158. [CrossRef] [PubMed]
26. Brady, S.; Andersen, E. An escape-room inspired game for genetics review. *J. Biol. Educ.* **2019**, *55*, 406–417. [CrossRef]
27. Jin, G.; Tu, M.; Kim, T.; Heffron, J.; White, J. Game based Cybersecurity Training for High School Students. In Proceedings of the 49th ACM Technical Symposium on Computer Science Education, Minneapolis, MN, USA, 27 February–2 March 2018. [CrossRef]
28. Hodhod, R.; Hardage, H.; Abbas, S.; Aldakheel, E. CyberHero: An Adaptive Serious Game to Promote Cybersecurity Awareness. *Electronics* **2023**, *12*, 3544. [CrossRef]
29. Kletenik, D.; Butbul, A.; Chan, D.; Kwok, D.; LaSpina, M. Cyber Secured: A Serious Game for Cybersecurity Novices. In Proceedings of the 51st ACM Technical Symposium on Computer Science Education, Portland, OR, USA, 11–14 March 2020. [CrossRef]
30. Nagarajan, A.; Allbeck, J.; Sood, A.; Janssen, T. Exploring game design for cybersecurity training. In Proceedings of the 2012 IEEE International Conference on Cyber Technology in Automation, Control, and Intelligent Systems (CYBER), Bangkok, Thailand, 27–31 May 2012; pp. 256–262. [CrossRef]
31. Hamman, S.; Hopkinson, K.; Markham, R.; Chaplik, A.; Metzler, G. Teaching Game Theory to Improve Adversarial Thinking in Cybersecurity Students. *IEEE Trans. Educ.* **2017**, *60*, 205–211. [CrossRef]
32. Jin, G.; Tu, M.; Kim, T.; Heffron, J.; White, J. Evaluation of Game-Based Learning in Cybersecurity Education for High School Students. *J. Educ. Learn.* **2018**, *12*, 150–158. [CrossRef]
33. Löffler, E.; Schneider, B.; Zanwar, T.; Asprion, P.T. CySecEscape 2.0—A Virtual Escape Room to Raise Cybersecurity Awareness. *Int. J. Serious Games* **2021**, *8*, 59–70. [CrossRef]
34. Beguin, E.; Besnard, S.; Cros, A.; Joannes, B.; Leclerc-Istria, O.; Noel, A.; Roels, N.; Taleb, F.; Thongphan, J.; Alata, E.; Nicomette, V. Computer-Security-Oriented Escape Room. *IEEE Secur. Priv.* **2019**, *17*, 78–83. [CrossRef]
35. Salahdine, F.; Kaabouch, N. Social Engineering Attacks: A Survey. *Future Internet* **2019**, *11*, 89. [CrossRef]
36. Brooke, J. SUS: A quick and dirty usability scale. In *Usability Evaluation in Industry*; CRC Press: Boca Raton, FL, USA, 1995; p. 189.
37. IJsselsteijn, W.A.; de Kort, Y.A.W.; Poels, K. *The Game Experience Questionnaire*; Technische Universiteit Eindhoven: Eindhoven, The Netherlands, 2013.
38. Law, E.L.-C.; Brühlmann, F.; Mekler, E.D. Systematic Review and Validation of the Game Experience Questionnaire (GEQ)—Implications for Citation and Reporting Practice. In Proceedings of the 2018 Annual Symposium on Computer-Human Interaction in Play, Melbourne, Australian, 28–31 October 2018; Association for Computing Machinery: New York, NY, USA, 2018; pp. 257–270.
39. Fotaris, P.; Mastoras, T. Escape rooms for learning: A systematic review. *Res. Pract. Technol. Enhanc. Learn.* **2019**, *14*, 235–243.
40. Cain, J. Exploring the benefits of using gamification and video games for adult learners. *J. Contin. High. Educ.* **2019**, *67*, 45–54.
41. Clarke, S.; Peel, D. Escape the norm! Using escape room activities to support experiential learning in undergraduate business education. *Int. J. Manag. Educ.* **2020**, *18*, 100425.
42. Eukel, H.N.; Frenzel, J.E.; Cernusca, D. Educational gaming for pharmacy students—Design and evaluation of a diabetes-themed escape room. *Am. J. Pharm. Educ.* **2017**, *81*, 6265. [CrossRef]
43. Adams, V.; Burger, S.; Crawford, K.; Setter, R. Can you escape? Creating an escape room to facilitate active learning. *J. Nurses Prof. Dev.* **2018**, *34*, 60–63. [CrossRef] [PubMed]

Disclaimer/Publisher's Note: The statements, opinions and data contained in all publications are solely those of the individual author(s) and contributor(s) and not of MDPI and/or the editor(s). MDPI and/or the editor(s) disclaim responsibility for any injury to people or property resulting from any ideas, methods, instructions or products referred to in the content.

Article

Stealth Literacy Assessments via Educational Games

Ying Fang [1], Tong Li [2], Linh Huynh [3], Katerina Christhilf [3], Rod D. Roscoe [4] and Danielle S. McNamara [3,*]

[1] Faculty of Artificial Intelligence in Education, Central China Normal University, Wuhan 430079, China; fangying@ccnu.edu.cn
[2] School of Journalism and Strategic Communication, Ball State University, Muncie, IN 47306, USA
[3] Department of Psychology, Arizona State University, Tempe, AZ 85281, USA
[4] Human Systems Engineering, Arizona State University, Mesa, AZ 85212, USA
* Correspondence: dsmcnamara1@gmail.com

Abstract: Literacy assessment is essential for effective literacy instruction and training. However, traditional paper-based literacy assessments are typically decontextualized and may cause stress and anxiety for test takers. In contrast, serious games and game environments allow for the assessment of literacy in more authentic and engaging ways, which has some potential to increase the assessment's validity and reliability. The primary objective of this study is to examine the feasibility of a novel approach for stealthily assessing literacy skills using games in an intelligent tutoring system (ITS) designed for reading comprehension strategy training. We investigated the degree to which learners' game performance and enjoyment predicted their scores on standardized reading tests. Amazon Mechanical Turk participants (n = 211) played three games in iSTART and self-reported their level of game enjoyment after each game. Participants also completed the Gates–MacGinitie Reading Test (GMRT), which includes vocabulary knowledge and reading comprehension measures. The results indicated that participants' performance in each game as well as the combined performance across all three games predicted their literacy skills. However, the relations between game enjoyment and literacy skills varied across games. These findings suggest the potential of leveraging serious games to assess students' literacy skills and improve the adaptivity of game-based learning environments.

Keywords: literacy; assessment; reading comprehension; educational games; intelligent tutoring system

Citation: Fang, Y.; Li, T.; Huynh, L.; Christhilf, K.; Roscoe, R.D.; McNamara, D.S. Stealth Literacy Assessments via Educational Games. *Computers* **2023**, *12*, 130. https://doi.org/10.3390/computers12070130

Academic Editors: Carlos Vaz de Carvalho, Hariklia Tsalapatas and Ricardo Baptista

Received: 23 May 2023
Revised: 18 June 2023
Accepted: 20 June 2023
Published: 25 June 2023

Copyright: © 2023 by the authors. Licensee MDPI, Basel, Switzerland. This article is an open access article distributed under the terms and conditions of the Creative Commons Attribution (CC BY) license (https://creativecommons.org/licenses/by/4.0/).

1. Introduction

Literacy can be broadly defined as "the ability to understand, evaluate, use, and engage with written texts to participate in society, to achieve one's goals, and to develop one's knowledge and potential" [1]. Literacy skills are critical to academic success and in life; however, large-scale reading assessment data reveal that many students and adults struggle with reading comprehension. The most recent National Assessment of Educational Progress reported that 27% of 8th grade students in the United States performed below the basic levels of reading comprehension and 66% did not reach proficient levels. Similarly, 30% and 63% of 12th graders did not reach basic and proficiency reading levels, respectively [2]. As might be expected, deficits in reading skills often continue into adulthood. According to the 2017 Program for the International Assessment of Adult Competencies assessment, 19% of U.S. adults aged 16 or older performed at or below the lowest literacy level [3].

Literacy assessments are a key component of any effort to improve students' literacy or remediate potential gaps in education. A good understanding of learners' current skills reveals the types and amounts of instruction they will need to grow. However, traditional literacy assessments (e.g., standardized tests) typically require a significant amount of time to administer, score, and interpret. Additionally, these tests usually occur before or after learning, making it difficult to provide timely feedback to guide teaching and learning [4–6]. Moreover, traditional assessments may cause stress and test anxiety, which may in turn negatively impact students' test-taking experiences [5,7], and change how students respond to the assessments [8].

In contrast to traditional literacy assessments, stealth assessment offers an innovative approach by implementing literacy assessments in computer-based learning environments. These assessments take place *during* learning activities, instead of summative or "checkpoint" assessments. In addition, the assessments are based on students' natural behaviors and performance rather than being presented as "tests." As such, stealth literacy assessments can evaluate student reading skills unobtrusively and dynamically, and provide timely feedback throughout the learning process. In this innovative context, serious games have strong potential to be more motivating and enjoyable than traditional reading assessments. Thus, in this study, we investigated the feasibility of game-based stealth assessment to predict literacy skills, specifically reading comprehension and vocabulary knowledge.

1.1. Stealth Assessment within Games

Stealth assessment refers to performance-based assessments that are seamlessly embedded in gaming environments without the awareness of students who are being evaluated [4,9]. Stealth assessment was initially proposed and explored to assess higher-order competencies such as persistence, creativity, self-efficacy, openness, and teamwork, primarily because these competencies substantially impact student academic achievement, but also because traditional methods of assessment often neglect these abilities [10,11]. As such, stealth assessments that analyze how students use knowledge and skills during gameplay have been embedded in serious games to unobtrusively assess those competencies [12–15]. Game environments may make assessment less salient or less visible, and thus students feel that they are merely "playing" rather than "being tested".

In stealth assessment, traditional test items are replaced by authentic, real-world scenarios or game tasks. Since stealth assessment items can be contextualized and potentially connected to the real world, students' skills, behaviors, and competencies may be more validly demonstrated through these game activities than in traditional assessments [16,17]. Within stealth assessments, students generate rich sequences of performance data (e.g., choices, actions, and errors) when they perform the tasks, which can serve as evidence for knowledge and skills assessment. When students are assessed without the feeling of being tested, it can reduce their stress and anxiety, which can in turn increase the reliability of the assessment [18,19].

Serious games in education are designed to enhance students' learning experience by providing a more fun way to acquire knowledge [20,21], which could be ideal for stealth assessments because they further separate students' experiences of play and enjoyment from experiences of testing and measurement. In a well-designed game, students are immersed in game scenarios and motivated to proceed through the challenges and meet learning goals, which might not feel like a learning or testing experience at all [22]. For example, Physics Playground is a game that emphasizes 2-D physics simulations. The game implements stealth assessments to evaluate students' physics knowledge, persistence, and creativity [15,23]. When students interact with the game, they produce a dense stream of performance data, which is recorded by the system in a log file and analyzed using Bayesian networks to infer students' knowledge and skills. The system then provides ongoing automated feedback to teachers and/or students, based on the assessment, to support student learning. Another example of stealth assessment is a game-based learning environment named ENGAGE that was designed to promote computational thinking skills. Students' behavioral data were collected during their gameplay and then analyzed using machine learning methods to infer their problem-solving skills and computational knowledge [13,24].

1.2. Stealth Reading Assessment via Games

Prior studies have explored stealth assessment via games to assess students' higher-order skills and competencies, such as problem solving, creativity, and persistence, along with scientific knowledge [14,17,23,25]. Only a few studies have investigated using stealth assessment to assess *literacy* [26–28]. These studies primarily leveraged natural language processing (NLP) techniques to extract linguistic properties of constructed responses (e.g., essays and explanations) to make predictions about students' reading skills. For example, Allen and McNamara analyzed the lexical properties of students' essays to predict students' vocabulary test scores. Two lexical indices associated with the use of sophisticated and academic word use accounted for 44% of the variance in vocabulary knowledge scores [26]. Fang et al. predicted students' comprehension test scores using the linguistic properties of the self-explanations generated during their practice in a game-based learning environment. Five linguistic features accounted for around 20% of the variance in comprehension test scores across datasets. The studies collectively demonstrate linguistic features at multiple levels of language (e.g., lexical, syntactic, and semantic) have strong potential to serve as proxies of reading skills, supporting the feasibility of stealth reading assessment using NLP [27].

However, the use of NLP for reading assessment is not without challenges. These methods rely on machine learning algorithms, which require a large amount of data to train and test to improve the analysis accuracy [29]. Additionally, some NLP methods rely on extensive computational resources to support the complex calculations, which might be difficult to adopt by development teams without those resources [29]. Most importantly, NLP is language-specific, rendering it challenging to generalize algorithms across languages.

An alternative to NLP is the analysis of students' performance data from games implementing multiple-choice questions. Multiple-choice questions provide a shortlist of answers for students to choose from, which do not require complex data analysis to evaluate students' answers. For example, Fang et al. investigated the association between students' reading skills and their performance in a *single* vocabulary game (i.e., Vocab Flash). The analysis was based on students' performance data from the game implementing what are essentially multiple-choice questions in disguise. The results of the study supported the value of using a simple vocabulary game to assess reading comprehension [30].

The current study examines both vocabulary and main idea games. In the following section, we introduce how literacy skills are reflected by the ability to identify word meaning and text main ideas, providing the theoretical grounds for leveraging vocabulary and main idea games to assess reading skills.

1.3. Assessing Reading Skills through Vocabulary and Main Idea Games

Reading skills are at least partially reflected by students' vocabulary knowledge and ability to identify main ideas of passages. Readers must be able to process the basic elements of the text, including the individual words and the syntax, to understand and gain meaning from texts. From those elements, the reader can construct an understanding of the meanings behind phrases and sentences. Readers with more vocabulary knowledge tend to have better reading comprehension skills [31,32]. When a reader is unfamiliar with certain keywords in a text, this can slow down or fully impair the processing of key points in the text. Although in some cases the meaning of words can be understood via contextual cues, comprehension becomes more challenging for readers with lower vocabulary knowledge [33]. This effect is exacerbated when the reader also has insufficient prior knowledge about the topic through which to build context [34]. Hence, when assessing texts for readability, educators and researchers have found that texts containing many low-frequency and sophisticated words make them more difficult to read [35]. In second-language learners, vocabulary is one of the most critical factors in determining how well students can comprehend texts [36,37].

Comprehending text requires not only knowledge of individual word meanings, but also the skills required to deduce relations between ideas and, in turn, the main ideas [38]. Identifying topic sentences is a comprehension strategy that requires students to recognize the main ideas within a text while dismissing information that is irrelevant or redundant [39]. Being able to distinguish main versus supporting ideas can foster deep comprehension because it encourages students to attend to the higher-level meaning and the global organization of information across texts [40]. Consequently, Bransford et al. suggested that identifying main ideas within a text can lead to enhanced comprehension and retention of the text content [40]. According to Wade-Stein and Kintsch [41], this task not only promotes students' construction of factual knowledge, but also of conceptual knowledge, as the process of identifying main ideas within a text reinforces students' memory representations of its content.

Deducing the gist of a text can be challenging [42,43]. Low-knowledge readers can find it hard to differentiate between the main arguments and supporting ideas of a text [44]. Likewise, Wigent found that students with reading difficulties focused more on details rather than identifying main ideas, subsequently recognizing and recalling fewer topic sentences compared to average readers [45]. By contrast, strategic and skilled readers are more likely to grasp the main ideas from text compared to less skilled readers [46–49].

Students' reading comprehension skills, vocabulary knowledge, and ability to recognize main ideas are closely associated. Therefore, this study employed three distinct games that emphasized vocabulary and main idea identification. First, Vocab Flash is a game that requires players to select appropriate synonyms for words. It is an adaptive game designed to measure vocabulary for variously skilled participants, making it ideal for stealth assessment of vocabulary. Second, Adventurer's Loot is a game that asks players to read a text and select the main ideas. Participants must be careful to select only the main ideas, and not any extraneous details. Finally, Dungeon Escape asks players to read a passage, and imagine they are about to write a summary of the passage. They must select the best topic sentence for the summary. To pick the best topic sentence, participants must locate and integrate the main ideas of the passage. Such integration may require further reading comprehension skill than Adventurer's Loot, which is why two main idea games are included.

1.4. Current Study

The goal of the current study is to investigate the feasibility of using games for stealth assessment of reading skills. Specifically, this study examines the predictive value of *three* distinct games, namely one game that targets vocabulary knowledge and two games that target main idea identification. We not only examine students' performance in each game individually, but also explore the value of combining performance data across all three games. The goal is to assess the extent to which students' performance in the three games is indicative of their reading skills as measured by standardized reading tests.

In addition to game performance, we consider students' subjective game experience. Based on the literature regarding serious games [50–52], we expect that most students will have overall positive attitudes toward playing the games. However, it is unknown whether students' game enjoyment will influence the validity of the stealth assessment. To that end, we address the following research questions in this study:

1. To what extent does students' performance in the three games predict their reading skills (i.e., vocabulary knowledge and reading comprehension)?
2. Does students' enjoyment of the games moderate the relations between game performance and reading skills?

2. Methods

2.1. Experimental Environment: iSTART

The Interactive Strategy Training for Active Reading and Thinking (iSTART) is a game-based intelligent tutoring system (ITS) designed to help students improve reading skills through adaptive instruction and training. iSTART was developed based upon Self-Explanation Reading Training (SERT) [53], a successful classroom intervention that taught students to explain the meaning of texts while reading (i.e., self-explain) through the use of comprehension strategies (i.e., comprehension monitoring, paraphrasing, predicting, bridging, and elaborating). The current version of iSTART includes three training modules focusing on self-explanation, summarization, and question asking [54].

iSTART learning materials consist of video lessons and two types of practice: regular and game-based practice. Video lessons provide students with information about comprehension strategies and prepare them for the practice. During the regular practice, students complete given tasks, and the system provides immediate feedback on students' performance. For example, when students generate a self-explanation on a target sentence, the NLP algorithms implemented in iSTART automatically analyze the self-explanation and provide real-time feedback. The feedback includes a holistic score on a scale of 0 ("poor") to 3 ("great") and actionable feedback to help students improve the self-explanation when the score is below a certain threshold [55]. Studies that investigate the effectiveness of iSTART indicate that iSTART facilitates both comprehension strategy learning and comprehension skills [53,54,56].

2.1.1. iSTART Games

iSTART implements two forms of game-based practice to increase learners' motivation and engagement: generative and identification games [54,56]. In generative games, students are asked to construct verbal responses such as self-explanations. NLP-based algorithms assess these constructed responses to determine the quality and/or the use of specific strategies. In contrast, identification games ask students to review short example stimuli or prompts, and then to choose one or more responses that correctly identify strategy use or follow from the prompts. For example, students might read an example self-explanation and then indicate whether the excerpt demonstrates "paraphrasing" or "elaborating." In a vocabulary game, students may be given a prompt term and then must choose a correct synonym from several choices. Importantly, alternatives typically comprise carefully generated foils, such that incorrect answers are diagnostic of student misunderstanding.

iSTART games also include narrative scenarios and other challenges to further motivate reading strategy practice. For instance, students are rewarded with "iBucks" during gameplay, which can be "spent" to unlock additional game backgrounds or to customize personal avatar characters [54]. In addition, students receive immediate feedback during or after gameplay to support their self-monitoring and engagement [52]. For example, in Showdown, students compete against a computer-controlled player to explain target sentences in given texts. At the end of each round, the system evaluates students' answers and informs them of their performance ("poor", "fair", "good", or "great"). Meanwhile, the performance scores of students are compared with the computer-controlled player to determine who wins the round (see Figure 1).

2.1.2. Adaptivity Facilitated by Assessments in iSTART

iSTART implements both inner-loop and outer-loop adaptivity to customize instruction to individual students. Inner-loop adaptivity refers to the immediate feedback students are given when they complete an individual task, and outer-loop adaptivity refers to the selection of subsequent tasks based on students' past performance [57].

Regarding inner-loop adaptivity, the generative games utilize NLP (e.g., LSA) and machine learning algorithms to assess constructed responses, and then provide holistic scores and actionable, individualized feedback [55,56]. Within identification games, the assessment of the answers matches students' selection with predetermined answers. The system then provides timely feedback including response accuracy, explanations of why the responses are correct or incorrect, and game performance scores.

To further promote skill acquisition, iSTART complements the inner-loop with outer-loop adaptations, which select practice texts based on the student model and the instruction model. An ITS typically employs three elements to assess students and select appropriate tasks: the domain model, the student model, and the instructional model [57–59]. The domain model represents ideal expert knowledge and may also address common student misconceptions. The domain model is usually created using detailed analyses of the knowledge elicited from subject matter experts. The student model represents students' current understanding of the subject matter, and it is constructed by examining student task performance in comparison to the domain model. Finally, the instructional model represents the instructional strategies. It is used to select instructional content or tasks based on inferences about student knowledge and skills. iSTART creates student models using students' self-explanation scores and scores on multiple-choice measures. The instructional model then determines the features of each presented task (i.e., text difficulty and scaffolds to support comprehension) using the evolving student model. For example, subsequent texts become more difficult if students' self-explanation quality on prior texts is higher. Conversely, when students' self-explanation quality is lower, the subsequent texts become easier [54].

Figure 1. Showdown is a practice game within iSTART. The first image (**top**) shows the game interface where the human player and computer-controlled player are about to start the competition. The second image (**bottom**) shows both players' explanations and the system-generated evaluations of the explanations.

Current adaptation facilitated by assessments in iSTART is at the micro-level. Specifically, the assessments are task-specific and may not transfer to the games implementing different types of tasks. For example, students' scores on self-explanation games may not reflect their performance in question-asking games. As such, students' self-explanation scores are not leveraged to guide the adaptivity (e.g., learning material selection) in the question-asking module. We anticipate that task-general assessments, such as reading skill assessments, have strong potential to supplement current assessments to guide the macro-level adaptation across modules.

2.2. Participants

Participants were 246 adults recruited on Amazon Mechanical Turk (an online platform). Thirty-five participants were excluded from the study due to failing an attention check, which resulted in the final sample of 211 participants (98 female, 113 male). In the final sample, 77.2% of participants identified as Caucasian, 10.9% as African American, 6.6% as Hispanic, 4.3% as Asian, and 1.0% as another race/ethnicity. Participants were 37.2 years old, on average, with a range of 17 to 68. Most participants (81.5%) reported holding a Bachelor's or advanced degree.

2.3. Procedure, Materials, and Measures

Participants first responded to a demographic questionnaire, and then played three iSTART games: Vocab Flash, Dungeon Escape, and Adventurer's Loot. Game order was counterbalanced. After every game, participants completed a brief questionnaire regarding their enjoyment. In the final step of the study, participants completed the Gates–MacGinitie Reading Test (GMRT), which included a vocabulary and a comprehension subtests.

Gates–MacGinitie Reading Test (GMRT). Participants' reading skills were measured by the Gates–MacGinitie Reading Test (GMRT) level 10/12 form S. The GMRT is an established and reliable measure of reading comprehension (α = 0.85–0.92) [60], which comprises both vocabulary and comprehension subtests. The vocabulary subtest (10 min) includes 45 multiple-choice questions that ask participants to choose the correct definition of target words in the given sentences. The comprehension subtest (20 min) consists of a series of textual passages with two to six multiple-choice comprehension questions per passage. There are a total of 48 questions.

Vocabulary and reading comprehension skills were operationally defined as the total number of correct answers on the vocabulary and reading comprehension GMRT subtests, respectively.

Vocab Flash. In this game, students read a target word and must choose a synonym out of four alternatives (i.e., one correct choice and three incorrect foils). Students are allotted 5 min to respond to as many terms as possible. For each target word, students are only allowed one attempt to select the answer. After students submit the answer, they receive feedback that (a) indicates whether their answer is correct and (b) clearly highlights the correct response. One key feature of the game is its adaptivity. The target words are classified into nine different levels of difficulty based on their frequency rating in Corpus of Contemporary American English (COCA) [61]. Uncommon words are typically more challenging. The game begins with the easiest words and progresses to higher levels of difficulty as students answer correctly. However, students can also return to easier levels after repeated errors. As in computer-adaptive testing [62], students can fluctuate between levels of difficulty, but more skilled students will generally encounter more difficult items. Game performance in Vocab Flash was measured by the proportion of correctly answered questions.

Dungeon Escape. Dungeon Escape is an iSTART game in which students are knights trapped in a dungeon. The way to escape it is to earn points by selecting topic sentences of given texts. Each student must complete six texts that are randomly selected from the game's science text pool. Four alternative sentences (i.e., one correct answer and three incorrect foils) are provided for each text. Students are allowed multiple attempts for each question, and they proceed to the next text by selecting the correct answer. Performance in Dungeon Escape was measured by the proportion of correct answers only in students' first attempts, because students may potentially game the system (e.g., try all of the answers sequentially).

Adventurer's Loot. Adventurer's Loot is an iSTART game in which students are asked to discover the hidden treasures on a map by selecting the main ideas of given texts. There are eight sites on the map, and each site corresponds to a specific text. Students can select a site from the map to explore and work on the corresponding text. Students are allowed multiple attempts on a text. The only way to proceed to the next text is by answering the question correctly, namely, selecting all the main ideas. Importantly, the number of correct answers (i.e., main ideas) in this game varies between texts. For the texts with multiple correct answers, the incorrect answers may be missing main ideas or selecting distractors. To be sensitive to different error types and potential user attempts to game the system, d prime was used as the performance measure of Adventure's Loot. It was based on (1) the proportion of correctly identified main ideas in the first attempt, which was computed using the number of correctly selected answers divided by the number of correct answers, and (2) the proportion of incorrectly selected distractors in the first attempt, which was computed using the number of incorrectly selected answers (i.e., selected distractors) divided by the number of distractors. D prime was calculated using the z score of (1), subtracting the z score of (2).

Game Survey. After each game, participants responded to six items pertaining to their subjective game enjoyment. These questions were derived from measures implemented in prior studies: (1) This game was fun to play; (2) This game was frustrating; (3) I enjoyed playing this game; (4) This game was boring; (5) The tasks in this game were easy; and (6) I would play this game again. Participants rated their agreement with these statements on a 6-point Likert scale ranging from "1" (strongly disagree) to "6" (strongly agree). Student enjoyment of the game was operationalized as the average score of the six items.

2.4. Statistical Analyses

Internal consistency between survey items (i.e., reliability) was measured using Cronbach's alpha calculated with the following formula:

$$\alpha = \frac{N \times \bar{c}}{\bar{v} + (N-1) \times \bar{c}}$$

where N = number of items, $\bar{c}$ = mean covariance between items, and $\bar{v}$ = mean item variance. Two items "This game was frustrating" and "This game was boring" were reverse coded before the calculation such that all of the items indicated positive attitudes toward the games.

Hierarchical linear regressions were conducted to determine whether student game performance and enjoyment predict reading test performance. More specifically, game performance scores and enjoyment scores for each game (Vocab Flash, Dungeon Escape, and Adventurer's Loot) were used to predict vocabulary test scores and comprehension test scores.

Finally, hierarchical linear regressions were conducted to examine whether participants' performance and enjoyment *combined* across all three games were better predictors of reading skills than performance and enjoyment of each *individual* game.

3. Results

3.1. Survey Item Internal Consistency

Cronbach's alpha of the six survey items for Vocab Flash, Dungeon Escape, and Adventurer's Loot were 0.68, 0.73, and 0.82, respectively. A general accepted rule is that alpha of 0.6–0.7 indicates an acceptable level of reliability, and 0.8 or greater a good level [63]. Therefore, the scores indicated acceptable to good internal consistency between the survey items.

3.2. Descriptive Statistics of Predictor and Predicted Variables

Table 1 provides descriptive statistics of participants' performance on the vocabulary and comprehension subtests, as well as their performance and enjoyment scores for the three games. Reading tests performance scores were calculated using the proportion of correct answers in the subtests. Game performance scores were calculated using the proportion of correct answers or the proportion of correct and incorrect answers, depending on the game. Game enjoyment scores were calculated using the sum of participants' ratings on the game enjoyment survey. As is shown in Table 1, participants' vocabulary and comprehension test scores were strongly and positively correlated ($r = 0.76$). The correlation between game performance scores and reading test scores were different for each game. Students tended to enjoy playing the games, particularly Vocab Flash ($M = 4.34$). However, the strength and direction of the correlations between game enjoyment and participants' reading test scores varied between games.

Table 1. Descriptive statistics and correlations between predictor and predicted variables.

Measure	M	SD	Vocabulary	Comprehension	VF Correct	DE Correct	AL Correct	AL Incorrect	VF Enjoyment	DE Enjoyment
Vocabulary	0.43	0.29								
Comprehension	0.36	0.22	0.76 **							
VF Correct	0.48	0.24	0.76 **	0.60 **						
DE Correct	0.41	0.23	0.46 **	0.50 **	0.38 **					
AL Correct	0.61	0.25	0.09	0.08	0.11	0.16 *				
AL Incorrect	0.55	0.26	−0.27 **	−0.26 **	−0.21 *	−0.07	0.67 **			
VF Enjoyment	4.34	0.84	0.13	0.16 *	0.24 **	0.07	0.03	−0.10		
DE Enjoyment	4.11	0.99	−0.26 **	−0.23 *	−0.13	0.00	−0.04	0.07	0.45 **	
AL Enjoyment	3.75	1.16	−0.60 **	−0.52 **	−0.41 **	−0.38 **	−0.10	0.12	0.31 **	0.63 **

Note. M = mean, SD = standard deviation, VF = Vocab Flash, DE = Dungeon Escape, AL = Adventurer's Loot, ** $p < 0.01$, * $p < 0.05$.

3.3. Predicting Vocabulary Knowledge with Individual Game Performance

Using hierarchical linear regression analyses, we explored whether vocabulary test scores could be predicted based on game performance and enjoyment for each game. Game performance measures were entered as predictors in Model 1, and then both game performance and enjoyment were entered as predictors in Model 2. More specifically, performance measures refer to the proportion of correct answers in Vocab Flash, proportion of correct answers for the first attempts in Dungeon Escape, and d prime scores for the first attempts in Adventurer's Loot (the calculation of d prime scores is introduced in Section 2.3). Enjoyment refers to the sum of participants' self-reported scores on the survey items. As is shown in Table 2, participants' performance in game Vocab Flash was a strong predictor and explained 57% of the variance in their vocabulary test scores. Their enjoyment of the game did not account for additional variance. For Dungeon Escape and Adventurer's Loot, participants' performance scores were again significant predictors of their vocabulary test scores. Their performance scores accounted for 20% of the variance in both games. The additional variance explained by game enjoyment was higher in Adventurer's Loot (24%) than in Dungeon Escape (6%).

3.4. Predicting Comprehension Test Scores with Individual Game Performance

As with vocabulary, hierarchical linear regression analyses sought to predict comprehension test scores based on game performance and enjoyment for Vocab Flash, Dungeon Escape, and Adventurer's Loot. In Model 1, game performance was entered as the sole predictor of comprehension test scores. Specifically, performance was measured by participants' proportion of correct answers in Vocab Flash, proportion of correct answers for the first attempts in Dungeon Escape, and d prime scores for the first attempts in Adventurer's Loot. In Model 2, both performance and enjoyment were entered as predictors of comprehension. For Vocab Flash, participants' performance scores explained 36% of the variance in comprehension test scores, with enjoyment adding no extra variance. For Dungeon Escape and Adventurer's Loot (see Table 3), participants' performance scores were significant predictors, which accounted for 25% and 18% of the variance in the comprehension test scores, respectively. The additional variance for which enjoyment accounted was 2% in Dungeon Escape and 18% in Adventurer's Loot.

Table 2. Regression analysis predicting vocabulary test scores with individual game performance and enjoyment.

Variable	Standardized Coefficient	t	R^2	R^2 Change
Vocab Flash				
Model 1			0.57	0.57 ***
Performance	0.76	16.68 ***		
Model 2			0.57	0.00
Performance	0.77	16.46 ***		
Enjoyment	−0.05	1.07		
Dungeon Escape				
Model 1			0.20	0.20 ***
Performance	0.45	6.95 ***		
Model 2			0.26	0.06 ***
Performance	0.45	7.20 ***		
Enjoyment	−0.25	−4.05 ***		
Adventurer's Loot				
Model 1			0.20	0.20 ***
Performance	0.44	7.18 ***		
Model 2			0.44	0.24 ***
Performance	0.31	5.72 ***		
Enjoyment	−0.51	−9.51 ***		

Note. *** $p < 0.001$.

Table 3. Regression analysis predicting comprehension test scores with individual game performance and enjoyment.

Variable	Standardized Coefficient	t	R^2	R^2 Change
Vocab Flash				
Model 1			0.36	0.36 ***
Performance	0.60	8.86 ***		
Model 2			0.36	0.00
Performance	0.60	8.49 ***		
Enjoyment	−0.01	0.16		
Dungeon Escape				
Model 1			0.25	0.25 ***
Performance	0.60	6.60 ***		
Model 2			0.27	0.02 ***
Performance	0.48	6.36 ***		
Enjoyment	−0.17	−2.31 *		
Adventurer's Loot				
Model 1			0.18	0.18 ***
Performance	0.43	5.69 ***		
Model 2			0.36	0.18 ***
Performance	0.31	4.48 ***		
Enjoyment	−0.44	−6.27 ***		

Note. *** $p < 0.001$, * $p < 0.05$.

3.5. Predicting Vocabulary Knowledge from Performance Combined across Games

In addition to the analyses examining each game separately, a hierarchical linear regression was conducted to predict vocabulary test scores based on their performance and enjoyment across all three games combined. The performance measures in Vocab

Flash, Dungeon Escape, and Adventurer's Loot were predictors of vocabulary test scores in Model 1. More specifically, the predictors were participants' proportion of correct scores in Vocab Flash, proportion of correct scores of the first attempts in Dungeon Escape, and d prime scores of the first attempts in Adventurer's Loot. Participants' performance in all three games were significant predictors of their vocabulary test scores and explained 65% of the variance. The explained variance was higher than that explained by performance measures in any individual game. Model 2 predicted vocabulary test scores with both performance and enjoyment scores in the three games. Game enjoyment scores added 9% of explained variance in vocabulary test scores (see Table 4).

Table 4. Regression analysis predicting vocabulary test scores with combined game performance.

Variable	Standardized Coefficient	t	R^2	R^2 Change
Model 1			0.65	0.65 ***
Performance (VF)	0.65	12.84 ***		
Performance (DE)	0.17	3.51 **		
D-prime (AL)	0.16	3.18 **		
Model 2			0.74	0.09 *
Performance (VF)	0.51	10.54 ***		
Performance (DE)	0.08	1.66		
Performance (AL)	0.11	2.55 *		
Enjoyment (VF)	0.11	2.33 *		
Enjoyment (DE)	0.03	0.50		
Enjoyment (AL)	−0.40	6.53 ***		

Note. VF = Vocab Flash, DE = Dungeon Escape, AL = Adventurer's Loot, * $p < 0.05$, ** $p < 0.01$, *** $p < 0.001$.

3.6. Predicting Comprehension Test Scores from Performance Combined across Games

A second hierarchical linear regression sought to predict comprehension test scores based on participants' game performance and enjoyment across all three games. Model 1 only included game performance measures in the three games as predictors. Results indicated that the performance scores of all three games were significant predictors of comprehension test scores and they accounted for 49% of the variance. Model 2 included both game performance and enjoyment measures in the three games as predicting variables. The significant predictors of comprehension test scores were participants' performances in Vocab Flash and Dungeon Escape and their enjoyment of Adventurer's Loot. The additional variance explained by game enjoyment beyond the performance scores was 6% (see Table 5).

Table 5. Regression analysis predicting comprehension test scores with combined game performance and enjoyment.

Variable	Standardized Coefficient	t	R^2	R^2 Change
Model 1			0.49	0.49 ***
Performance (VF)	0.43	5.86 ***		
Performance (DE)	0.30	4.35 ***		
Performance (AL)	0.18	2.59 *		
Model 2			0.55	0.06
Performance (VF)	0.30	3.66 ***		
Performance (DE)	0.20	2.77 **		
Performance (AL)	0.13	1.90		
Enjoyment (VF)	0.14	1.77		
Enjoyment (DE)	0.01	0.10		
Enjoyment (AL)	−0.35	3.24 **		

Note. VF = Vocab Flash, DE = Dungeon Escape, AL = Adventurer's Loot, * $p < 0.05$, ** $p < 0.01$, *** $p < 0.001$.

4. Discussion

In the current study, we investigated the feasibility of game-based stealth literacy assessment using games from the iSTART ITS. Specifically, we explored to what degree learners' game performance and enjoyment in three games (i.e., Vocab Flash, Dungeon Escape, and Adventurer's Loot) were able to predict their vocabulary knowledge and reading comprehension skills. In addition, we examined whether the associations between reading skills and game performance were moderated by participants' enjoyment of the games.

Our results suggest that game performance was predictive of participants' reading skills. Specifically, performance in Vocab Flash, Dungeon Escape, and Adventurer's Loot accounted for respectively 57%, 20%, and 20% of the variance of participants' vocabulary knowledge. The explained variance increased to 65% when the combined performance of all three games was used as a predictor. Performance in Vocab Flash, Dungeon Escape, and Adventurer's Loot explained respectively 36%, 25%, and 18% of the variance of participants' comprehension. The explained variance increased to 49% when using the combined performance across all three games as a predictor. These findings demonstrate that student performance in relatively well-designed reading games can provide valid measures of reading skills. As such, reading games may be a viable alternative to standardized reading tests, which can render the testing experience more enjoyable, motivating, and engaging [5,52,64]. Another benefit of using games for assessments is that students can be assessed during gameplay, without being interrupted or feeling "tested". This stealthy approach may reduce test anxiety, and in turn increase the reliability of the assessment [18,19,65].

One approach to assessing reading skills in prior research has been in the context of constructed responses wherein students generate self-explanations or essays. The linguistic features of those responses were found to be indicative of students' vocabulary knowledge and comprehension skills, which suggests the feasibility of stealth reading assessment using games that embed open-ended questions [26,27]. This study took a different approach by focusing on games implementing multiple-choice questions. Notably, students are less likely to consider the tasks to be multiple-choice questions because they are presented in the context of games. Our results indicate such games can also provide a means to stealthily assess reading skills, which complements the use of NLP methods for reading assessment. In the context of serious games and ITSs, stealth assessment affords ways to evaluate students' literacy skills and update student models as they naturally interact with the software. The stealth assessment of students' reading skills can augment the macro-level adaptation of ITS, such as guiding students' practice across modules. For example, the system may recommend students with lower reading skills to play more summarization games, but direct students with higher reading skills to engage in more difficult practice within the self-explanation module.

Another focus of this study was game enjoyment. Although participants tended to enjoy the games, game enjoyment was associated with reading skills differently in the three games. Reading skill and enjoyment of Vocab Flash were not correlated. However, reading skill was negatively correlated with enjoyment of Dungeon Escape and Adventurer's Loot: participants who enjoyed these two main idea games more had lower scores on the reading skills tests. This result supports the notion that those who are more likely to perform poorly and potentially be frustrated or anxious during a traditional test are also more likely to appreciate playing a game rather than taking a test. On the flip side, participants with higher reading skills may have had more positive experiences taking and succeeding on traditional tests, and thus had less appreciation for the games.

5. Conclusions and Implications

Literacy assessment is a key component of any effort to improve learners' literacy or remediate potential gaps in education. However, such assessments can be slow, disconnected from learning experiences, anxiety-inducing, or boring. The results from this study imply an important practical application as it provides a means to measure learners' literacy skills in real time via games. Stealth assessment via serious games can also inform adaptive instructional paths for students. Serious games and intelligent tutors may act as scaffolding for less skilled readers to receive more personalized instructions to enhance their skills. Furthermore, game-based assessment can replace traditional paper-based literacy measures, which are decontextualized, cause stress and anxiety for test takers, and in turn negatively impact the reliability of these assessments [66–68].

Notably, the games used for stealth assessment in this study were relatively simplistic. Thus, our findings indicate that simple, relatively inexpensive games can be leveraged to assess skills. Nonetheless, more elaborate, immersive games with comparable embedded pedagogical features have strong potential to augment the power of stealth assessment. Note that most participants in this study had Bachelor's or advanced degrees. Future research will involve a broader range of participants with respect to prior education, which will enable assessment of the generalizability of current findings.

Author Contributions: Conceptualization: D.S.M. and Y.F.; Methodology: D.S.M., R.D.R. and Y.F.; Formal analysis: Y.F.; Investigation: Y.F.; Resources: D.S.M.; Data curation: Y.F.; Writing—original draft: Y.F., T.L., L.H. and K.C.; Writing—review and editing: Y.F., T.L., L.H., K.C., R.D.R. and D.S.M.; Supervision: D.S.M. and R.D.R.; Funding acquisition: D.S.M. All authors have read and agreed to the published version of the manuscript.

Funding: This research was funded by The Office of Naval Research Grant N00014-20-1-2623 and Institute of Education Sciences Grant R305A190050.

Institutional Review Board Statement: This study was reviewed and approved by ASU's Institutional Review Board, and that the study conforms to recognized standards.

Data Availability Statement: Data may be accessed by emailing the first author at ying.fang07@gmail.com.

Conflicts of Interest: The authors declare no conflict of interest. The funders had no role in the design of the study; in the collection, analyses, or interpretation of data; in the writing of the manuscript, or in the decision to publish the results. The opinions expressed are those of the authors and do not represent views of the Office of Naval Research or Institute of Education Sciences.

References

1. Organization for Economic Cooperation and Development. *OECD Skills Outlook: First Results from the Survey of Adult Skills*; OECD Publishing: Paris, France, 2013.
2. NAEP Report Card: Reading. The Nations' Report Card. Available online: https://www.nationsreportcard.gov/reading/nation/achievement?grade=8 (accessed on 20 May 2022).
3. NCES. *Highlights of the 2017 U.S. PIAAC Results Web Report*; Department of Education, Institute of Education Sciences, National Center for Education Statistics: Washington, DC, USA, 2020. Available online: https://nces.ed.gov/surveys/piaac/current_results.asp (accessed on 20 May 2022).
4. Shute, V.J.; Ventura, M. *Measuring and Supporting Learning in Games: Stealth Assessment*; The MIT Press: Cambridge, MA, USA, 2013.
5. Kato, P.M.; de Klerk, S. Serious games for assessment: Welcome to the jungle. *J. Appl. Test. Technol.* **2017**, *18*, 1–6.
6. Francis, D.J.; Snow, C.E.; August, D.; Carlson, C.D.; Miller, J.; Iglesias, A. Measures of reading comprehension: A latent variable analysis of the diagnostic assessment of reading comprehension. *Sci. Stud. Read.* **2006**, *10*, 301–322. [CrossRef]
7. Petrovica, S.; Anohina-Naumeca, A. The adaptation approach for affective game-based assessment. *Appl. Comput. Syst.* **2017**, *22*, 13–20. [CrossRef]
8. Onwuegbuzie, A.J.; Leech, N.L. Sampling Designs in Qualitative Research: Making the Sampling Process More Public. *Qual. Rep.* **2007**, *12*, 238–254. [CrossRef]
9. Kim, Y.J.; Ifenthaler, D. Game-based assessment: The past ten years and moving forward. In *Game-Based Assessment Revisited*; Ifenthaler, D., Kim, Y.J., Eds.; Springer: Cham, Switzerland, 2019; pp. 3–11.
10. O'Connor, M.C.; Paunonen, S.V. Big Five personality predictors of post-secondary academic performance. *Personal. Individ. Differ.* **2007**, *43*, 971–990. [CrossRef]
11. Poropat, A.E. A meta-analysis of the five-factor model of personality and academic performance. *Psychol. Bull.* **2009**, *135*, 322. [CrossRef]
12. Ke, F.; Parajuli, B.; Smith, D. Assessing Game-Based Mathematics Learning in Action. In *Game-Based Assessment Revisited*; Springer: Cham, Switzerland, 2019; pp. 213–227.
13. Min, W.; Frankosky, M.H.; Mott, B.W.; Rowe, J.P.; Smith, A.; Wiebe, E.; Boyer, K.E.; Lester, J.C. DeepStealth: Game-based learning stealth assessment with deep neural networks. *IEEE Trans. Learn. Technol.* **2019**, *13*, 312–325. [CrossRef]
14. Shute, V.J.; Rahimi, S. Stealth assessment of creativity in a physics video game. *Comput. Hum. Behav.* **2021**, *116*, 106647. [CrossRef]

15. Shute, V.; Rahimi, S.; Smith, G.; Ke, F.; Almond, R.; Dai, C.P.; Kuba, R.; Liu, Z.; Yang, X.; Sun, C. Maximizing learning without sacrificing the fun: Stealth assessment, adaptivity and learning supports in educational games. *J. Comput. Assist. Learn.* **2021**, *37*, 127–141. [CrossRef]
16. Simonson, M.; Smaldino, S.; Albright, M.; Zvacek, S. Assessment for distance education. In *Teaching and Learning at a Distance: Foundations of Distance Education*; Prentice-Hall: Upper Saddle River, NJ, USA, 2000.
17. Shute, V.J.; Leighton, J.P.; Jang, E.E.; Chu, M.W. Advances in the science of assessment. *Educ. Assess.* **2016**, *21*, 34–59. [CrossRef]
18. De-Juan-Ripoll, C.; Soler-Domínguez, J.L.; Guixeres, J.; Contero, M.; Álvarez Gutiérrez, N.; Alcañiz, M. Virtual reality as a new approach for risk taking assessment. *Front. Psychol.* **2018**, *9*, 2532. [CrossRef]
19. De Rosier, M.E.; Thomas, J.M. Establishing the criterion validity of Zoo U's game-based social emotional skills assessment for school-based outcomes. *J. Appl. Dev. Psychol.* **2018**, *55*, 52–61. [CrossRef]
20. Salen, K.; Zimmerman, E. *Rules of Play: Game Design Fundamentals*; The MIT Press: Cambridge, UK, 2004.
21. Tsikinas, S.; Xinogalos, S. Towards a serious games design framework for people with intellectual disability or autism spectrum disorder. *Educ. Inf. Technol.* **2020**, *25*, 3405–3423. [CrossRef]
22. Annetta, L.A. The "I's" have it: A framework for serious educational game design. *Rev. Gen. Psychol.* **2010**, *14*, 105–112. [CrossRef]
23. Wang, L.; Shute, V.; Moore, G.R. Lessons learned and best practices of stealth assessment. *Int. J. Gaming Comput. Mediat. Simul.* **2015**, *7*, 66–87. [CrossRef]
24. Akram, B.; Min, W.; Wiebe, E.; Mott, B.; Boyer, K.E.; Lester, J. Improving stealth assessment in game-based learning with LSTM-based analytics. In Proceedings of the 11th International Conference on Educational Data Mining, Buffalo, NY, USA, 15–18 July 2018.
25. DiCerbo, K.E.; Bertling, M.; Stephenson, S.; Jia, Y.; Mislevy, R.J.; Bauer, M.; Jackson, G.T. An application of exploratory data analysis in the development of game-based assessments. In *Serious Games Analytics*; Springer: Cham, Switzerland, 2015; pp. 319–342.
26. Allen, L.K.; McNamara, D.S. You Are Your Words: Modeling Students' Vocabulary Knowledge with Natural Language Processing. In Proceedings of the 8th International Conference on Educational Data Mining, Madrid, Spain, 26–29 June 2015; Santos, O.C., Boticario, J.G., Romero, C., Pechenizkiy, M., Merceron, A., Mitros, P., Luna, J.M., Mihaescu, C., Moreno, P., Hershkovitz, A., et al., Eds.; International Educational Data Mining Society, 2015; pp. 258–265.
27. Fang, Y.; Allen, L.K.; Roscoe, R.D.; McNamara, D.S. Stealth literacy assessment: Leveraging games and NLP in iSTART. In *Advancing Natural Language Processing in Educational Assessment*; Yaneva, V., Davier, M., Eds.; Routledge: New York, NY, USA, 2023; pp. 183–199.
28. McCarthy, K.S.; Laura, K.A.; Scott, R.H. Predicting Reading Comprehension from Constructed Responses: Explanatory Retrievals as Stealth Assessment. In Proceedings of the International Conference on Artificial Intelligence in Education, Ifrane, Morocco, 6–10 July 2020; Springer: Cham, Switzerland, 2020; pp. 197–202.
29. Li, H. Deep learning for natural language processing: Advantages and challenges. *Natl. Sci. Rev.* **2018**, *5*, 24–26. [CrossRef]
30. Fang, Y.; Li, T.; Roscoe, R.D.; McNamara, D.S. Predicting literacy skills via stealth assessment in a simple vocabulary game. In Proceedings of the 23rd Human-Computer Interaction International Conference, Virtual Conference, 24–29 July 2021; Sottilare, R.A., Schwarz, J., Eds.; Springer: Cham, Switzerland, 2021; pp. 32–44.
31. Freebody, P.; Anderson, R.C. Effects of vocabulary difficulty, text cohesion, and schema availability on reading comprehension. *Read. Res. Q.* **1983**, *18*, 277–294. [CrossRef]
32. Bernhardt, E. Progress and procrastination in second language reading. *Annu. Rev. Appl. Linguist.* **2005**, *25*, 133–150. [CrossRef]
33. Cain, K.; Oakhill, J. Reading comprehension and vocabulary: Is vocabulary more important for some aspects of comprehension? *L'Année Psychol.* **2014**, *114*, 647–662. [CrossRef]
34. Cromley, J.G.; Azevedo, R. Testing and refining the direct and inferential mediation model of reading comprehension. *J. Educ. Psychol.* **2007**, *99*, 311–325. [CrossRef]
35. Chen, X.; Meurers, D. Word frequency and readability: Predicting the text-level readability with a lexical-level attribute. *J. Res. Read.* **2018**, *41*, 486–510. [CrossRef]
36. Masrai, A. Vocabulary and reading comprehension revisited: Evidence for high-, mid-, and low-frequency vocabulary knowledge. *Sage Open* **2019**, *9*, 2158244019845182. [CrossRef]
37. Jun Zhang, L.; Bin Anual, S. The role of vocabulary in reading comprehension: The case of secondary school students learning English in Singapore. *RELC J.* **2008**, *39*, 51–76. [CrossRef]
38. Kintsch, W.; Walter Kintsch, C. *Comprehension: A Paradigm for Cognition*; Cambridge University Press: Cambridge, UK, 1998.
39. Brown, A.L.; Campione, J.C.; Day, J.D. Learning to learn: On training students to learn from texts. *Educ. Res.* **1981**, *10*, 14–21. [CrossRef]
40. Bransford, J.D.; Brown, A.L.; Cocking, R.R. *How People Learn: Brain, Mind, Experience, and School: Expanded Edition*; National Academy Press: Washington, DC, USA, 2000.
41. Wade-Stein, D.; Kintsch, E. Summary Street: Interactive computer support for writing. *Cogn. Instr.* **2004**, *22*, 333–362. [CrossRef]
42. Fox, E. The Role of Reader Characteristics in Processing and Learning from Informational Text. *Rev. Educ. Res.* **2009**, *79*, 197–261. [CrossRef]
43. Williams, J.P. Teaching text structure to improve reading comprehension. In *Handbook of Learning Disabilities*; Swanson, H.L., Harris, K.R., Graham, S., Eds.; The Guilford Press: New York, NY, USA, 2003; pp. 293–305.

44. Anmarkrud, Ø.; Bråten, I.; Strømsø, H.I. Multiple-documents literacy: Strategic processing, source awareness, and argumentation when reading multiple conflicting documents. *Learn. Individ. Differ.* **2014**, *30*, 64–76. [CrossRef]
45. Wigent, C.A. High school readers: A profile of above average readers and readers with learning disabilities reading expository text. *Learn. Individ. Differ.* **2013**, *25*, 134–140. [CrossRef]
46. Lau, K.L. Reading strategy use between Chinese good and poor readers: A think aloud study. *J. Res. Read.* **2006**, *29*, 383–399. [CrossRef]
47. Shores, J.H. Are fast readers the best readers? A second report. *Elem. Engl.* **1961**, *38*, 236–245.
48. Johnston, P.; Afflerbach, P. The process of constructing main ideas from text. *Cogn. Instr.* **1985**, *2*, 207–232. [CrossRef]
49. Afflerbach, P.P. The influence of prior knowledge on expert readers' main idea construction strategies. *Read. Res. Q.* **1990**, *25*, 31–46. [CrossRef]
50. Chittaro, L.; Buttussi, F. Exploring the use of arcade game elements for attitude change: Two studies in the aviation safety domain. *Int. J. Hum. Comput. Stud.* **2019**, *127*, 112–123. [CrossRef]
51. Derbali, L.; Frasson, C. Players' motivation and EEG waves patterns in a serious game environment. In *International Conference on Intelligent Tutoring Systems*; Springer: Berlin/Heidelberg, Germany, 2010; pp. 297–299.
52. Jackson, G.T.; McNamara, D.S. Motivation and performance in a game-based intelligent tutoring system. *J. Educ. Psychol.* **2013**, *105*, 1036. [CrossRef]
53. McNamara, D.S. Self-explanation and reading strategy training (SERT) improves low-knowledge students' science course performance. *Discourse Process.* **2017**, *54*, 479–492. [CrossRef]
54. McCarthy, K.S.; Watanabe, M.; Dai, J.; McNamara, D.S. Personalized learning in iSTART: Past modifications and future design. *J. Res. Technol. Educ.* **2020**, *52*, 301–321. [CrossRef]
55. McNamara, D.S.; Boonthum, C.; Levinstein, I.B.; Millis, K. Evaluating self-explanations in iSTART: Comparing word-based and LSA algorithms. In *Handbook of Latent Semantic Analysis*; Landauer, T., McNamara, D.S., Dennis, S., Kintsch, W., Eds.; Erlbaum: Hillsdale, MI, USA, 2007; pp. 227–241.
56. McNamara, D.S. Chasing theory with technology: A quest to understand understanding. *Discourse Process.* **2021**, *58*, 442–448. [CrossRef]
57. VanLehn, K. The behavior of tutoring systems. *Int. J. Artif. Intell. Educ.* **2006**, *16*, 227–265.
58. Shute, V.J.; Psotka, J. Intelligent Tutoring Systems: Past, Present and Future. In *Handbook of Research on Educational Communications and Technology*; Jonassen, D., Ed.; Macmillan: New York, NY, USA, 1996; pp. 570–600.
59. Woolf, B.P. *Building Intelligent Interactive Tutors: Student-Centered Strategies for Revolutionizing E-Learning*; Morgan Kaufmann: San Francisco, CA, USA, 2010.
60. Phillips, L.M.; Norris, S.P.; Osmond, W.C.; Maynard, A.M. Relative reading achievement: A longitudinal study of 187 children from first through sixth grades. *J. Educ. Psychol.* **2002**, *94*, 3–13. [CrossRef]
61. Davies, M. The 385+ million word Corpus of Contemporary American English (1990–2008+): Design, architecture, and linguistic insights. *Int. J. Corpus Linguist.* **2009**, *14*, 159–190. [CrossRef]
62. Kimura, T. The impacts of computer adaptive testing from a variety of perspectives. *J. Educ. Eval. Health Prof.* **2017**, *14*, 1149050. [CrossRef]
63. Hulin, C.; Netemeyer, R.; Cudeck, R. Can a reliability coefficient be too high? *J. Consum. Psychol.* **2001**, *10*, 55–58.
64. McClarty, K.L.; Orr, A.; Frey, P.M.; Dolan, R.P.; Vassileva, V.; McVay, A. *A Literature Review of Gaming in Education*; Research Report; Pearson: Hoboken, NJ, USA, 2012; Available online: https://www.pearsonassessments.com/content/dam/school/global/clinical/us/assets/tmrs/lit-review-of-gaming-in-education.pdf (accessed on 22 May 2023).
65. Shute, V.J.; Rahimi, S. Review of computer-based assessment for learning in elementary and secondary education. *J. Comput. Assist. Learn.* **2017**, *33*, 1–19. [CrossRef]
66. Cassady, J.C.; Johnson, R.E. Cognitive test anxiety and academic performance. *Contemp. Educ. Psychol.* **2002**, *27*, 270–295. [CrossRef]
67. Segool, N.K.; Carlson, J.S.; Goforth, A.N.; Von Der Embse, N.; Barterian, J.A. Heightened test anxiety among young children: Elementary school students' anxious responses to high-stakes testing. *Psychol. Sch.* **2013**, *50*, 489–499. [CrossRef]
68. Von der Embse, N.P.; Witmer, S.E. High-stakes accountability: Student anxiety and large-scale testing. *J. Appl. Sch. Psychol.* **2014**, *30*, 132–156. [CrossRef]

Disclaimer/Publisher's Note: The statements, opinions and data contained in all publications are solely those of the individual author(s) and contributor(s) and not of MDPI and/or the editor(s). MDPI and/or the editor(s) disclaim responsibility for any injury to people or property resulting from any ideas, methods, instructions or products referred to in the content.

Article

Creating Location-Based Augmented Reality Games and Immersive Experiences for Touristic Destination Marketing and Education

Alexandros Kleftodimos [1,*], Athanasios Evagelou [1,2], Stefanos Gkoutzios [1], Maria Matsiola [1], Michalis Vrigkas [1], Anastasia Yannacopoulou [1], Amalia Triantafillidou [1] and Georgios Lappas [1]

1. Department of Communication and Digital Media, University of Western Macedonia, 52100 Kastoria, Greece; evagel@sch.gr (A.E.); sgoutzios@uowm.gr (S.G.); mmatsiola@uowm.gr (M.M.); mvrigkas@uowm.gr (M.V.); ayannacopoulou@uowm.gr (A.Y.); atriantafylidou@uowm.gr (A.T.); glappas@uowm.gr (G.L.)
2. The Centre for Environmental and Sustainability Education (CESE) of Kastoria, 52100 Kastoria, Greece
* Correspondence: akleftodimos@uowm.gr

Citation: Kleftodimos, A.; Evagelou, A.; Gkoutzios, S.; Matsiola, M.; Vrigkas, M.; Yannacopoulou, A.; Triantafillidou, A.; Lappas, G. Creating Location-Based Augmented Reality Games and Immersive Experiences for Touristic Destination Marketing and Education. *Computers* **2023**, *12*, 227. https://doi.org/10.3390/computers12110227

Academic Editors: Carlos Vaz de Carvalho, Hariklia Tsalapatas and Ricardo Baptista

Received: 19 September 2023
Revised: 30 October 2023
Accepted: 2 November 2023
Published: 7 November 2023

Copyright: © 2023 by the authors. Licensee MDPI, Basel, Switzerland. This article is an open access article distributed under the terms and conditions of the Creative Commons Attribution (CC BY) license (https:// creativecommons.org/licenses/by/ 4.0/).

Abstract: The aim of this paper is to present an approach that utilizes several mixed reality technologies for touristic promotion and education. More specifically, mixed reality applications and games were created to promote the mountainous areas of Western Macedonia, Greece, and to educate visitors on various aspects of these destinations, such as their history and cultural heritage. Location-based augmented reality (AR) games were designed to guide the users to visit and explore the destinations, get informed, gather points and prizes by accomplishing specific tasks, and meet virtual characters that tell stories. Furthermore, an immersive lab was established to inform visitors about the region of interest through mixed reality content designed for entertainment and education. The lab visitors can experience content and games through virtual reality (VR) and augmented reality (AR) wearable devices. Likewise, 3D content can be viewed through special stereoscopic monitors. An evaluation of the lab experience was performed with a sample of 82 visitors who positively evaluated features of the immersive experience such as the level of satisfaction, immersion, educational usefulness, the intention to visit the mountainous destinations of Western Macedonia, intention to revisit the lab, and intention to recommend the experience to others.

Keywords: augmented reality; virtual reality; location-based augmented reality; tourism marketing; educational tourism

1. Introduction

Mixed reality (MR) is being increasingly used in many fields for research and commercial purposes. Amongst these fields is tourism, where augmented reality is employed to enhance the touristic experience and educate visitors through applications that combine the promotion of destinations with education and entertainment. Hobson and William stated in their study [1] that travel itself can be regarded as a secondary reality, where tourists escape temporarily. Tourists often seek to escape into simulated experiences like amusement parks (e.g., Disneyland), totally absorbed into new alternate realities [2,3]. It can be argued that virtual and augmented reality (VR and AR) achieve similar experiences where tourists are immersed in alternate realities [4].

AR and VR applications that combine entertainment and education are being utilized to enhance the tourism experience and promote destinations. These applications can intensify the "presence" of the visitor at a destination [5], and create memorable tourism experiences through game elements such as challenges, rewards, competition, and role-playing [6]. Furthermore, they can increase visitor knowledge and awareness about a destination [7]. By experiencing such applications, the users can gain useful information about a destination, such as its history and geography, the natural environment and its flora and fauna, tangible and intangible cultural heritage [8], transport information, and gastronomy. Consequently, visitor satisfaction is enhanced [9], and that in turn has a positive impact on the destination's sustainability [10].

Augmented reality adds a digital information layer to the real world. Augmented reality (AR) applications are situated closer to the left end of the virtuality continuum as can be seen in Figure 1, where real environments lie on the left end and virtual environments on the right end. This space between the two ends contains various levels of mixing between virtual and real-world elements and is known as mixed reality [11].

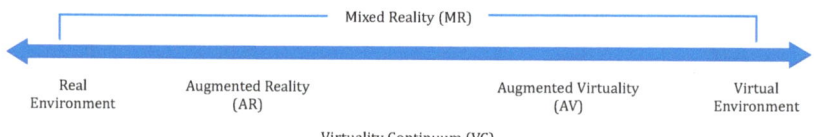

Figure 1. Representation of the "virtuality continuum".

Researchers and academics have proposed various definitions for AR [3]. One of the first definitions encountered in the literature was proposed by Azuma in 1997 [12], who defined AR as 3D objects integrated into a 3D real environment in real-time, highlighting three characteristics: (a) a combination of real and virtual, (b) interactive in real-time, and (c) registered in 3D. However, this definition does not encompass all the contemporary variations of augmented reality, and it is more aligned with image-based AR where specific images are needed to register the position of the 3D objects in the real environment [13]. The incorporation of Global Positioning Sensors (GPS) in mobile devices has opened new venues for augmented reality, resulting in a new subset of AR, the location-based or location-aware AR [14]. Location-based AR applications relying on the geolocation capabilities of mobile devices can present to users context-sensitive digital information that is related to their surroundings [3,13,15]. This section will further discuss the terms image-based (or marker-based) AR and location-based AR.

Given that AR, in the years to come after the first definition given by Azuma, evolved to include a variety of applications that blend real and virtual information, a broader and more encompassing definition of AR was needed [16]. By the end of the 2000 decade, many studies presented AR location-based applications that integrate different forms of digital information within real-world settings, such as videos, images, audio, and texts [15,17]. The present study focuses on mobile location-based AR applications and coincides with the broader definition provided by Fitzgerald et al. [16], which defines AR as including "the fusion of any digital information within real-world settings, i.e., being able to augment one's immediate surroundings with electronic data or information, in a variety of media formats that include not only visual/graphic media but also text, audio, video and haptic overlays".

Augmented reality can be achieved through wearable and non-wearable devices. Wearable devices include headsets and helmets, such as Microsoft HoloLens, Apple Vision Pro, and Magic Leap and non-wearable devices include mobile devices (smartphones, tablets, etc.) and stationary devices (TVs, PCs, etc.) [18]. Wearable devices are far more expensive than mobile devices, which is why mobile AR applications have played an important role in the significant popularity AR has gained over the last decade.

Quite a few categorizations have been presented throughout the evolution of AR, and one of these is based on the technologies that enable the AR experience (i.e., wearable and non-wearable devices) while another is based on the way that the digital content is triggered. In this categorization, we have marker-based, markerless, and location-based augmented reality.

Marker-based AR (or image-based AR) works by scanning a marker that triggers an augmented experience. Markers can be images such as drawings in art museums, street signs, and book pages, as well as real-world objects like statues and bridges. When a marker is viewed through the lens of a mobile device camera or an AR head-mounted display, it will activate digital content. The digital layer of information that superimposes the real world can come in many multimedia forms, such as text, images, videos, sound, 3D models, and 2D and 3D animations.

On the other hand, in markerless AR, digital content is chosen by the users and overlaid on physical surfaces on demand (floor, table, etc.). Finally, location-based AR applications, which can also be considered a type of markerless AR, present digital media to users when they reach specific locations as they move through the physical world. In these applications, GPS sensors constantly track users' location. Location-based AR has experienced significant growth after the advent of two popular augmented-reality location-based games, namely Pokémon Go and Ingress, created by Niantic in 2014 and 2016, respectively.

Location-based augmented reality games are currently used in several fields, such as entertainment, education, marketing, and tourism. An application can also be used to fulfill more than one purpose. For example, location-based AR games designed for tourism can be used to entertain tourists when they visit a destination, but at the same time, it can also educate them about various aspects of the destination, such as the cultural heritage and history of the place as well as aspects regarding its natural environment (e.g., mountains, rivers, wildlife). Having an experience that combines entertainment and learning about a destination is something that many tourists are looking for today, as acquiring new skills and learning something valuable while traveling is a current trend in tourism, which is especially popular amongst the representatives of Generation Y [19]. Location-based AR games attempt to create a deeper level of engagement between the users and the destination through an experience that combines educational entertainment, storytelling, personalized features, and social interaction [20].

Educational tourism (Edutourism) has been a trend in the global tourism industry for quite some time [21,22]. The main purpose of educational travel is to obtain knowledge and experience on certain topics rather than just enjoying the travel experience. Furthermore, AR and VR through wearable headsets can also promote destinations and educate individuals about these destinations before, during, and after their visit. Overall mixed reality (AR and VR applications) can be a significant asset in tourism marketing and educational tourism [23].

This paper aims to present a holistic approach that utilizes mixed reality technologies (AR and VR) to promote mountainous areas of Western Macedonia, Greece. Western Macedonia is a region in Greece with high unemployment rates, a region that could benefit economically from an increase in tourism. Furthermore, although mountainous regions possess exceptional beauty, they receive less tourism than other destinations in Greece, which are located near the sea.

This initiative is part of a broader project called Agrotour (the Agrotour project, https://agrotour.uowm.gr/, accessed on 10 September 2023), which aims at enhancing research and development in many areas that could strengthen the economy of Western Macedonia, such as agrotourism, smart farming, stock raising, bioeconomy and agri-food, new technologies in farming, etc.

The applications that have been developed include gamification features and aim at enhancing the travel experience by entertaining and educating the visitors.

More specifically, the paper aims to present the following:

(a) The methodology that was followed for identifying destinations in the mountainous regions of Western Macedonia that could benefit from touristic promotion.

(b) Three location-based applications with gamification features. Through these applications, the users visit places, and experience multimedia content when certain locations are reached. Users also gather points, virtual objects as well as prizes when certain tasks are accomplished. Various technologies have been used for developing these applications, which will be described in detail. Furthermore, by presenting these applications and the methodology followed to create them, the paper also aims to provide ideas and, to a certain degree, technical information for those wishing to develop similar applications.

(c) A mixed reality lab where visitors can experience 3D, virtual, and augmented reality content related to the mountainous areas of Western Macedonia, Greece. Many visitors have visited the lab so far, and the paper will also present evaluation results from these visits.

The contribution of this study to the existing literature is the presentation of a holistic approach that utilizes a range of AR and VR technologies in combination for touristic promotion and education. These technologies include augmented reality location-based games, an AR mobile application with virtual guides that combines location-based and image-based AR, and an immersive lab with content that can be viewed using virtual and augmented reality headsets as well as content that can be viewed through stereoscopic monitors. The equipment for producing the content for these devices is also presented. It is worth mentioning that most of the research initiatives encountered in the literature focus on specific applications that are either VR or AR rather than several AR and VR technologies that are used in combination. This is evident from literature reviews [3,24] that examine both VR and AR applications in tourism. For example, in the critical review by Wei [24] amongst the 60 papers that were retrieved and reviewed, 33 concerned VR, 25 AR, and only 2 combined both technologies.

Furthermore, this paper briefly presents how mobile AR games for educational tourism can be developed using an open source platform that is accessible and user-friendly to everyone who is interested in developing similar applications. This development platform is Taleblazer, an AR platform developed by MIT, a platform that is suitable also for people with no prior experience in programming. Therefore, it is expected that the paper will also provide ideas and guidance to people who do not have technical expertise and who are interested in developing location-based games for educational purposes.

Another contribution of this initiative is the use of the emerging generative artificial intelligence (AI) platforms for rapidly producing animated content for AR applications. More specifically, virtual talking characters were created and embedded in one of the mobile AR applications. Such a task typically requires both time and expertise, but generative AI tools can aid in producing such content fast and without any particular skills.

2. Related Work

There are many initiatives where virtual and augmented reality are used to enhance the destination experience by entertaining and educating the public. In order to spot these research initiatives, the literature was explored by searching Scholar Google and by conducting a bibliometric analysis using VOSViewer 1.6.20 [25] and publications from Scopus. The keywords used were virtual reality and augmented reality in tourism research. However, more focus is given to AR and more specifically location-based AR since a large part of the current paper concerns location-based augmented reality in tourism.

Starting from various initiatives that concern location-based AR games, the authors in [19] analyze how gamified augmented reality experiences impact tourist attractiveness. Their study focuses on the city of Bydgoszcz, which lacks the most popular attributes of sun, sand, and sea, a similar case to the Western Macedonia region in Greece. Their research aimed to analyze the potential of increasing the number of visitors in the city by creating a location-based AR game. The authors also explored whether the game created a memorable experience for the visitors, an experience that would bring them back to the city. The responses received using a survey led to the conclusion that a location-based mobile game using gamification techniques and AR is a memorable experience, and the development of such a game and its introduction on the market would increase the potential of foreign tourism in the destination.

The authors of [7] research through Structural Equation Modeling (SEM) the impact of location-based AR games on the users' intentions to visit a touristic destination, the role that gained knowledge plays in the experience, and factors driving the adoption of AR games. The study revealed that knowledge gained during gameplay has a statistically significant impact on intentions to visit.

To promote film-induced tourism in Macau, China, the authors of [26] proposed an AR application called "IfilmAR-tour-APP", which is a location-based AR mobile tour system related to film characters. The application provides film-related information (e.g., film shooting sites and transport to the attractions) as well as emotional attachment with film celebrities. Macau is a place where many film productions take place and this application is intended for tourists that visit film sites.

Location-based AR games are also used extensively in cultural heritage communication and education as this field holds an important role in tourism since many tourists are interested in exploring the history and cultural heritage of the places they visit. Heritage tourism is primarily concerned with the exploration of tangible (material) and intangible (immaterial) remnants of the past [27].

In the recent literature, there are many examples of initiatives that utilize the power of AR to present information regarding the history and cultural heritage of a destination.

For example, in [28], a mobile application called Korat Historical Explorer was developed and evaluated. Korat is the largest province in Thailand with many historical sites including more than 2000 temples. The application has three important AR usage modes consisting of (1) an AR map mode used for looking at the map in the form of print media to display 3D simulations of 10 tourist attractions in Korat, (2) AR landmark mode used for viewing actual objects in their actual locations such as ancient art paintings and mural paintings to view relevant 3D images, and (3) AR direction mode used for leading the user to the important points of the tourist attractions.

Museums and archaeological sites are also entities that employ AR to transmit information in a more engaging way. In these initiatives, when the user gets close to certain locations, the real environment is enriched with digital information, which can be in many multimedia forms such as text and sound, digital stories, animations, 3D reconstructions of monuments, etc.

For example, in [29,30], the researchers explored the enhancement of the visitor experience in outdoor archeological sites and indoor museums with the use of augmented reality technology. These AR applications managed to educate a broad and non-specialized audience with historical and archaeological content in the form of 3D virtual reconstructions. The visitors of these archaeological sites and museums may visualize these reconstructions by looking through the camera lens of their mobile devices at specific targets of the real world.

Jiang et al. [31] examined the efficacy of AR for enhancing the memorability of tourism experiences (MTE) at the heritage site of the Great Wall of China, using a smartphone app, equipped with four interrelated AR heritage tourism experiences. In [32], an AR location-based application that added a layer of 3D models of historical buildings in the real world was created. The 3D models presented how these constructions were in their past state. The authors of [33] developed an AR game called "The buildings speak about our city", which is a combination of location-based and marker-based augmented reality. The game aims to motivate primary school students to discover the buildings of tobacco warehouses in a city in western Greece, which have historical, architectural, and cultural value, and explore their relationship with the city's economic and cultural development.

Location-based AR can also be utilized in interior spaces (e.g., museums) where GPS signals are absent with the use of beacons and other tracking devices [30,34].

Furthermore, AR location-based games often harness the power of storytelling to transform an AR tour into a more exciting and engaging experience. Storytelling is known to be a powerful communication method. Storytelling is deeply embedded in human learning, as it provides an organizational structure for new experiences and knowledge [35]. People can mentally organize and memorize information better if it is communicated through stories. Any advancements in media technology that enable people to convey stories in new and innovative ways can have a profound impact. Storytelling is also an effective method of education and instruction. Stories can contain lessons, codified bits of wisdom that are passed on in a memorable and enjoyable form [36].

One of the earliest attempts of AR location-based games that utilize storytelling to entertain and educate tourists is REXplorer [37], an application created for tourists visiting the town of Regensburg, in Germany. REXplorer players encounter virtual characters and, more specifically, spirits of historical figures during gameplay. These virtual characters relate to historic buildings. The AR game aims to make the task of learning history a fun process. The spirits interact with the users, prompting them to go on quests at specific locations within the city center. By performing these quests, the players indirectly explore the historical city center and learn history in an entertaining and engaging way. Of course, there are also plenty of recent examples where AR and storytelling are jointly utilized in tourism to educate the public about a place's history and cultural heritage [38,39].

Furthermore, since this paper also utilizes virtual reality (VR), it is worth mentioning that there have been many studies in the past that research the effects of VR technologies in tourism promotion where it has been documented that VR can be used to promote a destination by generating a sense of presence. For example, in [40], the authors present an AR/VR application that presents and promotes shore excursions on cruise ships. The study investigated the effectiveness of different media—head-mounted display (HMD) or computer—used to view 360-degree tourism promotional videos as well as the impact of different information sources featured in a VR tourism video on destination image formation and intention to visit. The findings of many studies suggest that virtual reality can be an effective medium for tourism promotion (e.g., [40,41]).

The idea of "virtual portals" is presented in [42] for outdoor archaeological sites. "Virtual portals" is an idea inspired by many TV series, movies, and video games. Through these portals, the users perform transfers (or jumps) between the real world, AR, and VR reality. The authors explored the possibilities of a mixed reality that contains "shortcuts" (virtual portals) for performing "time travels" to bring the user from the real environment of the present to the virtual environment of the past.

And there are many more examples where VR is utilized for tourism promotion and such examples can be found in related literature reviews (e.g., [3,43,44]). Yung and Khoo-Lattimore [3] provide a systematic literature review on virtual reality and augmented reality in tourism research, while other reviews focus on either the use of VR or AR in tourism exclusively. For example, Beck et al. provide a state-of-the-art review on virtual reality in tourism [44], Theodoropoulos and Antoniou [43] provide a systematic review of VR games in cultural heritage, while Jingen and Elliot provide a systematic review on AR in tourism research [45]. In the Yung and Khoo-Lattimore review [3], virtual worlds were the most common focus (39% of the studies), and all studies of virtual worlds were based on the Second Life virtual world. The most common focus was studying the destination marketing potential of Second Life. Furthermore, the applications used for the enhancement of the touristic experience were exclusively AR applications. The authors argue that this could be explained with the higher mobile nature of AR when compared to VR, which typically demands the user to be stationary and requires more processing power. Similarly, Moro, Rita, Ramos, and Esmerado argue in their study [46] that although both AR and VR are progressively becoming more common in tourism experiences, VR is commonly designed as the basis of an experience, whilst AR is used to supplement an existing experience. In the study of Roman et al [47], more than 87% of the respondents that took part in a survey believed that VR tourism cannot substitute real-world tourism in the long run. However, the authors argue that VR tourism will be more beneficial for the citizens of developing countries who face difficulties in traveling to developed countries due to economic as well as other reasons (e.g., visas). Furthermore, virtual sightseeing may also constitute an alternative for people who cannot travel due to disabilities or other health conditions.

Moreover, the authors of [45] claim that while virtual reality might be a threat to travel and tourism as a potential substitute, augmented reality allows users to interact with the real environment that could potentially enhance visitors' experience. On the other hand, one of the findings of the Beck et al. literature review on virtual reality in tourism [44] is that research in this field has most commonly examined the pre-travel phase, using VR as a marketing tool for promotion and communication purposes and therefore end up investigating variables such as travel planning, behavioral intentions, or attitude. Study results suggest that VR, regardless of whether it is non-, semi-, or fully immersive, can positively influence the individual motivation to visit a place. Similarly, several studies that were analyzed in the review conducted by Yung and Khoo-Lattimore [3] found that the engagement and involvement participants felt when interacting with VR led to increased positive feelings toward the destination (e.g., [48–50]).

At this point, it has to be said that initiatives that combine both AR and VR are rather rare. The ArkaeVision project [51] is such an example that utilizes both AR and VR but also storytelling and gamification. More specifically, ArkaeVision introduces a game-like exploration of a 3D environment, virtually reconstructed, with elements of digital fiction and engaging storytelling, applied to two case studies: the exploration of the Hera II Temple of Paestum with virtual reality (VR) technology, and the exploration of the slab of the Swimmer Tomb with augmented reality (AR).

As far as education in tourism context is concerned, findings are contradicting amongst studies. In [44], it is stated that VR research in tourism in an educational context is rare, and there is a need for such VR applications. On the other hand, in the systematic review conducted by Yung and Khoo-Lattimore [3], VR and AR research in tourism education is found to be the second most common category.

3. Materials and Methods

As mentioned in the introduction, this study aims to present an approach that utilizes a range of mixed reality technologies (AR, VR, and stereoscopic technologies) for promoting mountainous areas of the region in Western Macedonia, Greece, areas that do not benefit from tourism as much as other places in Greece that are located near the sea. Furthermore, the mountain locations and villages of Western Macedonia may also be missed by tourists since more known places, such as the four municipality cities of the region (Kozani, Kastoria, Florina, and Grevena) are bound to attract the largest percentage of tourists visiting the area. Thus, the project's mission is the promotion of the mountainous destinations through mixed reality applications and games that would present to the visitors the most interesting locations and educate them about their history, cultural heritage, and the natural environment.

The project has four phases: Phase (1)—identifying the mountainous villages and sites of historical, cultural heritage and natural environment interest that can be part of application scenarios (AR games and mixed reality experiences), Phase (2)—designing and developing AR games that will intrigue users to visit the places, and learn important information about the areas, Phase (3)—creating an immersive lab for destination promotion and education, a lab where visitors can experience mixed reality content, and Phase (4)—evaluating the approach.

Regarding the first phase, the places that would be included in the augmented reality games would have to be identified. Western Macedonia is a region in the northwestern part of Greece that consists of 4 prefectures, Kozani, Grevena, Kastoria, and Florina. This large region has 486 cities, towns, and villages. Thus, the places and tasks that would be included in the AR game scenarios needed to be methodically determined. To select the places in the mountainous areas, a ranking method was developed that is based on data extracted from the relevant Wikipedia page articles for each of the 486 destinations. Page views were used as an indicator of the attractiveness of the destination and thus these were collected for candidate places during the same period of approximately a month's time (3 February 2022 until 2 March). Additional data related to altitude, population, and type of destination were also collected. The method provided an initial selection of mountain locations that are in the top ranks of the 486 destinations.

Three AR applications/games have been developed in order to prompt the users to visit a number of destinations in the region, and to educate them about different aspects of the destinations, such as historical facts, cultural heritage (monuments, etc.), the geography, and the natural environment.

To decide on the technologies that would be used for creating the AR games, the related literature as well as websites with software solutions or software reviews were investigated. It is worth mentioning that members of the project team have investigated authoring tools in the past, and performed a comparative analysis between tools, and the results of this research are presented in a previous publication [52]. Members of the team also created AR educational games for cultural heritage and evaluated the impact of these games [53–55].

Therefore, drawing from past experience, two of the AR games were built using Taleblazer http://taleblazer.org/, accessed on 1 September 2023), an open platform for developing location-based AR games for outdoor and interior spaces. Taleblazer is a product of the Massachusetts Institute of Technology (MIT). Taleblazer has a visual block-based programming environment similar to Scratch, a well-known product of MIT for introducing young children and novices to the basics of algorithmic thinking and programming. Furthermore, Taleblazer produces AR games that are available on both Android and IOS mobile devices.

Taleblazer was chosen for its simplicity in rapidly creating AR educational location-based games. As already mentioned, Taleblazer is suitable for people with no programming experience as long as they are willing to learn the basics of programming through the visual blocked-based programming environment that is suitable for beginners. Thus, the applications developed in this study can act as an example to people who wish to create similar games.

In these applications, the users are prompted to visit locations, be educated about these locations through information in multimedia form (text, images, and videos), answer questions, gather points and virtual objects, and receive prizes. The first application covers the whole region of Western Macedonia, and the second one the prefecture of Kastoria where the Digital Media and Strategic Communication laboratory of the Department of Communication and Digital Media is situated.

However, although Taleblazer can produce reliable location-based games, there are some features that cannot be achieved through this online platform. One of the features that cannot be achieved is augmentations that can be seen through the lens of the cameras of mobile devices (e.g., 3D graphics and animation). This feature is quite popular in AR applications and games, and it is amongst the features that impress users the most. Therefore, to overcome the lack of this feature in the first two applications, a third application was also created. The third application is a mobile application for Android devices developed using Android and ARCore (https://developers.google.com/ar, accessed on 1 September 2023) SDKs.

These SDK solutions were chosen for creating the third application, which would also incorporate 3D graphics and animation augmentations. This application utilizes both location-based and image-based AR. In the same fashion as the other two games, the users are prompted to visit places and get informed about the history and cultural heritage of these places. However, this time, information is often delivered through virtual characters. These are characters that act as guides for providing information to the tourists but also important historical figures that have lived in the region and characters from Greek mythology that come to life to tell stories to the users.

As mentioned in Section 2, storytelling is a powerful communication medium that has also proved to be efficient in AR location-based projects [36]. A certain procedure using generative AI online tools was followed to produce these virtual characters. To create such multimedia content in the past would require expertise and time. Today, these AI platforms can be used to create multimedia content quickly and easily. This work also contributes to the literature by presenting how such tools are utilized in the rapid development of content for AR applications.

A usability test was carried out for the mobile applications to test their functionality.

A mixed reality lab was also established for the purposes of the project consisting of VR, AR, and stereoscopic devices, as well as content to be viewed with these devices. The lab was equipped with 3D stereoscopic monitors, and VR and AR headsets. More specifically, the lab consists of one 3D TV where content can be viewed through 3D polarized glasses, an auto stereoscopic monitor for viewing 3D content without glasses, and plain 2D videos, two VR headsets, and an AR headset. To produce content for these devices, a 3D camera and a 3D 360° VR camera were used. Three-dimensional models were also built for the AR headset using an open source 3D computer graphics software tool.

The AR applications and the immersive lab were promoted through social media and the local press. Furthermore, a national TV channel dedicated a short broadcast on the lab and the staff involved in the project.

At this point, it is worth mentioning that several students from the Department of Communication and Digital Media worked on the project and contributed to the production of the deliverables.

The lab experience was evaluated through a questionnaire with various Likert scale items that assess aspects such as user satisfaction, level of immersion, educational usefulness, intention to revisit, and intention to recommend the experience to other people. In total, 82 individuals completed the questionnaire right after the experience and the evaluation results are presented in Section 7.

4. Creating AR Location-Based Games Using Taleblazer

Taleblazer is an online platform that is becoming increasingly popular for creating AR location-based games. An advantage of this platform over other solutions is its user-friendly developers' environment (Figure 2). The block-based programming environment is similar to Scratch, another famous product of MIT. Taleblazer was chosen as an authoring tool for developing these applications for two main reasons that are listed below:

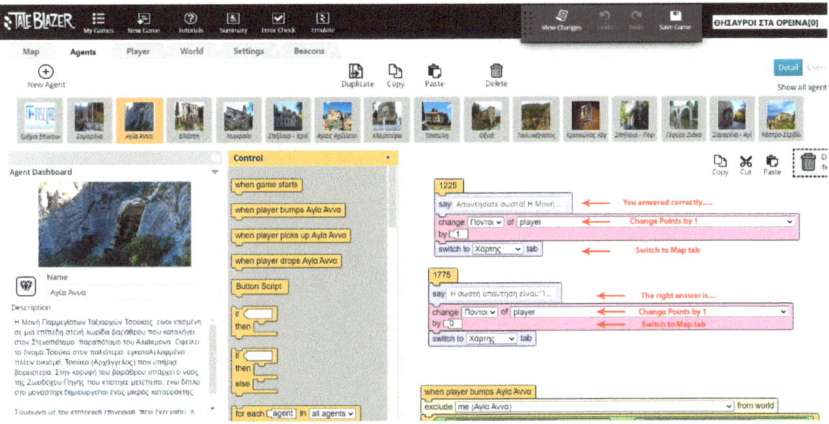

Figure 2. Taleblazer's visual block-based programming environment.

(a) Taleblazer is a low-code authoring tool that does not require advanced programming expertise. Therefore, it is suitable for many people who do not have development experience and would like to create AR location-based applications as long as they are willing to get acquainted with a programming environment that is similar to Scratch. It is anticipated that this paper will also provide ideas and some technical tips for many people who would like to create similar applications but do not have programming experience. Furthermore, Taleblazer can achieve every functionality required using a location-based application apart from image-based augmentations that appear in the real environment through the camera lens of the mobile devices. Since this is a feature that most people expect from an AR application, a third application using different tools was also developed, to provide this feature amongst others. This application will be covered in the next section.

(b) As mentioned in Section 3, members of the research team conducted a comparative analysis in the past [52] between low-code and no-code authoring tools for developing AR location-based applications, and Taleblazer proved to be a more suitable tool amongst other solutions such as ARIS and Metaverse Studio. Research team members have also developed several applications using this platform [53–55], so there is now substantial experience in building applications with Taleblazer.

In this initiative, two applications were created with the aim to gamify the travel experience of tourists visiting the region of Western Macedonia, Greece. The applications are called "Treasures in the mountainous areas of Western Macedonia" and "Mountainous destinations of Kastoria". The second application concentrates on the prefecture of Kastoria, where the Department of Communication and Digital Media is situated.

In the scenario of the game "Treasures in the mountainous areas of Western Macedonia", the users are presented with interesting destinations that can be visited during their stay in the area. On the Taleblazer map, these destinations are depicted with stars and diamond icons (Figure 3). The most significant places are depicted as stars. The visitor is prompted to visit at least four places, two of which would have to be "stars", in order to receive prizes. Points are also gathered during the visits.

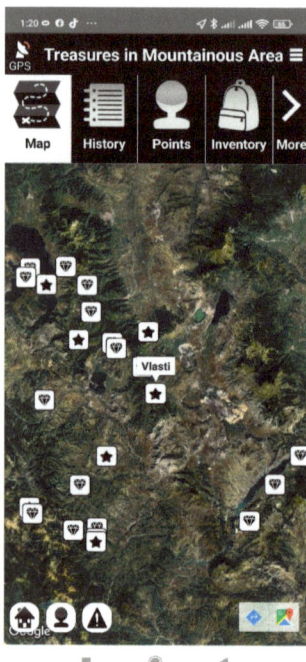

Figure 3. Destinations are depicted with star and diamond icons. The users have to visit four places (two of which must be stars) in order to receive prizes.

The prizes that the players will receive if they complete the mission of the game are souvenir gifts from a local store that has agreed to cooperate with the Digital Media and Strategic Communication lab for this project and special mixed reality content that can be viewed during a visit to the immersive lab. It is worth mentioning that the game does not end when the four places are visited. The player can continue to visit places, get informed about these places, and gather even more points.

The users of the applications are guided to the destinations through the Google Map API (in the same manner as the Google Map mobile application), which has been incorporated into Taleblazer. The icons shown at the bottom of the application (Figure 3) can enable Google Map navigation when clicked.

The basic elements of the Taleblazer programming environment are "Regions" and "Agents". Regions are the physical areas on the map where the game (or tour) will evolve. Using a selection tool, the designer can determine the game's region on a digital map. After the region is set, "Agents" can be introduced (Figure 4). Agents are multimedia elements associated with GPS locations and are activated when a learner "bumps" into these locations (Figure 4). This multimedia content can be in the form of images, multiple-choice questions, sounds, narration, and video.

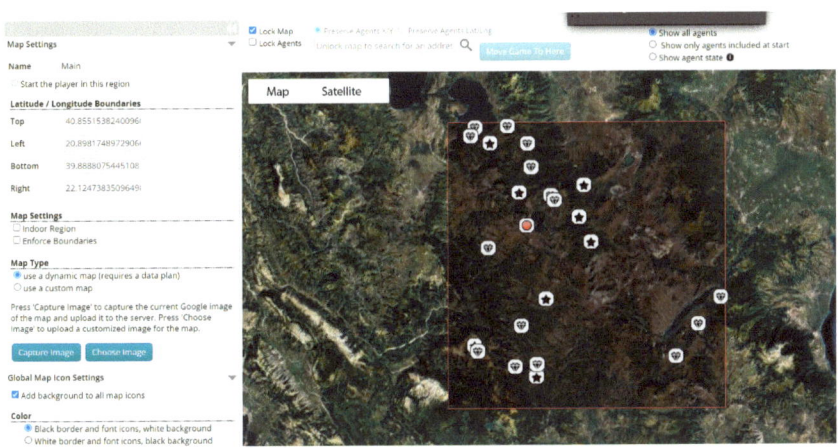

Figure 4. Regions and agents. The stars and diamonds depict Agents. Multimedia content is activated when the user reaches these GPS locations. The red dot is the user's position.

Two kinds of maps can be used in Taleblazer. The designer can either choose to utilize the Google Map API for user navigation or custom maps, that is, maps created by the designer (e.g., with image editing tools). Custom maps are very useful in small areas that can be walked on foot as well as applications for indoor spaces. With custom maps, the designers can also add their own details to a map. If custom maps are used, the designer would have to match the real-world locations with the custom map locations. However, for the applications of this initiative, the Google Map API would have to be used since the game would evolve in a large geographical area. The users would have to reach the destinations by vehicle and the Google Map API functionality that is embedded into Taleblazer is more suitable in this case to guide them on this tour.

To experience the application, the users would have to download Taleblazer from "Google Play" for Android mobile devices or the "App Store" for IOS operating systems, respectively, and then insert the unique code in the "Game Code" tab to install the game. The game codes of both applications are included in the webpage that was designed for distributing the two applications (the application webpage: https://dmsclab.uowm.gr/projects/serious-games/, accessed on 1 September 2023).

When the user reaches a point of interest (that is, when the blue dot of the user gets close to a star or diamond icon), informative multimedia content in the form of text and images is activated. This content appears on the screen of the mobile device, and the users are also often prompted to observe their surroundings (e.g., a bridge, church, etc.) to answer a multiple-choice question. Feedback is then given to the users by telling them if the answer is correct and by giving them more detailed information regarding the question topic (Figure 5). If the answer is correct, then a point is also given as a reward.

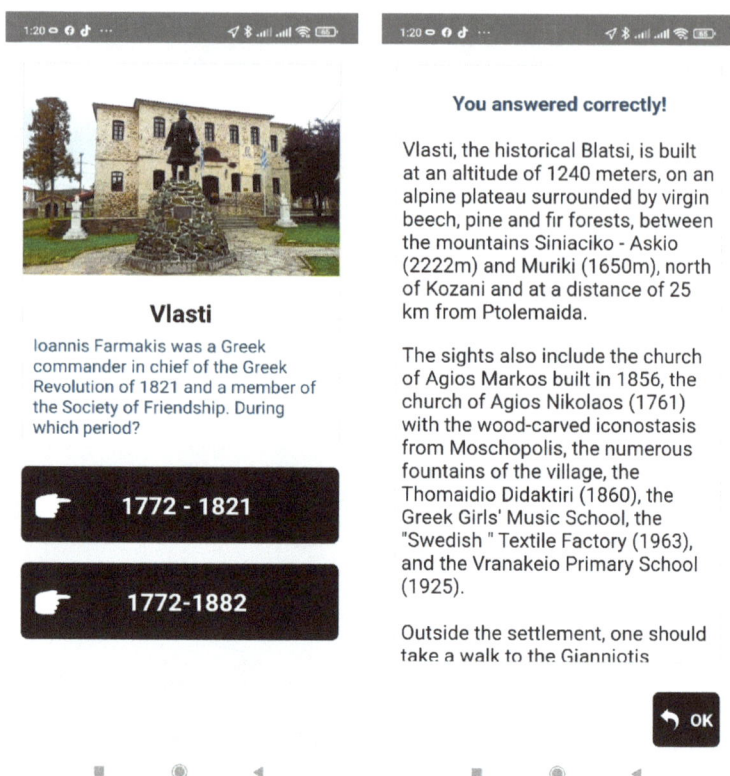

Figure 5. Indicative questions and answers.

In some places, the users are also prompted to watch short videos with information regarding the destination or to visit a webpage via a link that is included in the informational text. If the user decides to follow a link to a webpage, then this webpage is opened within the Taleblazer environment.

The visited places can be viewed at any time in the History tab, and the points gathered in the Points b. tab. Virtual objects (stars and diamonds) are collected in the Inventory tab (Figure 6).

The second AR game "Mountainous Destinations of Kastoria" is similar to the first one. In this game, the users are prompted to visit interesting places in the Prefecture of Kastoria, which is part of Western Macedonia. Again, four places would have to be visited in order to receive prizes. The places were equally important this time, so there were no stars or diamonds (Figure 7).

In both applications, when the users visit four of the places on the map, they receive a message telling them that they have succeeded and that they can now visit the immersive lab to have a mixed reality experience, and a local store in order to receive souvenir gifts. The red and yellow dots that appear on the application map when the task is accomplished indicate the locations of the immersive lab and the local store, respectively (Figure 8). The users can either choose to go directly to the lab or to the local store to receive their gift or to continue to visit other places and leave their visit to the lab and store for a later time. The users who managed to fulfill the games' mission will be able to receive the prizes by showing the places visited in the History tab or the virtual objects collected.

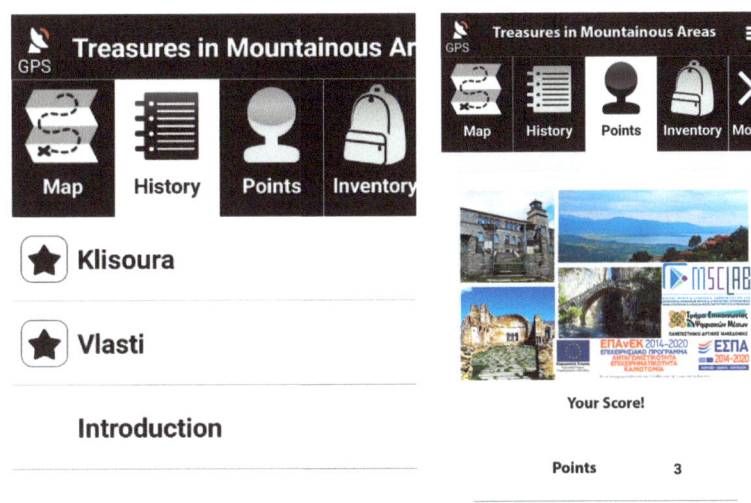

Figure 6. History and Points tabs.

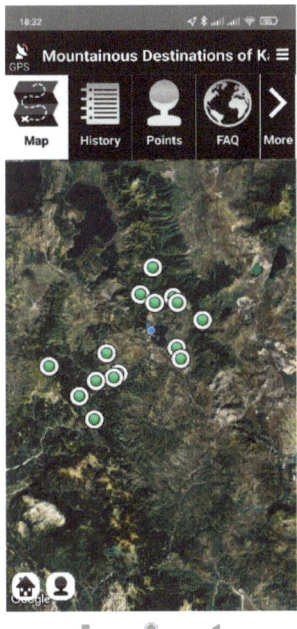

Figure 7. Mountainous destinations of Kastoria. The blue dot indicates the current position of the player and the green dots the places to visit.

Figure 8. Location of the immersive lab (red dot) and the local store (yellow dot). Blue dot indicates the GPS location of the user.

5. Creating an AR Application for Tourism Using AR SDK and Generative AI Multimedia Production Tools

As mentioned in Section 3, a third application was also created to introduce some engaging content through augmentations that appear through the camera lens of mobile devices. The name of the application is "Virtual Guide for mountainous areas of Western Macedonia".

This application is independent of the AR games described in the previous section and can be used alone or in conjunction with these games. The purpose of this application is to

(a) educate the users about the mountainous areas of Western Macedonia. The informative material of the application can be viewed at any time (e.g., before, during, or after the visits to the mountainous areas).

(b) entertain the users while they are visiting the places through virtual characters that provide information and tell stories. This content can only be viewed when the users are at specific locations.

Using this application, the users can be guided with Google Map API to visit places in the four regional units of Western Macedonia, Kozani, Kastoria, Florina, and Grevena. When starting the application, the first screen is a menu with image buttons for the four regional units plus an image button for the Department of Communication and Digital Media. By pressing any of the buttons, the user is forwarded to another menu with image buttons of the most known locations in each regional unit that are situated in mountainous areas (Figure 9). The snapshot images of the application in Figure 9 were edited to include English labels. The locations are mainly villages and places of historical and cultural heritage interest. It is also worth saying that this application contains some interesting mountainous paths. Tourists of the area can use the applications as a guide to walk these paths.

Figure 9. Snapshots from the Virtual Guide for mountainous areas of Western Macedonia.

By choosing any of the location buttons, a new screen is presented with information about the selected point of interest, and this information can be related to its history, geography, economy, demography, or natural environment. This information is mainly in textual form, but the application also includes narration so that the user can hear the information rather than read it. Google's text-to-speech service is used for obtaining mp3 audio files from text (Figure 10).

Several images are also included for each location, together with information in text and audio form. Furthermore, a link to a webpage that contains more information and a link to an informative video is typically included for many locations. The user is also given the option to view locations with Google Street View within the application. Some interior spaces, such as museums, are also available in Street View. For example, in the application, users can view in the 360° mode the house of Pavlos Melas, an important historical figure of the past (Figure 11).

A difference between this application and the applications described in the previous section is that the users can experience a large part of the application content remotely and without visiting the places indicated. However, if they decide to visit the places, they are given the option to be navigated to these places via the Google Map API embedded in the application. Moreover, they will be rewarded with augmented reality storytelling experiences if they decide to visit these places. More specifically, through the application, the users are informed about AR spots that exist in many locations. These AR spots are image-based markers that activate multimedia content. In many cases, this multimedia content is virtual characters that provide information or tell a story. As already mentioned for these augmentations, the AR core SDK was used together with the Sceneform API for inserting digital objects such as video and gITF (GL Transmission Format) 3D models into the real world.

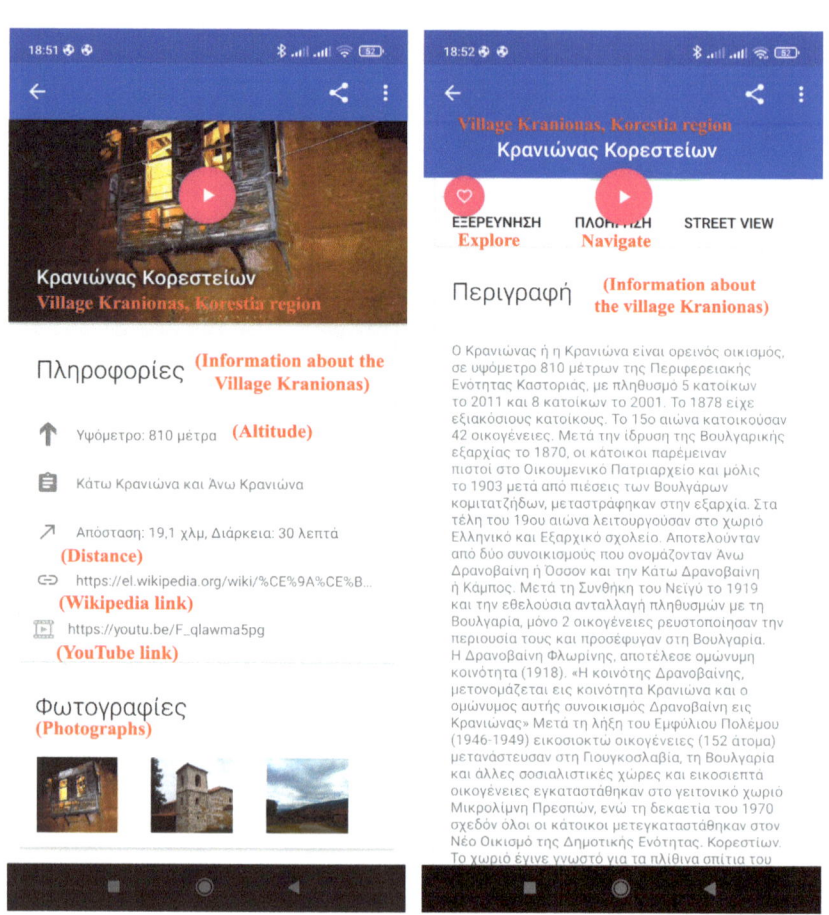

Figure 10. Information about the village Kranionas in the Korestia region.

For example, when the users get to a specific location in the village of Klisoura with the aid of the application, they are prompted to look for a specific street sign. When they find the street sign and view it through the camera lens of the application, a virtual guide appears and gives information about the village (Figure 12). This information is delivered with voice narration.

Similarly, in the village of Nimfeo in the regional unit of Florina, the users are guided to a specific location where they are asked to look for a sign. Nimfeo is also known for the Arctouros refuge for bears. This is a refuge for bears that were born or spent a long time in captivity and can no longer survive on their own in the wild. By scanning the sign with their mobile device, a character of mythology named Calisto (who was transformed by Hera, the wife of Zeus, into a bear) appears and tells her whole story. While listening to her story, the users can also see other augmentations that surround them (e.g., a bear).

Figure 11. A 360° view of an interior of a museum (house of Pavlos Melas).

Figure 12. Virtual guide at the village of Klisoura. The street sign in this figure can be found at the entrance of the village. The sign welcomes the visitors to the village (the martyred community of Kleisoura) and the virtual guide provides information about the village.

In other cases, important people who are connected to the history of the villages appear and either tell a story or give information about the place. For example, in a village called Melas, which took its name after Pavlos Melas (a Greek revolutionary and artillery officer of the Hellenic Army who participated in the Greco-Turkish War of 1897 and was amongst the first Greek officers to join the Macedonian Struggle), the users are prompted to find an AR spot, a sign that will trigger multimedia content. When this sign is scanned, Pavlos Melas appears and reads a letter that he sent to his parents during the Macedonian Struggle (Figure 13).

Figure 13. Melas virtual character. The virtual character is actvated when a street sign is scanned with the camera of a mobile phone.

In another example, if the users decide to visit a village called Vogatsiko, another historical figure called Ionas Dragoumis (a Greek politician) comes to life when an AR spot (a street sign) is scanned with the mobile device. Dragoumis then narrates some personal political views that have been written in his memoirs and have been recorded in history books (Figure 14).

In another instance, Athanasios Christopoulos (a celebrated Greek poet, playwright, distinguished scholar, and jurist) comes to life and recites one of his poems.

To create this application, the following SDKs were used: (a) Android SDK platform tools, (b) Google Maps SDK for navigating to the points of interest, and (c) the AR Core SDK, and the Sceneform 3D framework/SDK for creating the image-based augmentations. ARCore is Google's augmented reality SDK offering cross-platform APIs to build immersive experiences on Android, iOS, Unity, and Web. ARCore offers points, plane detection, pose, light estimation, anchors, image tracking, face tracking, object occlusion, and cloud anchors. With ARCore's motion-tracking capabilities, developers can track the phone's position relative to the surroundings. Other important capabilities of ARCore are environmental understanding, including detecting the size and location of surfaces, and light estimation, including real-life lighting conditions. Sceneform is a 3D framework/SDK that makes it easy to build ARCore apps for Android without OpenGL.

The application that was implemented in the JAVA programming language includes the basic functions of a mobile guiding application, such as displaying points of interest (POI) on a map; locating the user's location; retrieving information, photos, and online video for points of interest (POI) from a database; and navigating to the points of interest (POI) with the Google Maps API. The application uses the geolocation capabilities of mobile devices (GPS sensor) and internet connection (4G/5G) to provide real-time information and navigation. The application data (images, audio, textual information, etc.) are stored in a local SQLlite database. Audio files were created with the Google text-to-speech service. The database is implemented in Android using the ORM (Object Relational Mapping) library. A web server was used for storing animated videos (e.g., the virtual characters) as well as images used as triggers in image-based AR. This method has two advantages. First, it saves space from the mobile application database, making the application lighter. Second, the videos and trigger images can be changed without demanding application updates from the end users.

Figure 14. A historical figure, Ionas Dragoumis, comes to life, and expresses political views. The virual character appears when the sign of the village Vogatsiko is scanned with a mobile phone.

The application uses responsive design techniques to adapt to various mobile devices (smartphones and tablets). Text-to-speech Google API was also utilized for converting textual information about the destinations into audio.

There are several SDKs for developing AR mobile applications, but the most popular ones are ARCore, ARkit, and Vuforia. ARKit is Apple's framework and SDK for developers who wish to build AR apps for the IOS platform. ARKit has similar functions to ARCore but supports only the IOS platform, while ARCore and Vuforia support both IOS and Android, and this is a disadvantage for ARKit since the Android market has a large share of mobile devices. ARKit and ARCore offer equivalent results and capabilities in lighting estimation using a different approach for achieving this. ARKit fares better in reliable tracking, and ARCore leads in mapping and reliable recovery.

Vuforia, on the other hand, has limited capabilities in comparison to ARKit and ARCore but it can utilize the benefits of both ARCore and ARKit through Vuforia Engine SDK (which is available for Android and IOS development). However, it must be mentioned at this stage that the member of the team that developed the application had experience in programming with Android and ARCore, so although Vuforia could also be used for this task, the ARCore solution was chosen. Furthermore, although the application is currently only available for Android operating systems due to the cross-platform SDKs used, it could also be adapted to IOS platforms in the future.

To create the virtual characters, generative AI tools were used. Generative artificial intelligence (AI) describes algorithms (such as ChatGPT and DALL·E 2) that can be used to create multimedia content, including audio, code, images, text, simulations, and videos. Recent breakthroughs in the field have a significant impact on the way that content is created. Today, there are several AI tools for creating multimedia content. In our case, the aim was to create virtual talking characters that are identical to existing pictures and drawings. By searching the web and trying various tools, it was decided to use Heygen, an AI video generator platform (Heygen: https://app.heygen.com/, accessed on 1 September 2023). Using the platform, first, the images of the characters were uploaded onto the platform. These images were either drawings or old photos of historical characters or characters from mythology. Some character images that were also available on the platform were used (e.g., Figure 12). After choosing the images, the background was removed, and the resolution was adjusted. Then, the animated video was created by adding the related text, which was converted into speech with the text-to-speech algorithms of the platform.

The application functionality is depicted in Figure 15.

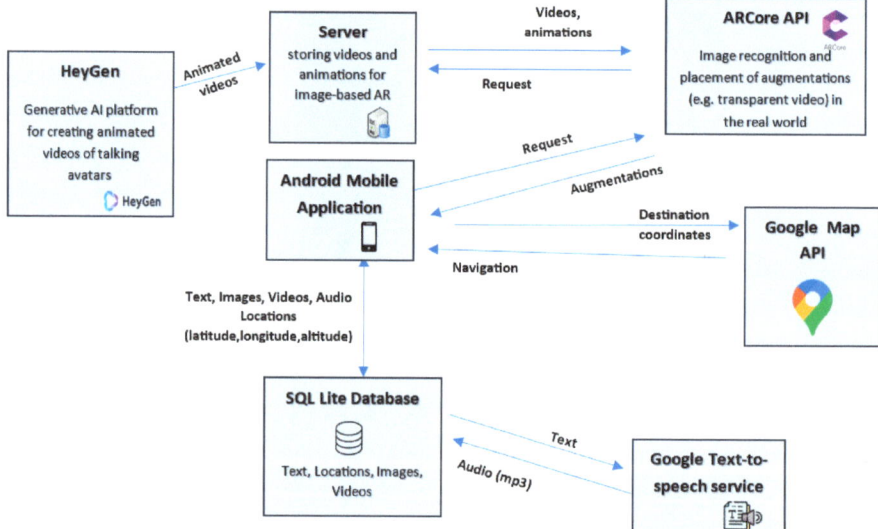

Figure 15. Application architecture.

6. The Immersive Lab

The immersive lab is a lab where people can come to experience immersive mixed reality content. As the lab encompasses many technologies that lie at different points on the mixed reality continuum (Figure 1), the term mixed reality will be used when talking about the lab experience as a whole.

This lab can be visited at any time either before or after visiting destinations of Western Macedonia. The lab aims to promote the mountainous places of Western Macedonia by providing visitors with impressive immersive content. The content that the visitors can experience while visiting the lab is 3D videos, 360° videos, 2D educational videos, and an AR game. Some impressive 360° content was reserved for the people who came to the lab after experiencing the location-based AR games described in Section 4.

The equipment of this immersive lab is the following:
- Virtual Reality Headsets, Oculus Quest 2, and HTC VIVE PRO;
- Augmented Reality Headset, HoloLens 2;

- Autostereoscopic Screen, Dimenco 65 model:DM654MAS;
- Three-dimensional TV, LG 47LM760S;
- Two computers, Intel Core i9-10900K Box 3.7 GHz, memory DDR4 64 Gb, Graphics Card Nvidia GeForce RTX 3080 10GB.

Additionally, 360° videos were captured using a Kandao Obsidian S 3D 360° VR camera and edited in Adobe Premiere PRO. This content is viewed through the virtual reality headsets (Oculus Quest 2 and HTC Vive). Three-dimensional videos were produced using a Panasonic AG-3DA1EJ 3D camera. Visitors can see these 3D videos in the 3D TV LG 47LM760S with the use of polarized glasses. Three-dimensional videos can also be viewed on the autostereoscopic screen. The autostereoscopic screen is also used for viewing plain 2D educational videos. Figure 16 depicts the mentioned technologies.

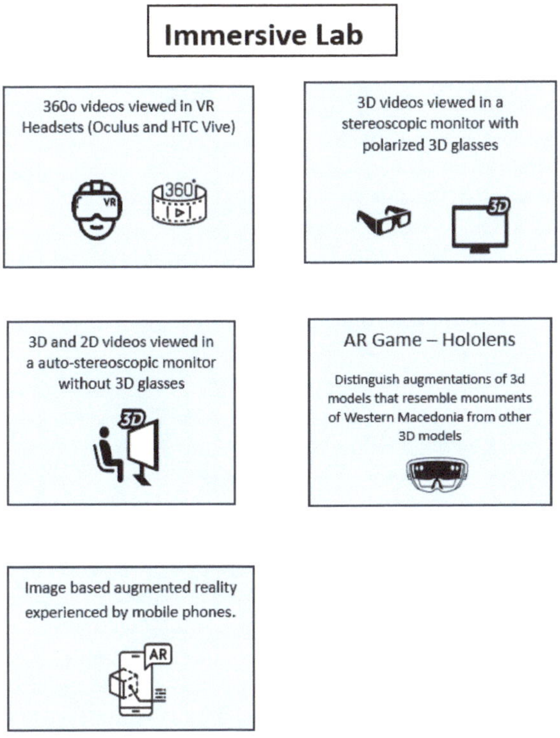

Figure 16. Immersive lab technologies.

A person who manages the visits is always present in the lab to welcome the visitors and guide them through the immersive experience. This person guides the visitors to experience the devices and the content in a specific order. This person was responsible for conveying more information regarding the destinations that were viewed through the immersive technologies (VR, AR headsets, and stereoscopic screens) in cases where the videos (360°, 3D, and 2D) contained only images.

Some images of the street signs that activate the virtual characters with the aid of the AR application described in the previous section are also placed on the lab walls.

The main aim of the lab was to impress the visitors with 3D images and videos. Around five people were employed for this task, which were mainly students (undergraduate and postgraduate) of the Department of Communication and Digital Media. These people were trained to operate the equipment and manage the visits (Figures 17 and 18).

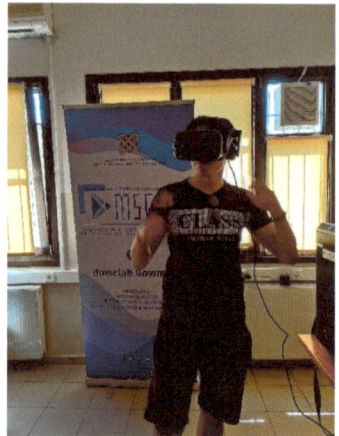

Figure 17. Immersive lab.

Figure 18. Immersive lab.

The AR Game

AR content can be viewed through the Microsoft HoloLens 2 headset. The game is rather simple. The visitors wearing the headset are presented with 3D objects and animations that augment the lab's physical space and are literally floating around them. The users are then asked to identify the 3D objects that could be related to the Western Macedonia region and move them to a specific area, differentiating them from the rest of the objects. For this reason, 3D models were produced using Blender. The models resemble monuments that exist in the region (Figure 19). For example, two models are an arched bridge (such bridges are typically found in the region) and a clock tower situated in the main square of the capital city of Western Macedonia, Kozani (Agiou Nikolaou Clock Tower).

Furthermore, a 3D representation model of a known Byzantine church that is situated in Kastoria, called Koubelidiki church, was created by the photogrammetry method using multiple photographs and stitching software (Figure 20).

The rest of the objects included in this AR game were drawn from the Hololens 3D library.

The person who is responsible for the lab provides more information about the models that resemble (or are identical to) monuments of the region.

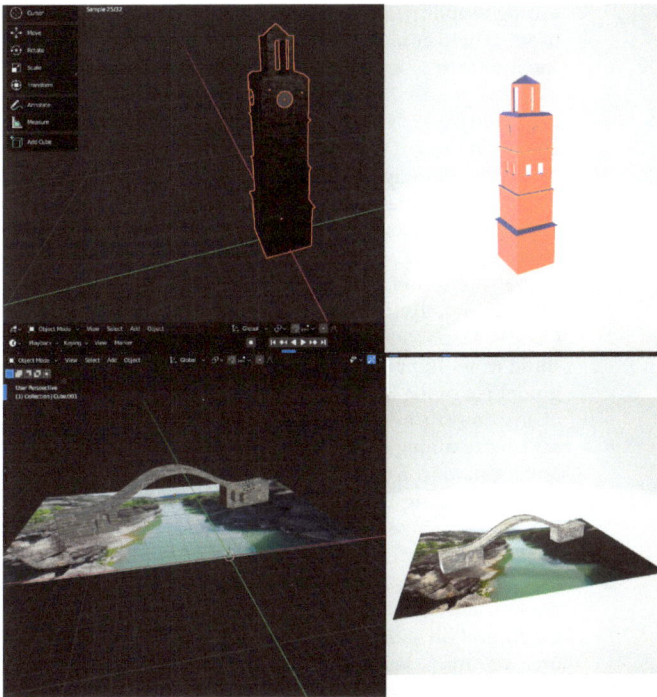

Figure 19. Three-dimensional models that resemble known monuments of the region.

Figure 20. Three-dimensional model of the Byzantine church "Koubelidiki".

7. Application Usability Testing of the Mobile Applications, Immersive Lab Experience Evaluation, and Results

A usability test was carried out to check the mobile applications' functionality and see if these applications are intuitive and user-friendly (the mobile apps described in Sections 4 and 5). Regarding the immersive lab experience, this was evaluated using a questionnaire that was distributed to 82 participants.

7.1. Application Usability Test

To test the usability of the mobile applications, a mixed approach was followed where the thinking-out-loud method was used in combination with the coaching method [56,57]. The coaching method allows participants to ask any system-related questions to an expert during usability testing. The main goal of this method is to define the information needs of users to deliver improved training and documentation in addition to probably redesigning the interface to eradicate the need for questions in the first place. The thinking-aloud protocol requires participants to articulate their thoughts, feelings, and opinions during a usability test. One goal of this approach is to empower the tester to obtain a better understanding of the participant's mental model during interaction with the interface.

Three participants and an expert who acted as an observer and a coach were employed for the usability test. During this test, certain problems arose while using the Taleblazer applications "Treasures in the mountainous areas of Western Macedonia" and "Mountainous destinations of Kastoria". These problems were then tackled by redesigning some aspects of the application. The problems are listed below:

(a) Initially, it was not obvious to the participants how to start their navigation using Google Maps. Although there are relevant icons, as we can see in Figure 3, Section 4, these icons skipped their attention. Therefore, this information was made clearer in the guidelines of the application, which are shown before the start of the game and on-demand at any time during gameplay. Furthermore, a text message was included on the initial map screen explaining to the users how to start their navigation with Google Maps (by pressing any location icon and then the Google Map icons at the bottom of the screen).

(b) Finding the GPS location that would activate the multimedia content was not always easy since GPS sensors might not work well in certain places. For the multimedia content to be activated, the users would have to get close to the GPS location of a place (e.g., a monument). This distance between the user and the place that is needed for the content activation is configured for all the agents (GPS locations) through the Taleblazer configuration settings ('autobump' and 'tap to bump' settings). Therefore, the activation area was made larger through the Taleblazer settings to solve this problem.

Besides those two problems, no other issues were encountered with the Taleblazer applications, and the users were able to navigate to the places by car using these applications without problems.

The third application, "Virtual Guide for mountainous areas of Western Macedonia", proved to be very easy to use. The participants were able to experience the application either from their home or by visiting the locations navigated with the Google Map API, which is incorporated into the application. The AR spots were also easy to locate once the users were close to the required locations, and the AR augmentations were activated without problems. The virtual guides were successful in directing the users to other 3D objects that were present in space in some of the locations.

7.2. Immersive Lab Evaluation

As already mentioned, a questionnaire consisting of Likert scale items was distributed to 82 participants, evaluating aspects such as satisfaction, positive feelings, level of immersion, educational usefulness, intention to revisit and recommend the experience to others, and motivation and intention to visit destinations in Western Macedonia. The visitors evaluated, with the questionnaire, the experience as a whole (i.e., all technologies and applications experienced in the lab).

The research was carried out according to all the guidelines provided by the Research Ethics Committee of the University of Western Macedonia, Greece. All participants were adults and were informed about the nature and purpose of the study. Informed consent was obtained from all subjects who completed the questionnaire. Furthermore, the questionnaire was anonymous, and the participants were informed both orally and in writing that the collected anonymous data would only be accessible to the researchers of the study and would be protected from any external access.

Below, in Figures 21 and 22, we can see the demographic statistics regarding the gender of the participants and their age. The participants were mainly in the age group of 18 to 25 since many students and their friends visited the lab.

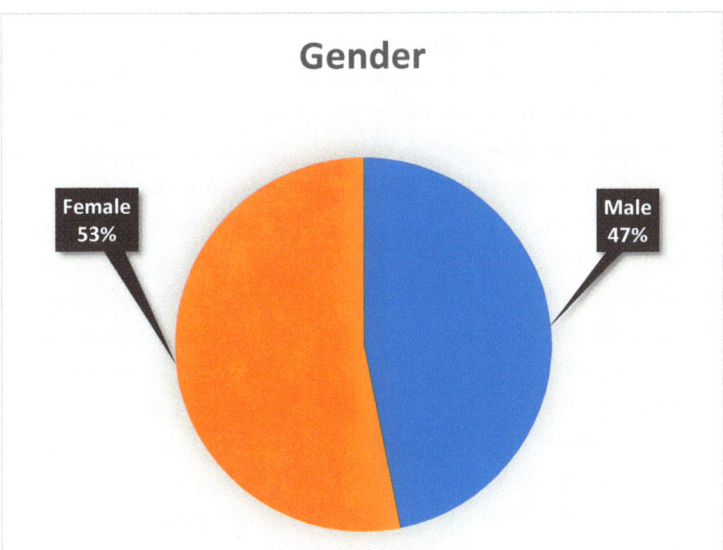

Figure 21. Gender.

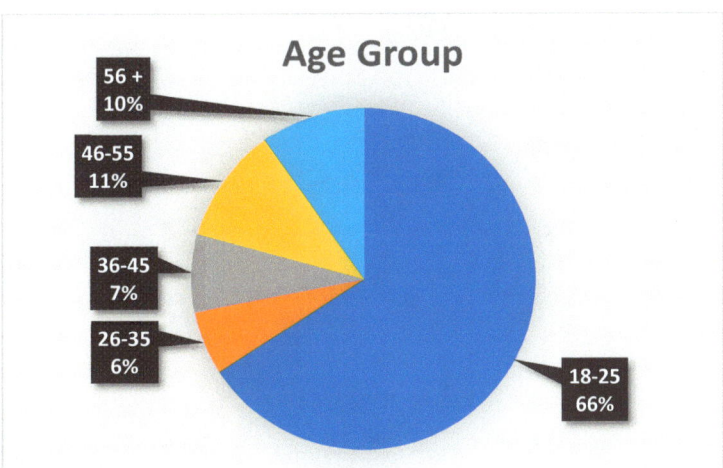

Figure 22. Age group.

As we can see in Figure 23, a large percentage of the participants were either totally unfamiliar with AR and VR technologies or had a small level of familiarity.

A large percentage of the visitors had visited destinations of Western Macedonia as it can be seen in Figure 24. This question includes all types of destinations and does not focus on mountainous destinations.

As can be seen from Figures 25 and 26, the visitors were highly satisfied with the experience that they had at the immersive laboratory and the quality of the content.

Figure 27 shows that the visitors were immersed in the experience, with their attention focused on the VR and AR content.

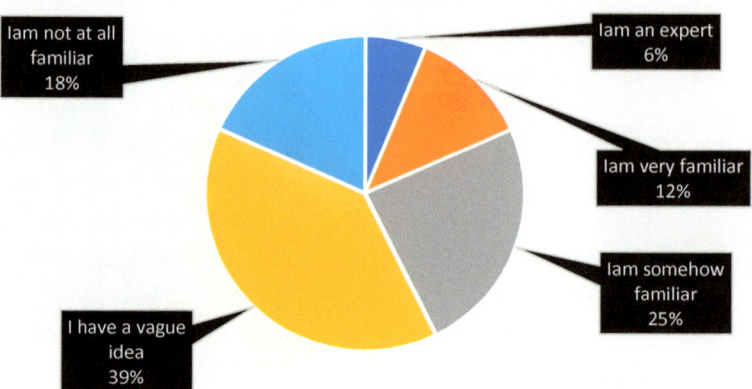

Figure 23. Familiarity with mixed reality technologies (AR and VR).

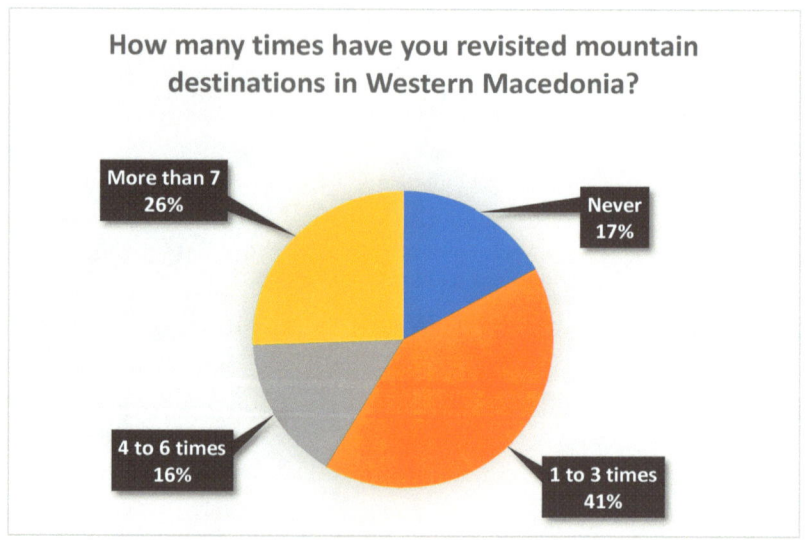

Figure 24. Number of visits to areas of Western Macedonia, Greece.

Figure 28 depicts the extent to which the visitors felt that the experience was educational and whether they gained knowledge regarding the mountainous regions of Western Macedonia. Based on the results, the experience was regarded as educational and helped strengthen the knowledge of participants.

Figure 29 depicts the intention to visit Western Macedonia's mountainous regions after the immersive lab visit. As results show, participants reported positive intentions to revisit the regions.

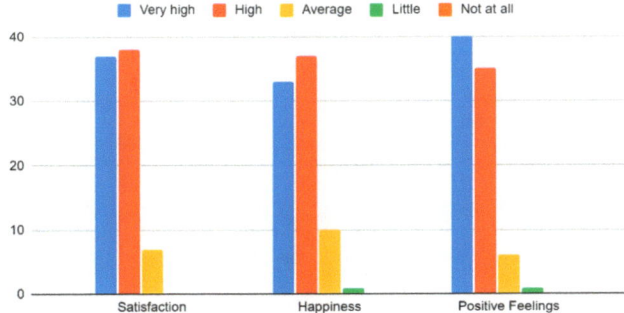

Figure 25. Measuring satisfaction and positive feelings after the visit to the immersive lab.

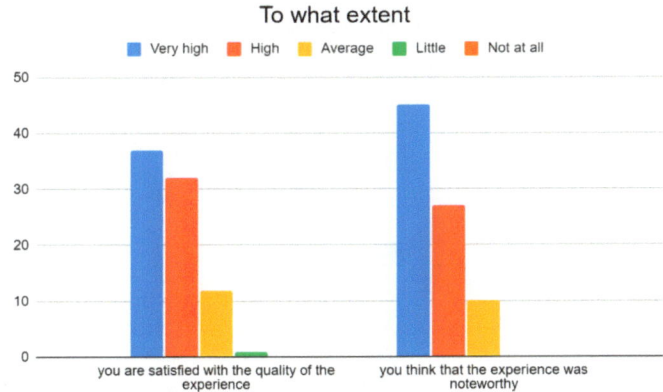

Figure 26. Assessing the quality of the experience.

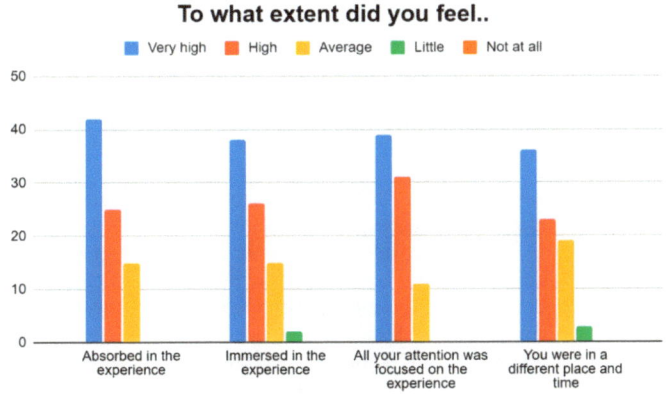

Figure 27. Measuring the level of immersion.

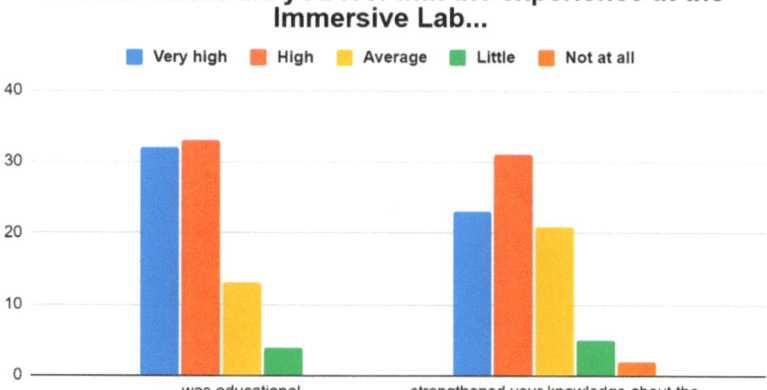

Figure 28. Educational usefulness.

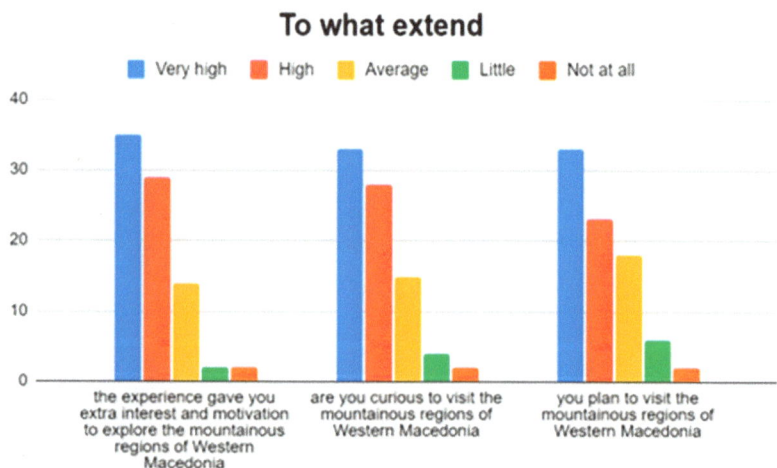

Figure 29. Interest initiated with the visit to the lab for visiting the mountainous areas of Western Macedonia.

As Figure 30 shows, participants exhibited positive intentions to revisit the lab and recommend the experience to others.

To explore whether the three dimensions of the virtual experience (e.g., entertainment, immersion, and knowledge) are related to participants' satisfaction, future intentions about the virtual experience, and future visit intentions to destinations of Western Macedonia, six summative scales were developed by adding the scores of the items of each construct and dividing them by the number of the items. All six scales exhibited adequate internal reliability based on the Cronbach's alpha values that exceeded the 0.70 criterion (entertainment: 0.859, immersion: 0.800, knowledge: 0.799, satisfaction: 0.837, future intentions of virtual experience: 0.780, future revisit intentions: 0.944).

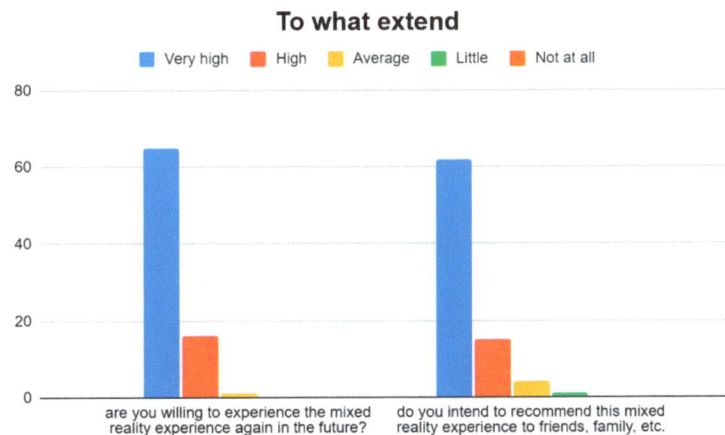

Figure 30. Intention to revisit the lab and to recommend the experience to other people.

Next, Pearson's correlation coefficient was calculated to test the impact of the three dimensions on participants' satisfaction and future intentions with the virtual experience and the destinations of Western Macedonia. Results are presented in Table 1.

Table 1. Correlations between constructs.

Relationship	Pearson's r (Significance)
Entertainment–Satisfaction	0.532 (0.000)
Entertainment–Lab Experience Intentions	0.436 (0.000)
Entertainment–Future Visit Intentions to Western Macedonia Destinations	0.246 (0.026)
Immersion–Satisfaction	0.517 (0.000)
Immersion–Lab Experience Intentions	0.323 (0.003)
Immersion–Future Visit Intentions to Western Macedonia Destinations	0.110 (0.325)
Knowledge–Satisfaction	0.446 (0.000)
Knowledge–Lab Experience Intentions	0.409 (0.000)
Knowledge–Future Visit Intentions to Western Macedonia Destinations	0.593 (0.000)

Based on the findings, the entertainment dimension of the virtual experience of the lab is significantly ($p < 0.05$) and positively related to visitors' satisfaction with the experience ($r = 0.532$, sig. = 0.000), future intentions with the immersive lab experience ($r = 0.436$, sig. = 0.000), and future visit intentions to destinations of Western Macedonia ($p = 0.246$, sig. = 0.026). In addition, the immersion dimension was found to be significantly ($p < 0.05$) and positively related to satisfaction ($r = 0.517$, sig. = 0.000) and virtual experience intentions ($r = 0.323$, sig. = 0.003). However, immersion did not influence in a significant way future revisit intentions ($r = 0.110$, sig. = 0.325).

Moreover, the educational dimension of the experience had a significant and positive relationship with visitors' satisfaction with the experience ($r = 0.446$, sig. = 0.000), future intentions with the mixed reality experience ($r = 0.409$, sig. = 0.000), and future revisit intentions ($p = 0.593$, sig. = 0.026).

Comparing the correlation coefficients of the significant relationships found, it can be argued that visitors' satisfaction can be influenced greatly by the entertainment and immersion dimensions of the experience, while future visit intentions to the destinations by the educational aspect of the experience. Intentions to re-experience the immersive lab tour were affected to the same degree by all three dimensions.

8. Discussion

The usability test revealed some aspects of the mobile applications that caused a certain level of confusion to the participants. Therefore, documentation was changed to include clearer instructions for the users, and a text message was added to show the users the appropriate buttons to click in order to start a Google Maps navigation (that will guide them to the locations indicated on the application map). Furthermore, a configuration setting was altered to broaden the necessary distance between the user's position and the agents' GPS position for the digital content to be activated. A possible limitation of this research is the lack of evaluation with a large sample of people visiting the area and experiencing the mobile AR applications. A questionnaire for evaluating the application experience and aspects such as engagement and satisfaction, ease of use, immersion, escapism, educational usefulness, and willingness to re-use or visit more destinations as a result of the experience was constructed, and incorporated in the applications. Users are asked to complete this questionnaire before exiting the applications. Responses are currently being collected, and an evaluation of the applications is pending.

Regarding the evaluation of the immersive lab experience, certain elements received high rankings, such as user satisfaction, immersion in the experience, intention to revisit the lab, and intention to recommend the experience to others. Furthermore, the experience influenced the users' interest to visit new places in the mountainous regions of Western Macedonia. The visitors perceived the experience as educational, and a significant percentage of these visitors feel that they gained knowledge regarding the destinations of Western Macedonia. The further analysis showed that entertainment and educational utility were factors that are positively related to the intentions of the survey participants to visit destinations of Western Macedonia. This is also in line with findings from previous research efforts. Although some early conceptual studies warned that VR will eventually threaten physical and corporeal travel [58], it was later proven through many studies that the engagement and involvement participants felt when interacting with VR led to increased positive feelings toward the destination (e.g., [3,48–50]). In a study conducted by Pantano and Servidio [59], participants who had a VR experience of a touristic destination expressed a desire to travel to the real tourism site in order to compare it to the one reconstructed in VR.

A possible limitation of this evaluation may be the fact that the visitors were predominately 18 to 25 years old, as many of the visitors were university students and their friends. However, the people responsible for the visits observed that older people were even more excited about their visit to the lab as most of them were completely unfamiliar with such technologies. Another observation was that most of the visitors showed particular interest and were impressed with the augmented reality content, which was experienced through the Microsoft HoloLens Headset.

As mentioned in the introduction, the contribution of this paper when compared to initiatives that are found in the literature (e.g., studies included in the systematic review in the use of AR and VR in tourism research by Yung and Khoo-Lattimore [3]) is the fact that it combines many technologies in order to promote touristic destinations, enhance the tourism experience, and educate users about the history, cultural heritage, demography, and natural environment of the destinations. To the best of our knowledge, there is a lack of studies and initiatives that use a combination of multiple technologies (AR and VR) for touristic promotion and education. Furthermore, there seems to be a shortage of research initiatives that focus on location-based (or location-aware AR) for tourism. This is true for educational applications as stated by the authors of a recent systematic mapping review of

STEM augmented reality applications in higher education [60] and it seems to be the case for tourism research as well. In the review conducted by Yung and Khoo-Lattimore [3], there seems to be only two papers that utilize this technology [61,62]. This is also the case with other review studies where image-based AR is more popular. For example, in [63], the authors examine twelve AR applications for tourism. Only four used location-based AR in comparison to 10 that used image-based AR (there were also applications that used both). Furthermore, as it is stated in the recent review conducted by Liang and Elliot [45], AR in tourism is currently an emerging topic, but journals that capture related articles are still limited. The authors also claim that the number of studies in this field is expected to grow significantly in the next decade. Therefore, the paper contributes to the emerging but still limited research literature on AR applications in tourism, concentrating on an even more limited niche of AR, that of location-based AR.

Another contribution of the paper is that it describes how a low-code platform such as Taleblazer, which is suitable for people who do not have programming experience, can be used to create AR games that enhance the tourist experience and knowledge about destinations. It is also worth mentioning that the game awards users who complete certain tasks with real prizes (content in the immersive lab and a gift from a local store) apart from points and virtual artifacts. This is another novelty of this initiative and as future work, we intend to examine the impact of this feature on user intentions (e.g., intentions to play the location-based AR games again by revisiting the region of Western Macedonia and recommend the games to others).

The paper also describes how more advanced technologies such as ARCore in combination with Google services (e.g., text-to-speech) and generative AI platforms can be used for creating mobile AR applications.

9. Conclusions

This paper aims to present an approach that utilizes a combination of AR and VR technologies and gaming applications for touristic destination marketing and education. Two location-based AR games have been developed with the aim to promote mountainous destinations in the region of Western Macedonia, Greece, and to educate users about various aspects of the destinations, including the history, cultural heritage, geography, and natural environment of the target destinations. Furthermore, a third AR mobile application was produced to inform and entertain visitors with the use of virtual guides. Besides information about the destinations, this application includes virtual characters that appear as augmentations when the mobile device's camera lens is pointed at various image targets (e.g., street signs). These characters either provide information about the destination or narrate stories. The characters that tell stories are mainly important people that are connected to the history or mythology of the target destinations. These characters were created using generative AI tools. Generative AI platforms can be utilized today to rapidly create computer graphics and animations, and these tools are bound to play an important role in content creation. Moreover, a usability test was carried out to test the mobile applications, and the problems revealed were tackled with changes to the user interface and the configuration settings.

Besides the mobile AR applications, an immersive lab was established with the aim to promote the mountainous destinations of Western Macedonia. The lab hosts various mixed reality technologies. By visiting the lab, the visitors can experience 360° videos with the use of VR headsets, 3D videos with the use of stereoscopic monitors, and an AR game with the use of an AR headset. The visitors can also view short 2D educational videos. The lab experience was evaluated using a questionnaire. The questionnaire measured the level of satisfaction, positive feelings associated with the experience, quality of the applications, immersion, educational usefulness, intention to visit the mountainous areas of Western Macedonia, intention to revisit the lab, and willingness to recommend the lab experience to family and friends. The evaluation results show that the recorded levels were high for all items and also revealed relations between these items. In the future, a similar evaluation will also take place for the mobile applications.

Author Contributions: Conceptualization, A.K., G.L.; methodology, A.K., G.L.; software, A.E., A.K., S.G.; validation, A.K., A.E., G.L.; data analysis, A.T., A.K.; Evaluation, A.T., A.K.; writing—original draft preparation, A.K.; writing—review and editing, A.K., M.M., M.V., A.Y., G.L.; visualization, A.K.; supervision, G.L., A.K.; project administration, G.L., A.K.; funding acquisition, G.L., A.K.; All authors have contributed equally. All authors have read and agreed to the published version of the manuscript.

Funding: This research has been co-funded by the European Union and Greek national funds through the Priority axis 'Development of Entrepreneurship Support Mechanisms' of the Operational Program 'Competitiveness, Entrepreneurship, and Innovation', Proposal in the framework of the action entitled 'Supporting Regional Excellence' (project code: 5047196).

Data Availability Statement: Data available on request due to privacy restrictions.

Conflicts of Interest: The authors declare no conflict of interest.

References

1. Perry Hobson, J.S.; Williams, A.P. Virtual reality: A new horizon for the tourism industry. *J. Vacat. Mark.* **1995**, *1*, 124–135. [CrossRef]
2. Cohen, E. Rethinking the sociology of tourism. *Ann. Tour. Res.* **1979**, *6*, 18–35. [CrossRef]
3. Yung, R.; Khoo-Lattimore, C. New realities: A systematic literature review on virtual reality and augmented reality in tourism research. *Curr. Issues Tour.* **2019**, *22*, 2056–2081. [CrossRef]
4. Williams, P.; Hobson, J.P. Virtual reality and tourism: Fact or fantasy? *Tour. Manag.* **1995**, *16*, 423–427. [CrossRef]
5. Champion, E. *Critical Gaming: Interactive History and Virtual Heritage*; Routledge: London, UK, 2016; ISBN 978-1-317-15739-7.
6. Xu, F.; Tian, F.; Buhalis, D.; Weber, J.; Zhang, H. Tourists as Mobile Gamers: Gamification for Tourism Marketing. *J. Travel Tour. Mark.* **2016**, *33*, 1124–1142. [CrossRef]
7. Lacka, E. Assessing the impact of full-fledged location-based augmented reality games on tourism destination visits. *Curr. Issues Tour.* **2020**, *23*, 345–357. [CrossRef]
8. Mortara, M.; Catalano, C.E.; Bellotti, F.; Fiucci, G.; Houry-Panchetti, M.; Petridis, P. Learning cultural heritage by serious games. *J. Cult. Herit.* **2014**, *15*, 318–325. [CrossRef]
9. Prakasa, F.B.P.; Emanuel, A.W.R. Review of Benefit Using Gamification Element for Countryside Tourism. In Proceedings of the 2019 International Conference of Artificial Intelligence and Information Technology (ICAIIT), Yogyakarta, Indonesia, 13–15 March 2019; pp. 196–200.
10. Yoo, C.; Kwon, S.; Na, H.; Chang, B. Factors Affecting the Adoption of Gamified Smart Tourism Applications: An Integrative Approach. *Sustainability* **2017**, *9*, 2162. [CrossRef]
11. Milgram, P.; Kishino, F. A Taxonomy of Mixed Reality Visual Displays. *IEICE Trans. Inf. Syst.* **1994**, *77*, 1321–1329. Available online: https://www.semanticscholar.org/paper/A-Taxonomy-of-Mixed-Reality-Visual-Displays-Milgram-Kishino/f78a31be8874eda176a5244c645289be9f1d4317 (accessed on 20 August 2023).
12. Azuma, R.T. A Survey of Augmented Reality. *Presence Teleoperators Virtual Environ.* **1997**, *6*, 355–385. [CrossRef]
13. Cheng, K.-H.; Tsai, C.-C. Affordances of Augmented Reality in Science Learning: Suggestions for Future Research. *J. Sci. Educ. Technol.* **2013**, *22*, 449–462. [CrossRef]
14. Martin, S.; Diaz, G.; Sancristobal, E.; Gil, R.; Castro, M.; Peire, J. New technology trends in education: Seven years of forecasts and convergence. *Comput. Educ.* **2011**, *57*, 1893–1906. [CrossRef]
15. Dunleavy, M.; Dede, C. Augmented Reality Teaching and Learning. In *Handbook of Research on Educational Communications and Technology*; Spector, J.M., Merrill, M.D., Elen, J., Bishop, M.J., Eds.; Springer: New York, NY, USA, 2014; pp. 735–745, ISBN 978-1-4614-3184-8.
16. FitzGerald, E.; Ferguson, R.; Adams, A.; Gaved, M.; Mor, Y.; Thomas, R. Augmented Reality and Mobile Learning: The State of the Art. *Int. J. Mob. Blended Learn.* **2013**, *5*, 43–58. [CrossRef]
17. Klopfer, E. *Augmented Learning: Research and Design of Mobile Educational Games*; MIT Press: Cambridge, MA, USA, 2008; ISBN 978-0-262-11315-1.
18. Peddie, J. Types of Augmented Reality. In *Augmented Reality*; Springer International Publishing: Cham, Switzerland, 2017; pp. 29–46, ISBN 978-3-319-54501-1.
19. Enhancing Tourism Potential by Using Gamification Techniques and Augmented Reality in Mobile Games | International Business Information Management Association (IBIMA). Available online: https://ibima.org/accepted-paper/enhancing-tourism-potential-by-using-gamification-techniques-and-augmented-reality-in-mobile-games/ (accessed on 5 August 2023).

20. Weber, J. Augmented Reality Gaming: A new Paradigm for Tourist Experience. In *Information and Communication Technologies in Tourism 2014: Proceedings of the International Conference in Dublin, Ireland, 21–24 January 2014*; Springer: Cham, Switzerland, 2014; pp. 57–64.
21. McGladdery, C.A.; Lubbe, B.A. Rethinking educational tourism: Proposing a new model and future directions. *Tour. Rev.* **2017**, *72*, 319–329. [CrossRef]
22. Ritchie, B. *An Introduction to Educational Tourism*; Channel View Publications: Clevedon, UK, 2003; ISBN 978-1-873150-50-4.
23. Zarzuela, M.M.; Pernas, F.J.D.; Calzón, S.M.; Ortega, D.G.; Rodríguez, M.A. Educational Tourism through a Virtual Reality Platform. *Procedia Comput. Sci.* **2013**, *25*, 382–388. [CrossRef]
24. Wei, W. Research progress on virtual reality (VR) and augmented reality (AR) in tourism and hospitality: A critical review of publications from 2000 to 2018. *J. Hosp. Tour. Technol.* **2019**, *10*, 539–570. [CrossRef]
25. Leiden University. VOSViewer, Visualizing Scientific landscapes. Available online: https://www.vosviewer.com/ (accessed on 27 October 2023).
26. Wu, X.; Lai, I.K.W. The acceptance of augmented reality tour app for promoting film-induced tourism: The effect of celebrity involvement and personal innovativeness. *J. Hosp. Tour. Technol.* **2021**, *12*, 454–470. [CrossRef]
27. Gravari-Barbas, M. Heritage and tourism: From opposition to coproduction. In *A Research Agenda for Heritage Tourism*; Gravari-Barbas, M., Ed.; Edward Elgar Publishing: Cheltenham, UK, 2020; ISBN 978-1-78990-352-2.
28. Phithak, T.; Kamollimsakul, S. Korat Historical Explorer: The Augmented Reality Mobile Application to Promote Historical Tourism in Korat. In Proceedings of the 2020 3rd International Conference on Computers in Management and Business, Tokyo, Japan, 31 January–2 February 2020; pp. 283–289.
29. Cisternino, D.; Gatto, C.; De Paolis, L.T. Augmented Reality for the Enhancement of Apulian Archaeological Areas. In *Augmented Reality, Virtual Reality, and Computer Graphics*; De Paolis, L.T., Bourdot, P., Eds.; Lecture Notes in Computer Science; Springer International Publishing: Cham, Switzerland, 2018; Volume 10851, pp. 370–382, ISBN 978-3-319-95281-9.
30. Tsai, T.-H.; Shen, C.-Y.; Lin, Z.-S.; Liu, H.-R.; Chiou, W.-K. Exploring Location-Based Augmented Reality Experience in Museums. In *Universal Access in Human–Computer Interaction. Designing Novel Interactions*; Antona, M., Stephanidis, C., Eds.; Lecture Notes in Computer Science; Springer International Publishing: Cham, Switzerland, 2017; Volume 10278, pp. 199–209, ISBN 978-3-319-58702-8.
31. Jiang, S.; Moyle, B.; Yung, R.; Tao, L.; Scott, N. Augmented reality and the enhancement of memorable tourism experiences at heritage sites. *Curr. Issues Tour.* **2023**, *26*, 242–257. [CrossRef]
32. Panou, C.; Ragia, L.; Dimelli, D.; Mania, K. An Architecture for Mobile Outdoors Augmented Reality for Cultural Heritage. *ISPRS Int. J. Geo-Inf.* **2018**, *7*, 463. [CrossRef]
33. Koutromanos, G.; Styliaras, G. "The buildings speak about our city": A location based augmented reality game. In Proceedings of the 2015 6th International Conference on Information, Intelligence, Systems and Applications (IISA), Corfu, Greece, 6–8 July 2015; pp. 1–6.
34. National Slate Museum Application. Available online: https://appadvice.com/app/national-slate-museum-amgueddfa-lechi-cymru/1122132874 (accessed on 25 August 2020).
35. Mandler, J.M. *Stories, Scripts, and Scenes*; Psychology Press: New York, NY, USA; Taylor and Francis Group: London, UK, 2014; ISBN 978-1-317-76859-3.
36. Azuma, R. Chapter 11: Location-Based Mixed and Augmented Reality Storytelling. In *Fundamentals of Wearable Computers and Augmented Reality*, 2nd ed.; Barfield, W., Ed.; CRC Press: Boca Raton, FL, USA, 2015; pp. 259–276, ISBN 978-1-4822-4350-5. [CrossRef]
37. Ballagas, R.A.; Kratz, S.G.; Borchers, J.; Yu, E.; Walz, S.P.; Fuhr, C.O.; Hovestadt, L.; Tann, M. REXplorer: A mobile, pervasive spell-casting game for tourists. In Proceedings of the CHI '07 Extended Abstracts on Human Factors in Computing Systems, San Jose, CA, USA, 28 April—3 May 2007; pp. 1929–1934.
38. Nobrega, R.; Jacob, J.; Coelho, A.; Weber, J.; Ribeiro, J.; Ferreira, S. Mobile location-based augmented reality applications for urban tourism storytelling. In Proceedings of the 2017 24o Encontro Português de Computação Gráfica e Interação (EPCGI), Guimaraes, Portugal, 12–13 October 2017; pp. 1–8.
39. Spierling, U.; Winzer, P.; Massarczyk, E. Experiencing the Presence of Historical Stories with Location-Based Augmented Reality. In *Interactive Storytelling*; Nunes, N., Oakley, I., Nisi, V., Eds.; Lecture Notes in Computer Science; Springer International Publishing: Cham, Switzerland, 2017; Volume 10690, pp. 49–62, ISBN 978-3-319-71026-6.
40. Arlati, S.; Spoladore, D.; Baldassini, D.; Sacco, M.; Greci, L. VirtualCruiseTour: An AR/VR Application to Promote Shore Excursions on Cruise Ships. In *Augmented Reality, Virtual Reality, and Computer Graphics*; De Paolis, L.T., Bourdot, P., Eds.; Lecture Notes in Computer Science; Springer International Publishing: Cham, Switzerland, 2018; Volume 10850, pp. 133–147, ISBN 978-3-319-95269-7.
41. Adachi, R.; Cramer, E.M.; Song, H. Using virtual reality for tourism marketing: A mediating role of self-presence. *Soc. Sci. J.* **2022**, *59*, 657–670. [CrossRef]
42. Cisternino, D.; Gatto, C.; D'Errico, G.; De Luca, V.; Barba, M.C.; Paladini, G.I.; De Paolis, L.T. Virtual Portals for a Smart Fruition of Historical and Archaeological Contexts. In *Augmented Reality, Virtual Reality, and Computer Graphics*; De Paolis, L.T., Bourdot, P., Eds.; Lecture Notes in Computer Science; Springer International Publishing: Cham, Switzerland, 2019; Volume 11614, pp. 264–273, ISBN 978-3-030-25998-3.

43. Theodoropoulos, A.; Antoniou, A. VR Games in Cultural Heritage: A Systematic Review of the Emerging Fields of Virtual Reality and Culture Games. *Appl. Sci.* **2022**, *12*, 8476. [CrossRef]
44. Beck, J.; Rainoldi, M.; Egger, R. Virtual reality in tourism: A state-of-the-art review. *Tour. Rev.* **2019**, *74*, 586–612. [CrossRef]
45. Jingen Liang, L.; Elliot, S. A systematic review of augmented reality tourism research: What is now and what is next? *Tour. Hosp. Res.* **2021**, *21*, 15–30. [CrossRef]
46. Moro, S.; Rita, P.; Ramos, P.; Esmerado, J. Analysing recent augmented and virtual reality developments in tourism. *J. Hosp. Tour. Technol.* **2019**, *10*, 571–586. [CrossRef]
47. Roman, M. Virtual and Space Tourism as New Trends in Travelling at the Time of the COVID-19 Pandemic. *Sustainability* **2022**, *14*, 628. [CrossRef]
48. Huang, Y.; Backman, S.J.; Backman, K.F. Exploring the impacts of involvement and flow experiences in Second Life on people's travel intentions. *J. Hosp. Tour. Technol.* **2012**, *3*, 4–23. [CrossRef]
49. Huang, Y.C.; Backman, K.F.; Backman, S.J.; Chang, L.L. Exploring the Implications of Virtual Reality Technology in Tourism Marketing: An Integrated Research Framework. *Int. J. Tour. Res.* **2016**, *18*, 116–128. [CrossRef]
50. Kim, J.; Hardin, A. The Impact of Virtual Worlds on Word-of-Mouth: Improving Social Networking and Servicescape in the Hospitality Industry. *J. Hosp. Mark. Manag.* **2010**, *19*, 735–753. [CrossRef]
51. Bozzelli, G.; Raia, A.; Ricciardi, S.; De Nino, M.; Barile, N.; Perrella, M.; Tramontano, M.; Pagano, A.; Palombini, A. An integrated VR/AR framework for user-centric interactive experience of cultural heritage: The ArkaeVision project. *Digit. Appl. Archaeol. Cult. Herit.* **2019**, *15*, e00124. [CrossRef]
52. Kleftodimos, A.; Lappas, G.; Vrigkas, M. Taleblazer vs. Metaverse: A comparative analysis of two platforms for building AR location-based educational games. *Int. J. Entertain. Technol. Manag.* **2022**, *1*, 290. [CrossRef]
53. Kleftodimos, A.; Evagelou, A.; Triantafyllidou, A.; Grigoriou, M.; Lappas, G. Location-Based Augmented Reality for Cultural Heritage Communication and Education: The Doltso District Application. *Sensors* **2023**, *23*, 4963. [CrossRef] [PubMed]
54. Kleftodimos, A.; Moustaka, M.; Evagelou, A. Location-Based Augmented Reality for Cultural Heritage Education: Creating Educational, Gamified Location-Based AR Applications for the Prehistoric Lake Settlement of Dispilio. *Digital* **2023**, *3*, 18–45. [CrossRef]
55. Kleftodimos, A.; Evagelou, A. Educational Location-Based Augmented Reality Applications for Indoor Spaces: Creating the Application "Exploring the Aquarium of Kastoria". Presented at the CORETA 2023, Advances on Core Technologies and Applications. September 2023. pp. 1–6. Available online: https://www.thinkmind.org/index.php?view=article&articleid=coreta_2023_1_10_10007 (accessed on 15 October 2023).
56. Nielsen, J. *Usability Engineering*; *Morgan Kaufmann*; Elsevier: Amsterdam, The Netherlands, 1993; ISBN 978-0-12-518406-9. [CrossRef]
57. St. Stephen's College; Gupta, S. A Comparative study of Usability Evaluation Methods. *Int. J. Comput. Trends Technol.* **2015**, *22*, 103–106. [CrossRef]
58. Cheong, R. The virtual threat to travel and tourism. *Tour. Manag.* **1995**, *16*, 417–422. [CrossRef]
59. Pantano, E.; Servidio, R. An exploratory study of the role of pervasive environments for promotion of tourism destinations. *J. Hosp. Tour. Technol.* **2011**, *2*, 50–65. [CrossRef]
60. Mystakidis, S.; Christopoulos, A.; Pellas, N. A systematic mapping review of augmented reality applications to support STEM learning in higher education. *Educ. Inf. Technol.* **2022**, *27*, 1883–1927. [CrossRef]
61. Pedrana, M. Location-based services and tourism: Possible implications for destination. *Curr. Issues Tour.* **2014**, *17*, 753–762. [CrossRef]
62. Trojan, J. Integrating AR services for the masses: Geotagged POI transformation platform. *J. Hosp. Tour. Technol.* **2016**, *7*, 254–265. [CrossRef]
63. Özkul, E.; Kumlu, S.T. Augmented Reality Applications in Tourism. *Int. J. Contemp. Tour. Res.* **2019**, *3*, 107–122. [CrossRef]

Disclaimer/Publisher's Note: The statements, opinions and data contained in all publications are solely those of the individual author(s) and contributor(s) and not of MDPI and/or the editor(s). MDPI and/or the editor(s) disclaim responsibility for any injury to people or property resulting from any ideas, methods, instructions or products referred to in the content.

Article

Creating a Newer and Improved Procedural Content Generation (PCG) Algorithm with Minimal Human Intervention for Computer Gaming Development †

Lazaros Lazaridis and George F. Fragulis *

Department of Electrical and Computer Engineering, University of Western Macedonia, 501 50 Kozani, Greece; dece00049@uowm.gr
* Correspondence: gfragulis@uowm.gr
† This paper is an extended version of our paper published in the 4th International Conference, HCI-Games 2022, Virtual Event, 26 June–1 July 2022.

Abstract: Procedural content generation (PCG) algorithms have become increasingly vital in video games developed by small studios due to their ability to save time while creating diverse and engaging environments, significantly enhancing replayability by ensuring that each gameplay experience is distinct. Previous research has demonstrated the effectiveness of PCG in generating various game elements, such as levels and weaponry, with unique attributes across different playthroughs. However, these studies often face limitations in processing efficiency and adaptability to real-time applications. The current study introduces an improved spawn algorithm designed for 2D map generation, capable of creating maps with multiple room sizes and a decorative object. Unlike traditional methods that rely solely on agent-based evaluations, this constructive algorithm emphasizes reduced processing power, making it suitable for generating small worlds in real time, particularly during loading screens. Our findings highlight the algorithm's potential to streamline game development processes, especially in resource-constrained environments, while maintaining high-quality content generation.

Keywords: procedural content generation (PCG); game development; replayability; algorithmic map generation; computer games; resource management; dynamic content creation

Citation: Lazaridis, L.; Fragulis, G.F. Creating a Newer and Improved Procedural Content Generation (PCG) Algorithm with Minimal Human Intervention for Computer Gaming Development. *Computers* **2024**, *13*, 304. https://doi.org/10.3390/computers13110304

Academic Editors: Carlos Vaz de Carvalho, Hariklia Tsalapatas and Ricardo Baptista

Received: 3 September 2024
Revised: 23 October 2024
Accepted: 1 November 2024
Published: 20 November 2024

Copyright: © 2024 by the authors. Licensee MDPI, Basel, Switzerland. This article is an open access article distributed under the terms and conditions of the Creative Commons Attribution (CC BY) license (https://creativecommons.org/licenses/by/4.0/).

1. Introduction

Procedural content generation (PCG) is known for its algorithmic generation of data, a method used to create random and streamlined content for several purposes such as maps, loot, item attributes, occasionally lore, etc. [1], in contrast with manual and static creation. Games are often evaluated in terms of their replayability, how elaborate their content is, play time, etc. Game content of high quality usually requires manual generation and significant effort by a large team that includes designers and developers, which considerably increases expenses and is also very time-consuming. Both wealthy studios and publishers have the required resources to support and invest in such concepts. On the other hand, this sumptuousness cannot be afforded to independent (indie) developers [2], so an alternative path must be found. PCG content has deep history in electronic gaming, and it has been relied upon by many games, particularly those heavily based on replayability to maintain the player's interest. Several popular games that utilize PCG methods are *The Binding of Issac* [3], which randomly generates rooms (see Figure 1), and *Minecraft* [4], in which its world is procedurally generated, with each component is uniquely arranged every time a new game is started, ensuring that no two players' worlds are alike. In *APEX Legends* [5], the weapons' spawn locations are completely randomized, every map is divided into subareas, and each subarea has a different spawn ratio of special or powerful equipment and power-ups, making some zones more desirable than others. In the case of generated maps in a 2D space, the same ones can also be reused in the game itself; for example, in a real-time strategy (RTS) game, the generated map can be used both as a real-scale map and as minimap with fewer details. The same concept can also be used for facial creation purposes based on players' preferences. In most RPGs (role-playing games), which allow a player to create their own unique hero, this process is performed manually, and, in rarer cases, a player can choose a randomly generated character based on a basic feature, e.g., race or clan. Modern games such as Grand Theft Auto Online [6] and Dark Souls 3 [7] offer very detailed character customization, from face details to body parts, by the user, but this procedure is considered laborious and time-consuming if the user wants a completely customized character. Recent works suggest methods for automatic face creation either by setting some features, e.g., skin color, nationality, class, etc., or by inputting a single photo [8,9]. Although faces cannot be produced by strict PCG algorithms, as the player needs their character to remain the same throughout a playthrough, there are situations where the hero is wounded, so a random scar can be depicted on a random part of the hero's face, or the hero may become older after several game years [10]. A positive effect of PCG methods is the fact that the demanded disk capacity requirements of the final game are significantly reduced, as the content generated by its game engine is not stored anywhere on the disk but is created on the fly, leading to improved resource management. However, content created on the fly demands more processing power, ideally before the game loads, but it turns out that this concession is worth the effort.

This paper's contribution is the presentation of an improved algorithm [11] that can create a top-down outdoor-level map filled with rooms of three basic sizes, which can make it quite congested, with a fountain in a random arrangement that changes every time the map is loaded. The three basic rooms can also be slightly changed during loading scenes in order to create a map with rooms that vary even more in size rooms. The same algorithm can be used in several other applications such as the inside of a dungeon or a room [12,13]. In other instances, it can be used to randomly generate loot and/or weapons with random range of properties or to add details such as vegetation, rocks, clouds, waterfalls, and so on [14]. Nevertheless, such methods are supposed to be used as helper applications rather than replacing jobs in the illustration field [15].

This paper is organized into five sections as follows: Section 2 provides an overview of fundamental methods and strategies related to procedural content generation (PCG). Section 3 addresses issues concerning content quality. Section 4 presents an analysis of the rules that the algorithm must adhere to, along with a detailed explanation of the algorithm itself. The results of our study are discussed in Section 5, followed by a discussion of the algorithm and its characteristics in Section 6. In Section 7, we offer a comprehensive conclusion on PCG, along with insights into future trends. The Appendix A contains a detailed presentation of the algorithm.

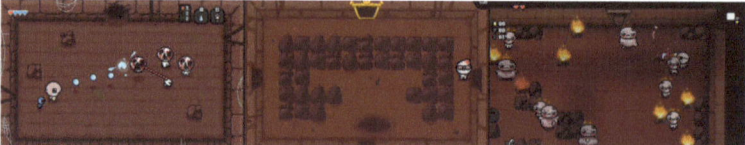

Figure 1. Binding of Isaac is a game that generates and decorates all of its rooms randomly.

2. Concepts of PCG

Over the years, multiple PCG methods have been developed, each differing majorly in its approach used to achieve content generation [16]. Some methods generate game content during the loading phase (*Offline*) [17], while others, such as *Online* ones, though less common, create content dynamically during gameplay based on miscellaneous factors, like the player's performance. The main difference between them is whether the implemented algorithm is classified as constructive or not. *Constructive* algorithms [18,19] do not require any evaluation as their outcome is considered unconditionally playable, in contrast to *generate-and-test* algorithms [20,21] where an agent must be present to test, for instance, if a game level leads to a dead-end. Nevertheless, special attention is needed for *softlocks* [22] as they lead to dead-end states even if they are not always undesirable, depending on game logic [23,24]. Consider a state in a Super Mario [25] game, as shown in Figure 2. If the player makes a successful jump, they can complete the level by reaching the pole where a flag is hung, whereas if they fail to avoid the intermediate gap, they will permanently become stuck between walls, a *softlock* state (see Figure 2).

The improved spawn algorithm belongs to *Offline* ones in which the map is being created just before the game level begins. In many cases, an agent should be present to examine in detail if the final map is playable. In this case, it is not necessary as it is a constructive one and it relies on a predefined ruleset that overcomes any undesirable states leading to dead-ends while the final content is being created in each playthrough literally from scratch [26]. Furthermore, although it relies heavily on its strict generative rules to construct diverse level maps every time a game level is loaded, it is not regarded as *random seed* [27] entirely as it can be accepted as a minor input from developers for complexity purposes by defining some *parameters* depending on how dense or sparse a level is desired to be.

Figure 2. A manually constructed level example in Super Mario that contains a possible softlock (if a player falls into the gap between walls, it is impossible to jump out, leading to a dead-end). Since there is nothing to kill the hero, the player must either manually reset the level or quit from it in order to abandon this state, abruptly losing any progress, as in this game there no saving points or autosaves.

2.1. The Role of Algorithms in PCG

PCG methods are capable of building a complete game, taking into consideration the needs of each asset (real-time difficulty adaption, special class loot rarity, map, room decoration, equipment etc.) that can be generated randomly in each load based on several rules by tracking the player's progress. Specifically, the most well-known methods they use are the following: (i) *Markov models* [28], which are considered particularly fast in PCG generation. (ii) *Cellular automata* [29,30], that consults a predetermined rule-set on a grid map area and explores if the adjacent cell can be occupied for a suitable asset or not. This method is commonly used for cave-like creations. (iii) *Generative grammar* [31] is essentially based on a grammatical rule system and then a parser undertakes the role to decide if an action can be applied or not. It is mostly used in games with complex lore where their progress is closely related to the user's actions and, depending on his choices, either a quest will be terminated or additional actions are required so as to be successfully accomplished, or a non-playable-character (NPC) team member will decide whether he will follow or reject the player, etc. (iv) *Machine learning* algorithms progressively learn and store all past actions, or they begin from a ready-to-use dataset [32] and extend with new data accordingly. Although they are very fast, they are characterized by their unreliability, as they do not guarantee that the final outcome will be playable, especially in narrow or dense areas such as a room interior, so an agent is necessary. However, a number of solutions have been proposed that subdue this behavior with specialized methods such as generative adversarial networks (GANs) [33,34], reinforcement learning (RL) [35–37], and deep learning [38]. (v) *Evolutionary Algorithms* [39]: although they are not yet widely used with PCG, they are preferred in 3D landscape modeling [40] to optimize game maps in strategy games that can accommodate massive armies and assist in dungeon modeling [41]. They present some failures to natural representation along with some minor conflicts between objects. Therefore, in specific operations they thrive with excellent results. Our algorithm is based on cellular automata (CA) with grid base as its kernel component.

2.2. PCG via Machine Learning

Procedural content generation via machine learning (PCGML) is considered as the generation of game content by using methods that have previously been trained on existing content from other instances [32,42]. It can be applied anywhere in a game where random content must be generated such as maps, items and their attributes, weapons and their features, character dialogues, cosmetics, etc. Machine learning can be effectively used to produce visual material that is closer to the user's preferences [43], for example, cosmetics for their equipment or building an initial character based on rudimentary questions or previous experience, but in case of levels, maps, or quests things become severely more complicated as other factors also take place. Especially on maps, machine learning methods should evaluate the final result to examine if the map is playable or if a character can jump on any permitted floor or be able to reach the exit. Such issues can be solved, but not entirely, by applying data augmentation methods [44], where the variety is increased in a given dataset, not by collecting more ready-to-use data but by adding modified versions of the already existing data [45]. In another case, a map can be represented as a set of puzzle pieces where each piece portrays a single element on a map. The puzzle pieces can also be shuffled to form another view of the same map or even to create a completely new one, while all pieces must fit seamlessly so as to avoid discontinuities (see Figure 3). The Bioshock collection [46] used this method to create myriad hacking puzzles with varying levels of difficulty.

Figure 3. A mini hacking puzzle game for a number of Bioshock alarm systems. The generated puzzle games are random, whereas the pieces of the puzzle are more than enough so that the puzzle can be solved in several ways.

2.3. Generative Adversarial Networks

This architecture is considered a special extension of machine learning. It can be assumed as an adversarial game between a generator and a discriminator. Initially, it generates synthetic data from a dataset and then exhibits similar characteristics to the real data; at the same time, a discriminator strives to classify if the generated data are considered fake or not. Generative adversarial networks (GANs), among all domains that have been applied, work better with image processing such as applications that involve human faces or handwritten characters. In terms of game levels, and the fact that GANs are based on machine learning methods [47], they encounter the same problems as they demand an initial set of known and functional maps in order to produce more of them. Additionally, due to the fact that such methods demand a lot of processing power, they are not suggested for real-time content generation [48]. Nevertheless, a study [49] managed to create several level maps for the well-known DOOM game from an initially created dataset suitable for training GAN from over than 1000 DOOM [50] levels by using two models—conditional and unconditional—where the conditional model uses several features as an input that are extracted from real levels, whereas the unconditional model uses only images from the given dataset (see Figure 4).

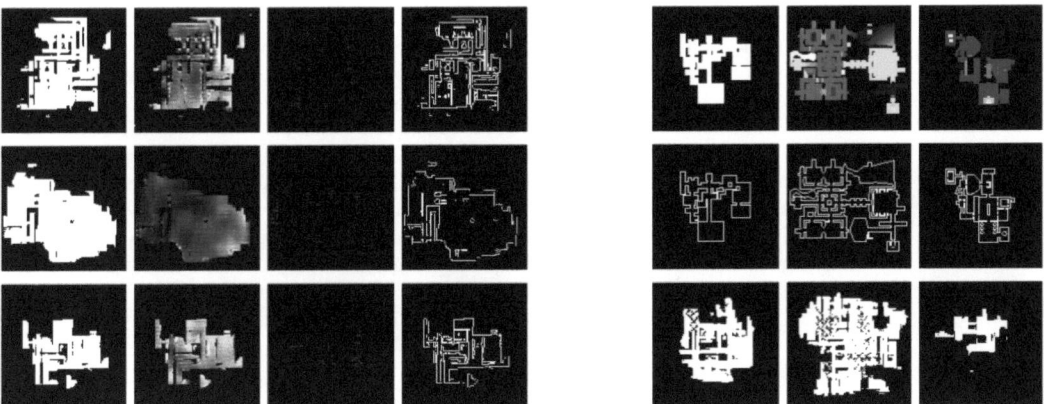

Figure 4. Left image shows maps that were created from an unconditional network while the maps in the right image were created from a conditional network.

2.4. Scenario Needs

Although there are a number of PCG algorithms to choose from, not all of them are suitable to implement any scenario. There is a big difference in creating random face portraits from scratch to produce playable terrains depending on different circumstances. Contemporary algorithms that use machine learning (ML) of GAN methods seem very promising but they are not used extensively for creating area maps of room interiors. In this paper, the proposed algorithm creates 2D area maps filled with varied-size rooms or caves and a decorative item, particularly a fountain. Each map is created from scratch without any previous experience and it can be configured so as to create maps with more or less room density depending on the desired difficulty, while the area around each room is considered as free roam state, and there are no specific paths. It uses the *cellular automata* technique, which fits perfectly for our cause as the whole map is based on a grid area where all objects are placed upon it, while at the same time the algorithm examines if there is enough space among them in order to avoid any collisions or blockages. The improved algorithm can generate a much larger number of varied-size rooms for better quality and a more diversified content. At this point, further analysis must be conducted for quality diversity and to determine how this content fits appropriately in a generated environment; either it is an open-air area with different climate conditions and terrain, or indoor areas such as caves or chambers.

3. Content Quality

Although random generated maps and environments are what we expect from a procedural content generation algorithm, the final outcome is not always as good as we would like it to be. This is not because of some failed object placements that overlap each other or an unexpected dead-end [51], but because the final scene does not seem natural or sensible, as the selected environmental objects do not match to a particular terrain or the co-existence of some objects does not make sense to the same place or map. Some practices define complex rules before the execution of an algorithm. Other solutions suggest a mixed approach where a PCG algorithm creates several templates [52], and at the end the designer chooses any or all of them that are compatible with their goals.

However, there are cases where some features are not disastrous; on the contrary, they enrich the complexity of a game or a level depending on what we would like to achieve. For example, roads on a strategic map should occasionally overlap each other, so crossroads, T-roads, or any other junction can be created. This method can convert a simple road network into quite a complicated one and it is very useful and easy to define the difficulty level of a game, e.g., if you are in early stages or in later ones. Also, the same technique can be used for defining the difficulty in real time, for instance, if a player's score is very high in one level, the next one will be much harder and vice versa. Another case is if a player is having a hard time passing a level so after a few failed tries on the same level, the next one may become easier. A study that uses roads to create new maps was applied to the *Kindom Rush: Frontiers* [53], a web-based tower defense game in which new maps were created with different road networks and random tower places with minimum distances between them all over the map, as overlap and close proximity in this situation must be avoided (see Figure 5). The Kullback–Leibler (KL) divergence was applied for the cover distribution:

$$KL(P||Q) = \sum_x P(x) \log(\frac{P(x)}{Q(x)}) \quad (1)$$

which defines a standard distribution for the whole map, which gives quite a natural appearance. At this point, for comparison purposes, our algorithm is based thoroughly on the asset number that is defined from the designer and it chooses if the selected asset can be placed or not on the map, based on the distance or location constraints giving a natural environment in a different way.

Other methods suggest a top-down approach where a game level has an entry point at the top of it and an exit point at the bottom; in other words, the player is always moving down. A known game that follows this strategy is Spelunky [54], where each map that represents a level is divided into 4 × 4 rooms and a path is planned throughout these rooms, as shown in Figure 6. At this point, it is important to mention that it is not necessary for all rooms to be used, which is something that easily defines how long or short a level can be. Each box is called a chunk and they are replaced by a random number; finally, the algorithm defines which number will be at the start and which one will be at the end [55]. The next step includes the room placement for each chunk, selected from quite a large set of templates, and the set of templates used depends on the area the level is in and whether the room falls on a path. For instance, according to Figure 6, beginning from the start chunk, a room with a corridor that has an exit point to the right of it is needed in order to continue in the room where the corresponding arrow indicates, so the suitable template set is the one that has exit points on the right side of the rooms, etc. Therefore, the room in Figure 7 fits in the second chunk as both entry points are located on the left of it and the exit point at the bottom. If we combine all of the above with decoration, obstacles, and monsters, this process leads to a game that generates unique levels in such a way that each game is always different and, at the same time, keeps the game fresh and exciting.

Based on the same principle is the game Diablo [56], which is separated into four stages where each stage includes four levels by using the top-down approach. Here, the player can access each level by finding an entrance to the lower one. In fact, all of the stages are dungeons with different names and design, and polished with appropriate themes.

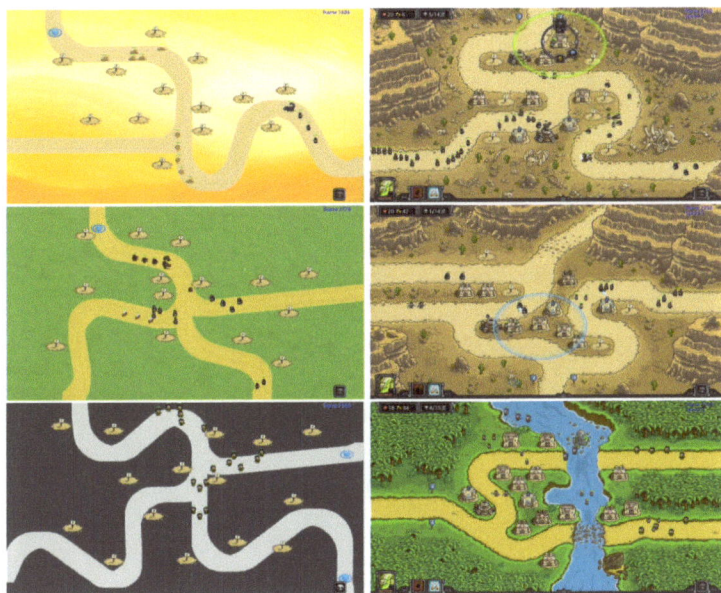

Figure 5. Three random levels are created in the *Kingdom Rush: Frontiers* with roads, tower places, and monster generation in consecutive waves. Also, in each wave, monsters are grouped based on their kind.

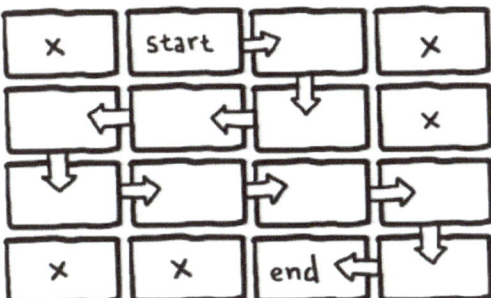

Figure 6. All levels in *Spelunky* game use this structure, with a preplanned top-down path, while all levels have the same size of 4×4 rooms. The start and end points are randomly generated in any room.

Figure 7. Example room design in *Spelunky* in its primary state, without any added decoration. Note that this room has two entry points, one from the left and another one from the right.

Everything in a level has a fully randomized arrangement at such a point that even the exit spots that lead to the lower levels are randomly placed throughout, but in this case an extra control is applied and the level is recreated from scratch if the exit spot cannot be placed due to the lack of space. On the other hand, extra attention is given both to the rooms that are located inside a level, which must be connected through corridors or other room entrances, and to the room entrances where their directions must not face a wall, something that we also check on our spawn algorithm. In Figure 8, a map is generated for the Cathedral level in Diablo, clearly showing the rooms' placement, corridors, and the exit point to the lower level.

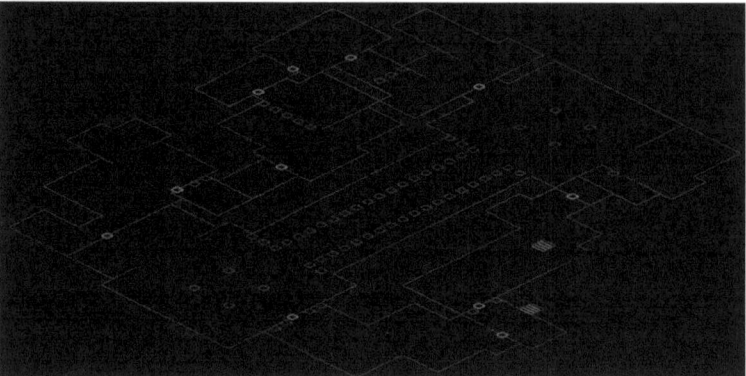

Figure 8. A randomly generated level of the Cathedral stage in *Diablo*.

3.1. Quality Diversity

Content quality not only refers to asset placement but also leads to quality diversity in order to make a game more enjoyable. As is well known, many games lack multifarious levels or stages so that a player can discover patterns from a point onward. Providing diverse levels to players adds extra value to entertainment since they need to deploy different strategies under specific circumstances [57]. Map diversity can also be combined with other game elements such as decoration objects, obstacles, statistics over offense/defense weapons, or rewards based on specific criteria. For example, in *Spelunky*, after the room selection, the next step comprises choosing obstacle and trap placement in certain points of each map; finally, object and monster placements are selected in random locations of each map, as shown in Figure 9.

Figure 9. After room selection in *Spelunky*, as a last step, randomized decorations, obstacles, and monster placement takes place.

Different parts of an environment need different manipulation from an algorithm. For example, in the case of decoration, the first thing needed is a very large object dataset; at the same time, this dataset should be divided into several categories depending on place, i.e., if it is outdoor, indoor, rooms, corridors, etc. Irrespective of the concept we would like a map to contain, it is of critical importance for a sequence to be kept. For instance, rooms are usually placed first on maps, then a road/corridor network to connect all or part of them or adjacent rooms that are connected with a door, and, finally, decoration comes last. Each step comprises different strategies and can be implemented by different algorithms, or one algorithm can also be used by configuring several settings. As long as a step belongs to the late stages in the chain of action, more restrictions are applied. For instance, it is important that the placement of decoration objects is predefined on certain spots that will be used as static objects in order for unnecessary obstacles to be avoided; otherwise, both motion and visual attributes will probably be hindered, except for cases where someone would like to hide a valuable object, treasure, contraption, etc. The improved Spawn algorithm keeps the same functionality as it is designed to be used in several scenarios that include both outdoor and indoor actions.

To keep the uniqueness of an environment, several techniques are developed, and this is where artificial intelligence (AI) shines, as special uses of it produce diversity with rich environments. One of the most reliable and fast techniques from the modern AI field used to create content on maps is reinforcement learning (RL), where environments are usually modeled by Markov decision processes (MDPs), a mathematical formulation that is used to study optimization problems. MDPs are often represented by the type (S, A, P, R, γ), where S is state space, A is the action space, $P_a(s, s') = P_r(s_{t+1} = s'|s_t = s, a_t = a)$ defines the transition probability from a state s to s' by executing an action a at time t, $R_a(s, s')$ consists of the immediate reward earned from the current transition, and, finally, $\gamma \in [0, 1]$ is the discount factor, which computes how many future cumulative rewards are used compared to the current one [58]. Due to the fact that RL is still adding complex computational tasks [47], it is somehow difficult to used in real-time implementations, but, on the other hand, it fits perfectly in turn-based games where the real-time actions exist at a minimum, mainly in nonbattle events. In addition, as Figure 10 shows, RL algorithms agents are necessary to use, and, in order to ensure that the quality will be high, a feedback system is required to monitor the overall process step by step [36] by comparing the previous state and the current updated state by using a reward calculator for a particular game. Depending on earned rewards, RL agents can generate random playable maps producing infinite unique designs, increasing both replayability and winning strategies. A recent study proposed stage creation in two phases, battle and nonbattle events [59]. Here, an evolutionary method is proposed, as in machine learning (ML) algorithms, a sufficient amount of content is required for training, but in this case, everything can be generated from scratch by learning *online* behaviors, either from the player or ready-to-use environments which represent levels or stages. In other words, it applies self-learning by interacting with the environment, something that can lead to unsupervised learning, which significantly reduces the overall processing time.

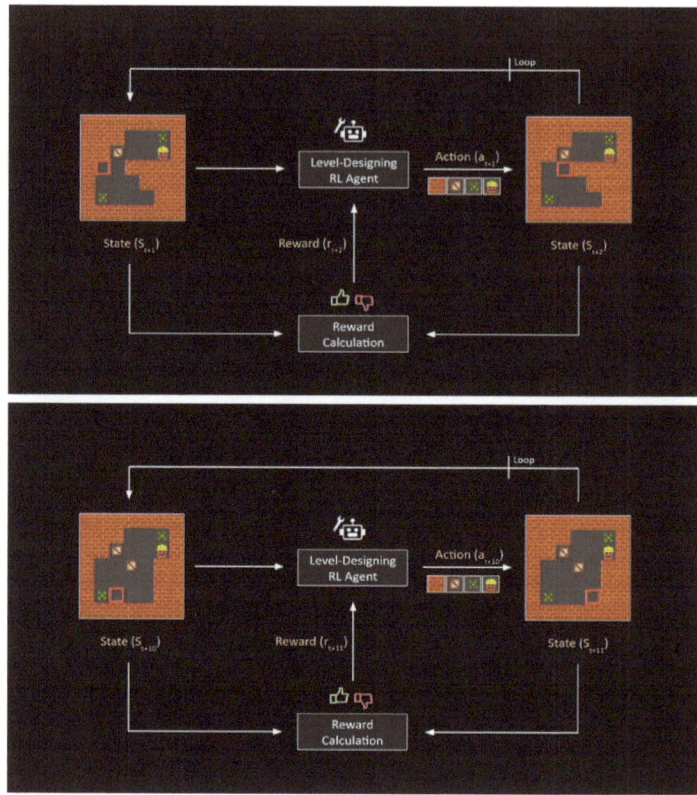

Figure 10. The first image shows the beginning of an RL agent and how it is interacting with the stage, while in the second image, the RL agent has finished its iterations. A reward calculator is present as it is needed to advise the agent if the proposed changes can lead to a dead-end or not, or if they can be accessed in general.

3.2. Intelligent Diversity

Except for environmental diversity, the replayability of a game also depends on its playstyle. Content diversity usually refers to static objects such as terrain, trees, roads, or weapons with stats that are randomly found in a treasure chest or after defeating enemies. The same static environment can also change the course of a battle by changing the location in which a fight takes place. For instance, for a boss that is vulnerable in ranged attacks and is easily defeated in one attempt, in another playthrough, they could be placed in a location that does not favor ranged attacks, so another approach must be developed by the player in order to defeat the same adversary. In other cases, bosses could be different. Along with their properties and stats, adding extra challenges in terms of the unknown of what one can encounter negates the same strategy being used every time a new game starts. In particular scenarios, if a difficulty option has been added, and a variety of heroes with completely different playstyles are present, e.g., characters who specialize in ranged attacks, melee combat, mages, summoners, etc., in the highest difficulty levels, every winning condition is totally different, as special combined strategies are required. In Diablo II [60], especially the *Hell* difficulty, levels, the playstyle of each player character must be changed as the player's resistances to all elements are dramatically dropped, even below zero, and all enemies have immunity at least to one kind of attack with different bosses. In Figure 11, a mini boss has two immunities, to physical and magic damage, so a player must use other kinds of attacks, e.g., poison, to defeat him.

Figure 11. *Hell* difficulty in Diablo II adds several random immunities to all foes; as a consequence, all of a character's properties must be reassigned in such a way to enhance a combination of two or three dissimilar attacks that can damage anyone in the battle field.

As a result, any player must redistribute any earned skill points in such a way so that his character becomes specialized in several attacks that better fit his playstyle, and if he belongs in a party with a specific role, such as a damage dealer or a defender. Furthermore, changing the stats only is not enough as it is mandatory for a player to also replace his weapons and armor as the game in this difficult level spawns even more advanced objects by piling up properties and giving the opportunity to players to customize their equipment even more depending on their playstyle. All of them inarguably add more challenge to the game by entirely changing its perspective, as if a different game is created.

To make a game more entertaining, a mechanism could be used to evaluate the diversity by measuring a possible satisfaction factor with the help of entropy, which is used as a base reward received by an agent. It is used to estimate the possible amount of information that a scenario has, for example, the number of segments that seem to display possible repetitions. The greater that number of repeated segments is, the lower the entropy value will be, and vice versa; this is how the diversity of a scenario is evaluated and measured. In [61]'s study, the entropy is calculated using the formula

$$H(x) = -\sum_{i}^{n} p(x_i) log(p(x_i)) \quad (2)$$

where $p(x_i)$ defines the probability occurrence for each event of the variable x, which in this case is represented by each segment, and the $log(p(x_i))$ calculates the amount of gathered information for each segment. Another study promoted a solution where it used the KL-divergence in order to quantify the similarity between the segments [62] in an RL algorithm to be able to generate endless playable levels in the Super Mario [25] game. In more detail, a level is divided into segments, then multiple RL agents evaluate the degree of diversity; finally, they are concatenated to form a full level. The KL-divergence was also used for the same Mario game, but in the study of [63], it was implemented asymmetrically, generating complex and rich environments, as shown in the Figure 12 level, but, on the contrary, a large dataset was required by the algorithm.

Figure 12. This level was generated by asymmetrical KL−Div with the use of a considerable size of a training sample where any novel pattern that did not exist in that sample is subjected as a candidate for use in the level generation.

In Table 1 we can see a summarized of the methods used to produce PCG environments.

Table 1. Methods used by applications to produce PCG content.

	Training Sample Required	Space	Predefined Levels	Asset Rotation
Minecraft	No	3D	No	No need
Binding of Isaac	No	2D	Merely	No
Spelunky	No	2D	Merely	No
Diablo	No	2D (map overlay)	Yes	No
Super Mario	Yes	2D	Merely	No need
Doom levels	Yes	3D	No	No
Kingdom Rush: Frontiers	No	2D	No	No
Our Spawn algorithm	No	2D	No	Merely

4. The Improved Spawn Algorithm

In general, the algorithm produces three room sizes that are placed randomly on a 2D grid-based map. Large and medium rooms have four standard entrance directions, specifically north, south, east, and west, while the small rooms' entrances can rotate everywhere in 360°. The entrance rotation of each room is also randomly generated and special attention is given to the fact that all entrances cannot face any wall side, as a minimum distance from it is estimated. The number of rooms that can be placed also remains the same, so the maximum number of each type is as follows: one large, two medium, and three small rooms. The fountain is considered as a decoration item so its size remains the same. Each item on the map has its own hitbox for collision detection purposes, where AABB [64] fits perfectly for rooms and a spheroidal for the fountain, as it is very important to distinguish boundaries to be set for future expansions.

The map uses the cellular automata (CA) technique, a method that is very suitable for building maps with ready-to-use places upon them. All grids are grouped in 5×5 groups to form larger ones and all rooms are placed into them without the obligation to fit exactly, as shown in Figure 13. The final result visually remains the same and comprises a minimap that can later be translated to a full-scale one. In terms of improvement, this version expands the overall number of room sizes in each type. In particular, large rooms in this version have 11 sizes, medium rooms have 31 sizes, and small rooms have 51 sizes. Each room size of each type cannot be scaled in a way that overlaps another type, e.g., a large room cannot be shrunk in a size equal to or less than a medium room. As a consequence, the end result becomes even more varied, with possible unique combinations based on type

$$C(n,r) = \binom{n}{r} = \frac{n!}{(r!(n-r)!)} \qquad (3)$$

where r represents the maximum number of rooms that are likely to exist on the map, while n is the possible size of a room. Therefore, the combinations are as follows:

- Large rooms: 11 ($r = 1$ and $n = 11$);
- Medium rooms: 465 ($r = 2$ and $n = 31$);
- Small rooms: 20,825 ($r = 3$ and $n = 51$).

The overall possible combinations exceed the 213 million, a number that cannot actually be achieved in reality, so the possibility of two maps totally matching is dramatically diminished. Table 2 summarizes the differences between the initial and improved algorithm.

Figure 13. The grid map is divided into larger grids and each one includes 5×5 small grids. The first room fits exactly in the large grid while the second one is placed in the corner, essentially occupying four large grids. This placement constitutes the worst-case scenario, something the algorithm takes into account and acts on accordingly.

Table 2. Key differences between two algorithm versions.

	Method	Trained Dataset Required	Room Sizes (Total)	Map Size
Initial algorithm	Cellular Automata	No	3	10×3 cells
Improved algorithm	Cellular Automata	No	93	10×3 cells

4.1. Algorithm Complexity

The algorithm divides the creation procedure into several steps, where each one executes a small part of the overall process; specifically, the first one is the creation of a blank grid map with rows and columns that are given before the algorithm's execution. For this instance, we decided that fifteen (15) rows and fifty columns (50) are enough for the algorithm to work, consisting of the smallest grids on the map. While rooms and any decorations occupy more than one tile, right afterwards, the algorithm divides the grid into larger areas containing part of the tiles; specifically, each large area contains five (5) rows and ten (10) columns of small tiles: a total amount of fifty (50). In this space, any object from small rooms to large ones and any of the decorations can fit from any angle, but it is decided that only small rooms can use this feature for simplicity reasons. This is a standard process that also demands a fixed time that takes about 45×10^{-11} seconds.

All other steps include loops, as their basic goal is to decide if and where a room will be placed as long as there is a free place. As is shown in Appendix A, large rooms are placed first as they are considered the most difficult because of their size. In general, bigger rooms are placed first, if selected, and then smaller ones, in order to minimize the exclusion possibility of bigger rooms not being placed at all due to lack of space. There are four loop stages in total, where each is used for creating a room type and the fourth is used for decoration placing. Each stage includes two nested loops; the outer one examines the large columns while the inner loops scrutinize each row of the selected column to determine if the chosen room can fit. All four loops are executed at least one time, and an extra compromise is taken into account where, if a large room is present in a large column, no other rooms in that specific column can be placed except for decoration. This compromise prevents the probability for a small part of the map to become populous and the rest of it being underpopulated, as, for vision purposes, it is optimal for all objects to be placed all over the map as much as possible.

In terms of computational load, all four loops execute almost the same calculations and they are executed only if a special condition is true. First of all, each loop chooses a random number that corresponds to an asset, e.g., if the chosen value is zero (0), that means a large room is selected. If the chosen value is within the correct loop, then the process of finding a random suitable place on the map begins by scanning all the large rows within the corresponding large column. Also, minor actions take place, such as if the maximum number of an asset is reached or determining which direction the entrance of a room faces; at this point, the loop ends. The maximum computation cost can be estimated as follows: ten loops are used for scanning the large columns, multiplied by three loops for each asset along with their simple computational costs; in other words, there are thirty loops. This means a total time based on $\log(n)$ calculations, specifically 9×10^{-11} seconds, translated in a few milliseconds.

4.2. Worst-Case Complexity

The maximum possible loop number cannot exceed thirty in this instance; in fact, it is impossible to reach this value because of the fact that the maximum number of assets that can be placed on a map is seven, as shown in Figure 14. A loop is only executed for a valid asset if it is chosen by a random generator, so the maximum loop number becomes seven, where in this case the algorithm searches for a free place to set it on. The worst-case scenario in this instance is the selection of all seven rooms and the fountain with their placement to take place at the end of the map, meaning the right part of the map, which adds a tiny fraction of time to the overall process. As a result, the total time becomes $\log(n)$ calculations, specifically 3×10^{-11} seconds, plus the time for the grid map creation.

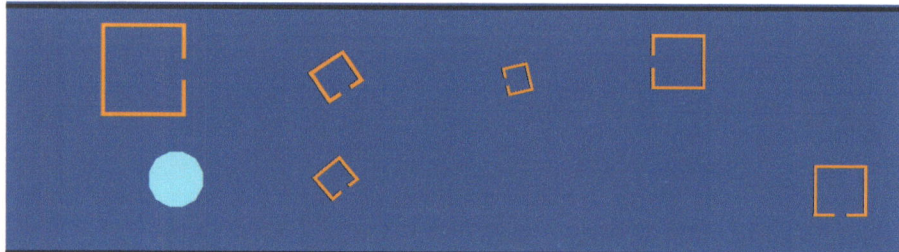

Figure 14. A random example with full rooms in the new improved algorithm.

5. Results

The overall modifications achieved a better result while the differences, in terms of size among rooms, are easily detected visually. As shown in Figure 14, all rooms, even those that belong to the same category, have obviously dissimilar sizes. The two medium rooms have few differences, while the small ones have more obvious varied sizes. The large room is an exception as the algorithm produces only one, so there is no comparison measure, but in the end, if we compare it with Figure 15, in this instance it is a little smaller. The newer algorithm was tested thoroughly by executing it multiple times, where each time a completely new map was created.

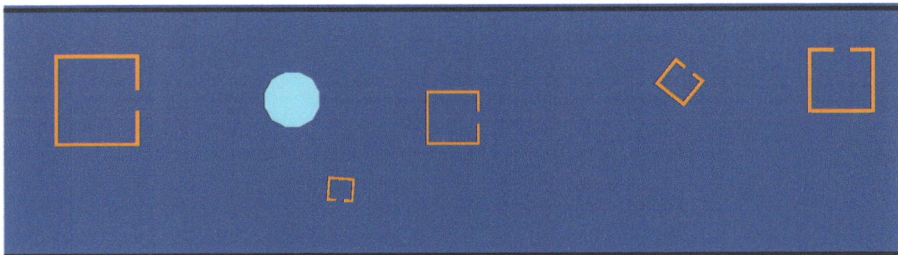

Figure 15. Another instance focusing on the large room which is a bit smaller.

Nevertheless, according to algorithm rules, there are situations where the fountain is not placed at all, producing an environment that lacks any decoration. Although this is not considered as a bug, a plain or barren map is not attractive at all except if it is used for special occasions like a small secret treasure room or a trap that leads into a room that the player must escape. On the other hand, despite the modifications, the spawn algorithm was executed seamlessly without showing any crushes or critical errors, and in terms of performance, the extra additions almost did not affect it at all. In the end, the final product was accomplished by producing a considerably more varied map without increasing the utilization of computing resources. These updates help an application to remain both replayable and lightweight.

6. Discussion

The algorithm was tested several times; specifically, 500 maps were created in a row, and all output results were within limits that were defined during the software development. Before the improvement, there was a rare situation in which two medium rooms collided by overlapping each other, but in this version, everything is corrected. The graphics engine remained the same, in particular Unity, and C# was used as the programming language as it is embedded with a very friendly development environment and is also very versatile, offering complete integration and full compatibility with any previous versions. The generated maps are quite different from each other, but because the room sizes can vary by a vast number, only half of them can be visually distinguished by the human eye. The algorithm generates a map in milliseconds, even for fully loaded ones, but it could become even faster if some conditions are removed, as in some cases they are not necessary. A case was observed in which a large room was placed at the top of a column, and while no other object can be placed underneath, the algorithm continues to check if something can be placed in the same column. This behavior was detected recently, and while it does add a negligible working load, in larger-scale maps this could become a bigger problem. But on the other hand, why should there be extra effort when it can be avoided? Because of the fact that the Spawn algorithm is rather lightweight, it can easily be used for web-based applications as it does not demand significant computational resources and its output is small in size so it can be downloaded without much effort, even over slow connections.

Its uniqueness relies on the fact that in each load a new map is generated from scratch, building a basic layout, but in the future this feature is going to change a bit as a road network will be added, at least among rooms. There are many games that use PCG methods for generating levels and maps but they use several tricks to display different outputs, and two of the best games in this matter are *Binding of Issac* [3] and *Spelunky* [54]. Especially in Spelunky, there is a large level dataset in which several of them are selected and then obstacles are placed in a random manner to change the overall view while all levels are generated from the beginning. On the other hand, *Kingdom Rush: Frontiers* [65] was used by a revolutionary RL algorithm to generate maps with several characteristics which were defined for use in such a way to add extra value in difficulty modes. Most of the 2D maps, despite the technique they use, have a common place as they are strictly grid-based, because this kind of arrangement offers a very convenient way to place everything in a determined fashion and it is also proven to be very fast if the video game relies on levels, especially if even the secret ones are considered distinct rooms. Grid-based methods are also capable of creating not only small-scale maps but huge worlds which represent a full game playthrough, but there are various methods that use other algorithms to achieve unique generated worlds, such as the *Terraria* [66], which generates uniquely random places by adding noise per pixel and then applies multiple scans each time the algorithm adds something, such as dirt, cavities, water, flora, etc.

7. Conclusions

In this study, we explored the development of an algorithm designed to enhance the procedural generation of game environments, particularly for role-playing games (RPGs). The aim was to create more dynamic and immersive game worlds by varying environmental structures and conditions, reducing the need for human intervention in the early stages of large-scale game projects. Our findings demonstrate that while many games, including dungeon crawlers, maintain a basic pattern across playthroughs, our algorithm can introduce significant variability by altering the placement of structures within environments, such as replacing a fountain with a market in a village. Additionally, the algorithm has the potential to create expansive open worlds with distinct regions characterized by unique environmental conditions. This ability to generate diverse and complex environments automatically enhances the depth and replayability of RPGs, pushing the boundaries of procedural content generation [54]. The core program is very lightweight as it does not demand a lot of resources, about 60 MB on disk and 70 MB on RAM as a final build in Unity, since all levels/maps are created on the fly during the loading stage and, based on worst-case complexity, the loading time is negligible.

The algorithm's capacity to modify game environments on a larger scale presents significant implications for game design, particularly in the context of RPGs. By automating the placement and variation of game structures, developers can focus on higher-level design elements while ensuring that each playthrough offers a unique experience. This approach not only increases replayability but also opens up new possibilities for creating more intricate and interactive game worlds. The potential to apply the algorithm to both outdoor and indoor settings, as well as to multilevel maps, further emphasizes its versatility and relevance in modern game development. Special attention was given to softlock state avoidance where a dead-end is revealed without implying that is a bug. This problem was not present in any cases of our tests, as we predicted, first of all, enough space between rooms so that a road (in future version) can be added. This behavior prevents a player from becoming trapped in adjacent rooms as the character is also placed randomly. All this effort is additionally enhanced by meticulous collision detection methods, especially those applied in small rooms that are rotated in 360 degrees, where a rotated corner can dangerously narrow an already tight space. In conclusion, the algorithm represents a significant step forward in the procedural generation of game environments, particularly for RPGs and dungeon crawlers. By minimizing human intervention and automating the creation of diverse and interactive game worlds, this approach has the potential to revolutionize game development.

Future Trends

Despite the advancements presented, the algorithm still lacks certain features necessary for practical application in real-world software. For example, while the algorithm effectively varies the placement and size of rooms, it currently does not address the integration of cosmetic assets or points of interest, such as hidden treasures. Future research should focus on these areas, particularly on connecting room entrances with paths that avoid overlapping with other map elements, and on decorating room interiors with a broader range of assets. Additionally, enhancing the algorithm to support the creation of multilevel maps with entrances and exits would further extend its capabilities. The final intention is for the same algorithm to also be used for decorating room interiors, as the asset arrangement between outdoor and indoor settings could be similar with similar restrictions, but in the case of the interior, the number and the variety of elements that will be chosen to input will be rather larger than outdoor ones. For replayability purposes, in order for the maximum result to be achieved, outdoor maps will vary in terms of environmental conditions while decorating them with proper assets. At the same time, the room interior will vary depending on randomly chosen themes that will be created on the fly by using suitable rules attached to the current one. Finally, a special addition will be the placement of an entrance or exit, or both, in each map in the form of a ladder, upwards or downwards, in order for multilevel maps to be created by moving back and forth, a feature that adds extra playable time and difficulty. The ultimate goal, though, is for the same algorithm, with minor adjustments for each case, to be able to create a playable game level from scratch without any human intervention.

Author Contributions: Conceptualization, L.L. and G.F.F.; methodology, L.L.; validation, L.L. and G.F.F.; writing—original draft preparation, L.L.; writing—review and editing, L.L. and G.F.F.; supervision, G.F.F. All authors have read and agreed to the published version of the manuscript.

Funding: This research received no external funding.

Data Availability Statement: No research data available.

Conflicts of Interest: The authors declare no conflicts of interest.

Appendix A

Input: 15 rows × 50 columns grid area, where maximum number of large, medium, and small rooms is set to 1, 2, and 3, respectively.

> **2 dimension vectors:** map[x, y] **Variables:** i, j as counters.
> **Random values:** 0 → large room, 1 → medium room, 2 → small room, 3 → no room
> **Maximum elements number:** large rooms → 1, medium rooms → 2, small rooms → 3, decorations (fountain) → 1

–Grid Map creation–
1. for i = 1 to maxColumns(50) do
2.   for j = 1 to maxRows(15) do
3.   map[i, j] = new Vector2(x, y)
4.   y = y + 1 (+1 tile in the row)
5.   end for (j)
6.   y = 0 (initiate the row tile)
7.   x = x + 1 (+1 tile in the column)
8. end for (i)

–Large column loop–
9. for i = 1 to largeColumns(10) do

–The nested "for j" loops chooses in which large row the rooms and decorations will be placed–

–Large room loop–
10.   for j = 1 to largeRows(3) do
11.   randomGenerator = randomValue 0 to 3

12.	if randomGenerator == 0 and maxNumberLargeRoom != 0 and noPresenceOfAnotherLargeRoom
13.		choose a direction other than no face wall
14.		create a large room as Vector3(map[i,j].x, map[i,j].y, direction)
15.		choose a random scale
16.		reduce the maxNumberLargeRoom by 1
17.	end if
18.	end for (j)

–Medium room loop–
19.	for j = 1 to largeRows do
20.		randomGenerator = randomValue 0 to 3
21.		if randomGenerator == 0 and maxNumberMediumRoom != 0 and noPresenceOfAnotherLargeRoom
22.			choose a random direction
23.			create a medium room as Vector3(map[i,j].x, map[i,j].y, direction)
24.			choose a random scale
25.			reduce the maxNumberMediumRoom by 1
26.		end if
27.	end for (j)

–Small room loop–
28.	for j = 1 to largeRows do
29.		randomGenerator = randomValue 0 to 3
30.		if randomGenerator == 2 and maxNumberSmallRoom != 0 and noPresenceOfAnotherLargeRoom
31.			choose a freely random direction
32.			create a small room as Vector3(map[i,j].x, map[i,j].y, direction)
33.			choose a random scale
34.			reduce the maxNumberSmallRoom by 1
35.		end if
36.	end for (j)

–Decoration (fountain) loop–
37.	for j = 1 to largeRows do
38.		randomFountainGenerator = randomValue 0 to 1
39.		if randomFountainGenerator == 1 and maxNumberFountain != 0
40.			create a fountain as Vector3(map[i,j].x, map[i,j].y, direction)
41.			reduce the maxNumberFountain by 1
42.		end if
43.	end for (j)

–Move to next map large column–
44. nextPointerLargeGridX = 0
45. nextPointerLargeGridY = nextPointerLargeGridY + stepY (10)
46. end for (i)

References

1. Viana, B.M.; dos Santos, S.R. Procedural Dungeon Generation: A Survey. *J. Interact. Syst.* **2021**, *12*, 83–101. [CrossRef]
2. Barriga, N.A. A short introduction to procedural content generation algorithms for videogames. *Int. J. Artif. Intell. Tools* **2019**, *28*, 1930001. [CrossRef]
3. Edmund M, F.H. The Binding of Isaac. Available online: https://bindingofisaac.fandom.com (accessed on 16 August 2024).
4. Persson, M. Minecraft. Available online: https://www.minecraft.net/en-us (accessed on 16 August 2024).
5. Electronic-Arts. APEX Legends. Available online: https://www.ea.com/games/apex-legends (accessed on 16 August 2024).

6. Games, R. Grand Theft Auto Online. Available online: https://www.rockstargames.com/gta-online (accessed on 16 August 2024).
7. Namco, B. Dark Souls III. Available online: https://en.bandainamcoent.eu/dark-souls/dark-souls-iii (accessed on 16 August 2024).
8. Shi, T.; Zou, Z.; Shi, Z.; Yuan, Y. Neural rendering for game character auto-creation. *IEEE Trans. Pattern Anal. Mach. Intell.* **2020**, *44*, 1489–1502. [CrossRef] [PubMed]
9. Shi, T.; Zuo, Z.; Yuan, Y.; Fan, C. Fast and robust face-to-parameter translation for game character auto-creation. In Proceedings of the AAAI Conference on Artificial Intelligence, New York, NY, USA, 7–12 February 2020; Volume 34, pp. 1733–1740.
10. Zhao, J.; Cheng, Y.; Cheng, Y.; Yang, Y.; Zhao, F.; Li, J.; Liu, H.; Yan, S.; Feng, J. Look across elapse: Disentangled representation learning and photorealistic cross-age face synthesis for age-invariant face recognition. In Proceedings of the AAAI Conference on Artificial Intelligence, Honolulu, HI, USA, 27 January–1 February 2019; Volume 33, pp. 9251–9258.
11. Lazaridis, L.; Kollias, K.F.; Maraslidis, G.; Michailidis, H.; Papatsimouli, M.; Fragulis, G.F. Auto Generating Maps in a 2D Environment. In Proceedings of the International Conference on Human-Computer Interaction, Virtual Event, 26 June 26–1 July 2022; pp. 40–50.
12. Freitas, V.M.R.d. Procedural Generation of Cave-Like Maps for 2D Top-Down Games. Bachelor's Thesis, Universidade Federal Do Rio Grande Do Sul Instituto De InformáTica Curso De Engenharia De ComputaçãO, Porto Alegre, Brazil, 2021.
13. Viana, B.M.; dos Santos, S.R. A survey of procedural dungeon generation. In Proceedings of the 2019 18th Brazilian Symposium on Computer Games and Digital Entertainment (SBGames), Rio de Janeiro, Brazil, 28–31 October 2019; pp. 29–38.
14. Minini, P.; Assuncao, J. Combining Constructive Procedural Dungeon Generation Methods with WaveFunctionCollapse in Top-Down 2D Games. In Proceedings of the SBGames, Recife, Brazil, 7–10 November 2020.
15. Lai, G.; Latham, W.; Leymarie, F.F. Towards friendly mixed initiative procedural content generation: Three pillars of industry. In Proceedings of the International Conference on the Foundations of Digital Games, Bugibba, Malta, 15–18 September 2020; pp. 1–4.
16. Gellel, A.; Sweetser, P. A hybrid approach to procedural generation of roguelike video game levels. In Proceedings of the International Conference on the Foundations of Digital Games, Bugibba Malta, 15–18 September 2020; pp. 1–10.
17. De Kegel, B.; Haahr, M. Procedural puzzle generation: A survey. *IEEE Trans. Games* **2019**, *12*, 21–40. [CrossRef]
18. Green, M.C.; Khalifa, A.; Alsoughayer, A.; Surana, D.; Liapis, A.; Togelius, J. Two-step constructive approaches for dungeon generation. In Proceedings of the 14th International Conference on the Foundations of Digital Games, San Luis Obispo, CA, USA, 26–30 August 2019; pp. 1–7.
19. Liapis, A. 10 Years of the PCG workshop: Past and Future Trends. In Proceedings of the International Conference on the Foundations of Digital Games, Bugibba, Malta, 15–18 September 2020; pp. 1–10.
20. Gisslén, L.; Eakins, A.; Gordillo, C.; Bergdahl, J.; Tollmar, K. Adversarial reinforcement learning for procedural content generation. In Proceedings of the 2021 IEEE Conference on Games (CoG), Copenhagen, Denmark, 17–20 August 2021; pp. 1–8.
21. Song, A.; Whitehead, J. TownSim: Agent-based city evolution for naturalistic road network generation. In Proceedings of the 14th International Conference on the Foundations of Digital Games, San Luis Obispo, CA, USA, 26–30 August 2019; pp. 1–9.
22. Mawhorter, R.; Smith, A. Softlock Detection for Super Metroid with Computation Tree Logic. In Proceedings of the 16th International Conference on the Foundations of Digital Games, Montreal, QC, Canada, 3–6 August 2021; pp. 1–10.
23. Cook, M.; Raad, A. Hyperstate space graphs for automated game analysis. In Proceedings of the 2019 IEEE Conference on Games (CoG), London, UK, 20–23 August 2019; pp. 1–8.
24. Chang, K.; Aytemiz, B.; Smith, A.M. Reveal-more: Amplifying human effort in quality assurance testing using automated exploration. In Proceedings of the 2019 IEEE Conference on Games (CoG), London, UK, 20–23 August 2019; pp. 1–8.
25. Nintendo Ltd. Super Mario Bros. Available online: https://www.nintendo.com/en-gb/Games/NES/Super-Mario-Bros-803853.html (accessed on 17 August 2024).
26. Bontrager, P.; Togelius, J. Learning to Generate Levels From Nothing. In Proceedings of the 2021 IEEE Conference on Games (CoG), Copenhagen, Denmark, 17–20 August 2021; pp. 1–8.
27. Summerville, A. Expanding expressive range: Evaluation methodologies for procedural content generation. In Proceedings of the Fourteenth Artificial Intelligence and Interactive Digital Entertainment Conference, Edmonton, AB, Canada, 13–17 November 2018.
28. Snodgrass, S.; Ontanón, S. Learning to generate video game maps using markov models. *IEEE Trans. Comput. Intell. AI Games* **2016**, *9*, 410–422. [CrossRef]
29. Adams, C.; Louis, S. Procedural maze level generation with evolutionary cellular automata. In Proceedings of the 2017 IEEE Symposium Series on Computational Intelligence (SSCI), Honolulu, HI, USA, 27 November–1 December 2017; pp. 1–8.
30. Flores-Aquino, G.O.; Ortega, J.D.D.; Arvizu, R.Y.A.; Muñoz, R.L.; Gutierrez-Frias, O.O.; Vasquez-Gomez, J.I. 2D Grid Map Generation for Deep-Learning-based Navigation Approaches. *arXiv* **2021**, arXiv:2110.13242.
31. Thompson, T.; Lavender, B. A generative grammar approach for action-adventure map generation in the legend of zelda. 2017. In Proceedings of the 7th International Symposium for AI & Games, Artificial Intelligence and Simulation of Behaviour, Bath, UK, 18–21 April 2017.
32. Summerville, A.; Snodgrass, S.; Guzdial, M.; Holmgård, C.; Hoover, A.K.; Isaksen, A.; Nealen, A.; Togelius, J. Procedural content generation via machine learning (PCGML). *IEEE Trans. Games* **2018**, *10*, 257–270. [CrossRef]

33. Gutierrez, J.; Schrum, J. Generative adversarial network rooms in generative graph grammar dungeons for the legend of zelda. In Proceedings of the 2020 IEEE Congress on Evolutionary Computation (CEC), Glasgow, UK, 19–24 July 2020; pp. 1–8.
34. Torrado, R.R.; Khalifa, A.; Green, M.C.; Justesen, N.; Risi, S.; Togelius, J. Bootstrapping conditional gans for video game level generation. In Proceedings of the 2020 IEEE Conference on Games (CoG), Osaka, Japan, 24–27 August 2020; pp. 41–48.
35. Sutton, R.S.; Barto, A.G. *Reinforcement Learning: An Introduction*; MIT Press: Cambridge, MA, USA, 2018.
36. Khalifa, A.; Bontrager, P.; Earle, S.; Togelius, J. Pcgrl: Procedural content generation via reinforcement learning. In Proceedings of the AAAI Conference on Artificial Intelligence and Interactive Digital Entertainment, Online, 19–23 October 2020; Volume 16, pp. 95–101.
37. Delarosa, O.; Dong, H.; Ruan, M.; Khalifa, A.; Togelius, J. Mixed-initiative level design with rl brush. In Proceedings of the International Conference on Computational Intelligence in Music, Sound, Art and Design (Part of EvoStar), Virtual Event, 7–9 April 2021; Springer: Cham, Switzerland, 2021; pp. 412–426.
38. Liu, J.; Snodgrass, S.; Khalifa, A.; Risi, S.; Yannakakis, G.N.; Togelius, J. Deep learning for procedural content generation. *Neural Comput. Appl.* **2021**, *33*, 19–37. [CrossRef]
39. Alvarez, A.; Dahlskog, S.; Font, J.; Togelius, J. Empowering quality diversity in dungeon design with interactive constrained map-elites. In Proceedings of the 2019 IEEE Conference on Games (CoG), London, UK, 20–23 August 2019; pp. 1–8.
40. Silva, R.C.; Fachada, N.; De Andrade, D.; Códices, N. Procedural generation of 3D maps with snappable meshes. *IEEE Access* **2022**, *10*, 43093–43111. [CrossRef]
41. Gravina, D.; Khalifa, A.; Liapis, A.; Togelius, J.; Yannakakis, G.N. Procedural content generation through quality diversity. In Proceedings of the 2019 IEEE Conference on Games (CoG), London, UK, 20–23 August 2019; pp. 1–8.
42. Yannakakis, G.N.; Togelius, J. *Artificial Intelligence and Games*; Springer: New York, NY, USA, 2018; Volume 2.
43. Juliani, A.; Berges, V.P.; Teng, E.; Cohen, A.; Harper, J.; Elion, C.; Goy, C.; Gao, Y.; Henry, H.; Mattar, M.; et al. Unity: A general platform for intelligent agents, 2018. *arXiv* **1809**, arXiv:1809.02627.
44. Risi, S.; Togelius, J. Increasing generality in machine learning through procedural content generation. *Nat. Mach. Intell.* **2020**, *2*, 428–436. [CrossRef]
45. Werneck, M.; Clua, E.W. Generating procedural dungeons using machine learning methods. In Proceedings of the 2020 19th Brazilian Symposium on Computer Games and Digital Entertainment (SBGames), Recife, Brazil, 7–10 November 2020; pp. 90–96.
46. Levine, K. Bioshock. Available online: https://2k.com/en-US/game/bioshock-the-collection/ (accessed on 23 August 2024).
47. Park, K.; Mott, B.W.; Min, W.; Boyer, K.E.; Wiebe, E.N.; Lester, J.C. Generating educational game levels with multistep deep convolutional generative adversarial networks. In Proceedings of the 2019 IEEE Conference on Games (CoG), London, UK, 20–23 August 2019; pp. 1–8.
48. Volz, V.; Schrum, J.; Liu, J.; Lucas, S.M.; Smith, A.; Risi, S. Evolving mario levels in the latent space of a deep convolutional generative adversarial network. In Proceedings of the Genetic and Evolutionary Computation Conference, Kyoto, Japan, 15–19 July 2018; pp. 221–228.
49. Giacomello, E.; Lanzi, P.L.; Loiacono, D. Doom level generation using generative adversarial networks. In Proceedings of the 2018 IEEE Games, Entertainment, Media Conference (GEM), Galway, Ireland, 15–17 August 2018; pp. 316–323.
50. id Software. Doom. Available online: https://www.idsoftware.com/en (accessed on 25 August 2024).
51. Alvarez, A.; Dahlskog, S.; Font, J.; Holmberg, J.; Johansson, S. Assessing aesthetic criteria in the evolutionary dungeon designer. In Proceedings of the 13th International Conference on the Foundations of Digital Games, Malmö, Sweden, 7–10 August 2018; pp. 1–4.
52. Alvarez, A.; Dahlskog, S.; Font, J.; Holmberg, J.; Nolasco, C.; Österman, A. Fostering creativity in the mixed-initiative evolutionary dungeon designer. In Proceedings of the 13th International Conference on the Foundations of Digital Games, Malmö, Sweden, 7–10 August 2018; pp. 1–8.
53. Liu, S.; Chaoran, L.; Yue, L.; Heng, M.; Xiao, H.; Yiming, S.; Licong, W.; Ze, C.; Xianghao, G.; Hengtong, L.; et al. Automatic generation of tower defense levels using PCG. In Proceedings of the 14th International Conference on the Foundations of Digital Games, San Luis Obispo, CA, USA, 26–30 August 2019; pp. 1–9.
54. Yu, D. Spelunky. Available online: https://spelunkyworld.com/original.html (accessed on 24 August 2024).
55. Lee, N.; Morris, J. A Procedural generation platform to create randomized gaming maps using 2D model and machine learning. In Proceedings of the CS & IT Conference Proceedings, Jakarta, Indonesia, 16 February 2023; Volume 13.
56. Entertainment, B. Diablo. Available online: https://us.shop.battle.net/en-us/product/diablo (accessed on 29 August 2024).
57. Pereira, L.T.; de Souza Prado, P.V.; Lopes, R.M.; Toledo, C.F.M. Procedural generation of dungeons' maps and locked-door missions through an evolutionary algorithm validated with players. *Expert Syst. Appl.* **2021**, *180*, 115009. [CrossRef]
58. Nam, S.; Ikeda, K. Generation of diverse stages in turn-based role-playing game using reinforcement learning. In Proceedings of the 2019 IEEE Conference on Games (CoG), London, UK, 20–23 August 2019; pp. 1–8.
59. Nam, S.G.; Hsueh, C.H.; Ikeda, K. Generation of game stages with quality and diversity by reinforcement learning in turn-based RPG. *IEEE Trans. Games* **2021**, *14*, 488–501. [CrossRef]
60. Entertainment, B. Diablo II. Available online: https://diablo2.blizzard.com/en-us/ (accessed on 27 August 2024).
61. Dutra, P.V.M.; Villela, S.M.; Neto, R.F. Procedural content generation using reinforcement learning and entropy measure as feedback. In Proceedings of the 2022 21st Brazilian Symposium on Computer Games and Digital Entertainment (SBGames), Natal, Brazil, 24–27 October 2022; pp. 1–6.
62. Shu, T.; Liu, J.; Yannakakis, G.N. Experience-driven PCG via reinforcement learning: A Super Mario Bros study. In Proceedings of the 2021 IEEE Conference on Games (CoG), Copenhagen, Denmark, 17–20 August 2021; pp. 1–9.

63. Lucas, S.M.; Volz, V. Tile pattern KL-divergence for analysing and evolving game levels. In Proceedings of the Genetic and Evolutionary Computation Conference, Prague, Czech Republic, 13–17 July 2019; pp. 170–178.
64. Lazaridis, L.; Papatsimouli, M.; Kollias, K.F.; Sarigiannidis, P.; Fragulis, G.F. Hitboxes: A survey about collision detection in video games. In Proceedings of the International Conference on Human-Computer Interaction, Virtual Event, 24–29 July 2021; Springer: Cham, Switzerland, 2021; pp. 314–326.
65. Ironhide. Kingdom Rush: Frontiers. Available online: https://www.kingdomrush.com/kingdom-rush-frontiers (accessed on 29 August 2024).
66. Re-Logic. Terraria. Available online: https://terraria.org/ (accessed on 29 August 2024).

Disclaimer/Publisher's Note: The statements, opinions and data contained in all publications are solely those of the individual author(s) and contributor(s) and not of MDPI and/or the editor(s). MDPI and/or the editor(s) disclaim responsibility for any injury to people or property resulting from any ideas, methods, instructions or products referred to in the content.

Article

FGPE+: The Mobile FGPE Environment and the Pareto-Optimized Gamified Programming Exercise Selection Model—An Empirical Evaluation

Rytis Maskeliūnas [1,*], Robertas Damaševičius [1], Tomas Blažauskas [1], Jakub Swacha [2,*], Ricardo Queirós [3,4] and José Carlos Paiva [4,5]

[1] Center of Excellence Forest 4.0, Faculty of Informatics, Kaunas University of Technology, 51423 Kaunas, Lithuania; robertas.damasevicius@ktu.lt (R.D.); tomas.blazauskas@ktu.lt (T.B.)
[2] Department of Information Technology in Management, University of Szczecin, 70-453 Szczecin, Poland
[3] uniMAD—ESMAD, Polytechnic of Porto, 4480-876 Vila do Conde, Portugal; ricardoqueiros@esmad.ipp.pt
[4] CRACS, INESC TEC, 4169-007 Porto, Portugal; jose.c.paiva@inesctec.pt
[5] Department of Computer Science, Faculty of Sciences, University of Porto, 4169-007 Porto, Portugal
* Correspondence: rytis.maskeliunas@ktu.lt (R.M.); jakub.swacha@usz.edu.pl (J.S.)

Abstract: This paper is poised to inform educators, policy makers and software developers about the untapped potential of PWAs in creating engaging, effective, and personalized learning experiences in the field of programming education. We aim to address a significant gap in the current understanding of the potential advantages and underutilisation of Progressive Web Applications (PWAs) within the education sector, specifically for programming education. Despite the evident lack of recognition of PWAs in this arena, we present an innovative approach through the Framework for Gamification in Programming Education (FGPE). This framework takes advantage of the ubiquity and ease of use of PWAs, integrating it with a Pareto optimized gamified programming exercise selection model ensuring personalized adaptive learning experiences by dynamically adjusting the complexity, content, and feedback of gamified exercises in response to the learners' ongoing progress and performance. This study examines the mobile user experience of the FGPE PLE in different countries, namely Poland and Lithuania, providing novel insights into its applicability and efficiency. Our results demonstrate that combining advanced adaptive algorithms with the convenience of mobile technology has the potential to revolutionize programming education. The FGPE+ course group outperformed the Moodle group in terms of the average perceived knowledge (M = 4.11, SD = 0.51).

Keywords: FGPE; pareto optimization; gamified programming; personalized learning; adaptive learning; progressive Web Applications (PWAs); mobile learning

Citation: Maskeliūnas, R.; Damaševičius, R.; Blažauskas, T.; Swacha, J.; Queirós, R.; Paiva, J.C. FGPE+: The Mobile FGPE Environment and the Pareto-Optimized Gamified Programming Exercise Selection Model—An Empirical Evaluation. *Computers* 2023, 12, 144. https://doi.org/10.3390/computers12070144

Academic Editors: Carlos Vaz de Carvalho, Hariklia Tsalapatas and Ricardo Baptista

Received: 24 June 2023
Revised: 18 July 2023
Accepted: 19 July 2023
Published: 21 July 2023

Copyright: © 2023 by the authors. Licensee MDPI, Basel, Switzerland. This article is an open access article distributed under the terms and conditions of the Creative Commons Attribution (CC BY) license (https:// creativecommons.org/licenses/by/ 4.0/).

1. Introduction

Traditional programming education techniques sometimes struggle to keep students engaged and motivated. Gamification models provide a solution by including interactive and game-like features that pique students' attention, improve their engagement, and create a better understanding of programming ideas. The events of the COVID-19 epidemic prompted higher education institutions around the world to quickly adjust to remote learning approaches [1,2]. The incorporation of e-learning platforms has been a popular technique, albeit the transition has not always been effective, frequently resulting in inferior performance when compared to traditional classroom-based training [3]. To fully realize the potential of gamified exercises and improve the e-learning experience, it became clear that these activities should be smoothly integrated into the existing course structure rather than given via separate platforms [4]. The Learning Tools Interoperability (LTI) standard developed as a feasible approach to support this integration. Students would be routed to a gamified learning environment within the e-learning platform itself if LTI was implemented, where they may engage in programming exercises. Furthermore, the outcomes of these activities might be quickly conveyed back to the e-learning platform, allowing for real-time feedback [5]. Gamified programming exercise selection models have received a lot of attention in recent years as a way to improve learning and engagement in programming education. These platforms use gamification aspects in conjunction with programming exercises to incentivize and encourage learners to actively engage in the development of their programming abilities. A significant body of research now suggests that gamification techniques and serious games play a crucial role in enhancing learning outcomes in programming education [6,7]. The underlying premise is that these approaches make learning more enjoyable and engaging, which in turn leads to improved knowledge retention and application [8,9].

Gamification has a positive impact on students' motivation [10,11]. Techniques such as points, badges, leaderboards, and challenges can stimulate students' competitive nature and foster collaboration [12,13]. Serious games, especially those addressing real-world issues, such as sustainable development [14] and healthcare [15,16], can foster deeper understanding and enhance programming skills. The game 'Eco JSity' is a successful example of integrating sustainability topics into programming education [14]. Gamification can improve self-efficacy in programming [11]. The perceived ability to perform a task often translates into better performance. Game-based learning approaches enhance student experience, knowledge gain, and usability in higher education programming courses [17]. Researchers also highlight different levels of gamification, from partial to full integration, into the course. For example, the SHOOT2LEARN project [18] and the game 'SQL Island' [19] mix gameplay with programming, enhancing student engagement. Meanwhile, 'SpAI War' is used for learning artificial intelligence algorithms [20]. These studies collectively suggest that gamification and serious games provide new, innovative ways to teach programming [21]. They transform the traditionally complex and abstract programming concepts into interactive, relatable scenarios for students. This helps to reduce the intimidation factor often associated with learning to program, which in turn can lead to improved motivation and self-efficacy [22]. However, it is essential to consider the varying degrees of success in implementing gamification, which is not always granted [23]. The effective integration of these techniques into a course requires thoughtful design and ongoing evaluation to ensure they are meeting the intended learning objectives [24]. Gamified programming systems rely heavily on engaging learners. The studies look at the motivating elements, pleasure, and immersion that learners have when using the mobile platform [25]. Analyzing the effectiveness of gamification features in creating intrinsic motivation and sustained engagement is part of this. Moreover, while game-based learning and gamification demonstrate promise, they are not a one-size-fits-all solution. They should be used alongside traditional teaching methods to cater to diverse learning styles and needs. More empirical studies are also needed to understand their long-term impacts and potential side effects (e.g., whether they lead to an over-reliance on reward structures) [26].

Technological advancements, such as the growth of mobile devices, online learning platforms, and instructional software, have opened the opportunity for gamification methods to be implemented in programming education. The availability of technologies and platforms to facilitate gamified learning experiences has fuelled the demand for these approaches. Although gamified exercises through LTI offer potential [27], there was always a need to improve the programming learning environment for mobile device users [28]. To solve this, a mobile gamified Programming Learning Environment (PLE) can be created utilizing Progressive Web Application (PWA) technology [29]. PWAs leverage new APIs and capabilities to deliver functionality comparable to native apps, such as push notifications, sensor integration, and the ability to have an access icon on the device's application launcher [30]. It is critical to assess the usability of mobile platforms to ensure that learners can easily explore, participate, and finish programming tasks. Examining characteristics such as ease of use, learnability, efficiency, and mistake prevention are all part of usability evaluations. Despite the potential benefits of PWAs in education, their use is restricted and their recognition in the area is minimal. The reasons for this limited acceptance are yet unknown [31]. PWAs tend to be underused, with the method going mostly unnoticed in the programming education sector [32]. This knowledge gap emphasizes the need for more PWA research and study in the context of educational applications. The Framework for Gamification in Programming Education (FGPE) was created to fill these shortcomings. This method sought to provide a comprehensive framework for the use of gamification techniques in programming education, spanning requirement design, the collection of gamified exercises, and the creation of supporting software [33]. The PWA design was chosen as part of its implementation to create mobile-like web apps with increased functionality and user experience.

This paper provides an in-depth examination of the integration of the Pareto-optimized gamified exercise selection model into e-learning courses using the LTI standard. It also emphasizes the potential of mobile gamified PLEs based on PWA technology to improve the programming learning environment for mobile device users. The FGPE method offers a framework for gamification in programming instruction, as well as the development of accompanying applications. Significant efforts were made during the installation of the FGPE strategy to optimize the performance of the mobile gamified PLE with a Pareto optimal exercise selection approach. This improved the efficiency of the app in providing programming training and the overall learning experience, as was confirmed by students utilizing these methodologies. Cross-country comparisons were also conducted to help understand the differences in user experience and educational results between areas and technical aspects that can impact user experience [34] in different nations.

We have aimed to assess the general attitude of mobile device users towards the PWA version of the FGPE PLE platform, as well as the suitability of the mobile version for different learning contexts. The evaluation used self-report measures, such as rating scales, to gather feedback from the users. Additionally, the study employed the User Experience Questionnaire (UEQ) to obtain a more detailed understanding of users' views on the mobile version of the platform. The responses from students in Poland and Lithuania were compared to identify any cross-country differences. Furthermore, a knowledge evaluation survey compared the effectiveness of the FGPE+ model with a classic Moodle course in teaching programming. The survey assessed the perceived knowledge of learners in each group. Lastly, the study evaluated the effectiveness of using the Sharable Content Object Reference Model (SCORM) in the FGPE+ model.

This paper is organized as follows. Section 2 reviews existing approaches to gamification in education. Section 3 describes the conducted experiment, including information on participants, the design of the gamified programming task selection model, and the implementation of such a model in FGPE. Section 4 presents the results of the evaluation of this implementation in various aspects. Section 5 discusses the outcomes of our study and its potential impact in STEM (Science, Technology, Engineering, and Mathematics) education. Finally, Section 6 summarizes the contributions of this work.

2. Overview of Approaches to Gamification in Educational Apps

According to research, well-designed gamification models can improve learning outcomes in programming education. They can boost student motivation, encourage long-term engagement, and improve information retention and skill development. As a result, educational institutions and companies are looking for efficient ways to improve learning outcomes in programming education. By bringing game design ideas and mechanics to programming learning environments, general gamification approaches for programming education attempt to improve student engagement, motivation, and learning results [35]. We frequently encounter recognizable elements such as:

- The Points, Badges, and Leaderboard (PBL) model entails allocating points to accomplished tasks or achievements, awarding badges for particular achievements, and keeping leaderboards to display student standings [36]. It uses competition and prizes to inspire students and promote a sense of accomplishment and growth.

- Quests and Levels model, in which learning activities are arranged into quests or missions, and learners move through a series of obstacles or levels [37]. Each level often provides new concepts or programming tasks, resulting in a logical learning path. Completing quests or levels opens additional content or features, giving the player a sense of advancement and success.
- Storytelling and narrative model that includes storytelling and narrative components into the programming learning process [38]. It develops immersive settings, characters, and plotlines to emotionally engage learners and relate learning information to real-world circumstances. Learners take on parts in the plot and embark on quests or missions, making learning more interesting and meaningful [39].
- The Achievements and unlockables model, as in video games, provides a system of achievements and unlockables [40]. Learners receive rewards when they complete particular programming assignments, grasp concepts, or reach milestones. Unlockables are extra challenges, added content, or special features that become available as learners progress and meet certain goals.
- The Companion or virtual pets model involves the incorporation of companion characters or virtual pets into the programming learning environment [41]. By completing programming assignments or demonstrating expertise, students can care for their virtual pets, earn rewards, and unlock additional capabilities. Throughout the learning process, the companions provide comments, direction, and encouragement [42].
- By integrating collaborative challenges and tournaments [43], the Collaborative challenge and tournament model stimulates collaboration and competitiveness among learners. Students create groups, collaborate to solve programming issues, and compete against other groups. It promotes camaraderie and pleasant rivalry while encouraging cooperation, communication, and problem-solving abilities [44].
- The Simulations and Virtual Environments Model [45] use simulations or virtual environments to provide realistic programming scenarios. Within the simulated environment, learners participate in hands-on coding activities such as constructing virtual apps or solving virtual programming tasks. Simulations provide a secure environment for experimentation and practice, allowing students to apply principles in a real-world setting [46].
- Gamification models almost always some include feedback and progress monitoring [47]. Learners immediately receive feedback on their coding performance, flagging problems and offering improvements. Progress tracking techniques such as progress bars, skill trees, or visual representations assist learners in seeing their progress and providing a sense of success [35].
- Gamified Learning Analytics combines learning analytics [48] with gamification [49]. It collects data from learners' interactions with gamified programming environments to gather insights into their learning habits, progress, and areas for growth. To improve learning outcomes, these insights enable tailored feedback, adaptive interventions, and instructional decision-making.

Empirical studies and scientific research are used to develop evidence-based gamification models for programming education in order to improve learning outcomes and engagement in programming education [50]. Below, we offer a rundown of some of the more popular research-based gamification approaches used in programming education:

- Goal-Structure Theory emphasizes the importance of defining specific objectives and creating a structured learning environment [51]. It underlines the significance of defining distinct, difficult, and attainable goals in programming assignments. Clear goals provide learners a feeling of direction and purpose, which promotes motivation and attention throughout the learning process.
- Learner autonomy, competence, and relatedness are all emphasized in Self-Determination Theory [52]. It demonstrates that learners are motivated when they feel in charge of their learning, recognize their competency in programming tasks, and feel connected

to others. Gamification components that encourage autonomy, skill development, and social interaction can boost student motivation and engagement.
- The goal of Flow Theory [53] is to generate a state of "flow" in which learners are completely absorbed and interested in programming tasks. Flow occurs when learners confront tasks that are appropriate for their ability level, receive fast feedback, and feel a sense of control and concentration. Flow experiences in programming education can be facilitated by gamification features that increase challenge, feedback, and focused attention.
- In programming education, Social Cognitive Theory stresses the importance of observational learning, social interaction, and feedback [54]. Learners may watch and learn from other people's programming techniques, participate in collaborative coding exercises, and receive constructive comments from peers or instructors. Based on this principle, gamification models stimulate social learning, give chances for knowledge exchange, and promote positive reinforcement [55].
- The Cognitive burden Theory [56] is concerned with controlling cognitive burden during programming activities. It implies that instructional design should aim to reduce external cognitive strain while increasing internal cognitive demand. Based on this principle, gamification models may include interactive components, step-by-step assistance, and scaffolded learning to minimize cognitive load and improve learning efficiency [57].
- Mastery Learning [58] advocates for a mastery-based approach to programming instruction. It pushes students to grasp basic programming concepts and abilities before moving on to more difficult topics. Gamification methods based on mastery learning give adaptive feedback, tailored learning routes, and chances for purposeful practice to assist learners in achieving mastery and establishing a solid programming foundation.
- Personalized and adaptive gamification models adjust the learning experience to the qualities, interests, and needs of the individual learner [59]. They use student data, such as performance history and learning styles, to modify the difficulty level, material sequencing, and feedback in programming exercises dynamically. Personalization and adaptability boost student engagement, improve learning efficiency, and accommodate a wide range of learning profiles [60].

To prevent possible downsides, it is critical to find a balance between gamification components and primary learning objectives [61]. Careful implementation, constant assessment, and adaptive design based on learner input may help maximize the benefits of research-based gamification models, while limiting the obstacles that come with them. Gamification models can enhance programming education by catching learners' attention and inspiring them to actively participate in learning sessions. The incorporation of game features can boost engagement and interest in programming by creating a sense of excitement and challenge [62]. By combining aspects such as incentives, accomplishments, and competition, gamification taps into both inner and extrinsic motivating factors. Learners are driven to complete goals, gain badges, and climb leaderboards, which might inspire them to continue studying and achieve higher outcomes. Gamification models may be tailored to meet the requirements and preferences of individual learners. Adaptive components and individualized feedback can create tailored learning experiences, allowing students to advance at their own speed while receiving focused assistance [63]. Personalization encourages a more effective and efficient learning process. Gamification models frequently include interactive coding exercises, simulations, and challenges, giving learners hands-on practice. Through active experimentation, learners may apply programming ideas in a practical setting, developing their problem-solving and coding abilities [64]. Gamified programming environments may provide learners with rapid feedback on their coding performance. Instant feedback enables learners to quickly detect and rectify errors, reinforcing their comprehension and enabling deeper learning.

However, there are various drawbacks to employing research-based gamification techniques [65]. There is a danger that learners will become more focused on obtaining points, badges, or rankings than on true learning. If the game aspects take precedence over the learning objectives, learners may participate in superficial engagement, prioritizing the gamification features above real comprehension and skill development. While competition may be inspiring, it can also create a climate that inhibits learners from collaborating and cooperating. Some students may become demotivated or disengaged if they believe they are slipping behind their peers, resulting in unpleasant learning experiences [66]. Gamification models should be carefully built to be compatible with the programming education setting and learning objectives. Not every game element may be appropriate or helpful for every programming topic or learning scenario. Gamification features should be relevant, meaningful, and connected with the targeted learning goals. Learners' learning methods, interests, and motivations vary [67]. Gamification methods may not appeal to all learners in the same way, and what drives one student may not inspire another. Individual variations must be considered, and gamification tactics must be tailored to different learner profiles. Maintaining and implementing gamified programming environments necessitates constant work and resources. Updating material, tracking learner progress, and fixing technological difficulties can be time-consuming and may need specialized staff or systems to support the gamification model's long-term viability [68].

In order to overcome all these drawbacks, several gamification design frameworks appeared in the last decades to provide a systematic and structured approach to avoid the ad hoc implementation of gamification components and learning theories, enabling organizations to effectively scaffold the development of a well-defined gamification strategy. Existing gamification design frameworks can be divided into process-based frameworks, component-based frameworks, and goal-based frameworks.

Process-based frameworks provide a structured approach to gamification design, guiding designers through a series of steps. Following these steps, designers ensure that game design elements are incorporated effectively. Some examples of these frameworks are Octalysis, 6D, and Game Thinking. The Octalysis framework defines and builds upon the eight core drives of human motivation [69]: Epic Meaning and Calling, Development and Accomplishment, Empowerment of Creativity and Feedback, Ownership and Possession, Social Influence and Relatedness, Scarcity and Impatience, Unpredictability and Curiosity, and Loss and Avoidance. The framework is based on a human-focused design, i.e., designed to motivate users, as opposed to function-focused design, i.e., designed assuming users complete their tasks in a timely manner, as they are required to do so. The 6D Framework [70] is a design process composed of six steps, namely "Define business objectives", "Delineate target behavior", "Describe yours players", "Devise activity loops", "Don't forget the fun" (i.e., ensure that there is fun while using with the system), and "Deploy appropriate tools". The Game Thinking Framework [71] combines game design, systems thinking, design thinking, and agile/lean practices into a recipe for gamification designers.

Component-based frameworks focus on specific game design elements, providing guidelines to apply these elements into non-game contexts. One of the most popular gamification design frameworks, the Mechanics–Dynamics–Aesthetics (MDA) framework [72], belongs to this group. The MDA framework breaks games down into three components: mechanics, dynamics, and aesthetics. Mechanics are the base game components such as rules and algorithms. Dynamics include the run-time behavior of the mechanics acting on player input, which involve other mechanics. Aesthetics are the player's emotional responses to the system, such as sensation, fantasy, narrative, challenge, fellowship, discovery, expression, and submission. The MDA framework provides precise definitions for these terms and seeks to explain how they relate to each other and influence the player's experience. Other component-based frameworks include the SAPS (Status, Access, Power, and Stuff) framework [73], which focuses on the four key components of gamification design: status, access, power, and stuff, and provides guidelines for incorporating these components into non-game contexts.

Goal-based frameworks capitalize on the desired outcomes of the gamification design process (e.g., increased engagement, motivation, and learning), providing guidelines to achieve these outcomes by using game design elements. Examples of frameworks in this category are RAMP (Relatedness, Autonomy, Mastery, and Purpose) [74] and ARCS (Attention, Relevance, Confidence, and Satisfaction) [75]. The RAMP framework focuses on four key outcomes of gamification design: retention, achievement, mastery, and progress. The framework provides guidelines for achieving these outcomes through game design elements such as points, levels/stages, badges, leaderboards, prizes and rewards, progress bars, storyline, and feedback. The ARCS model focuses on the four key components of motivation: attention, relevance, confidence, and satisfaction.

3. Materials and Methods

3.1. Materials

The 23 Lithuanian students from the Faculty of Informatics ranged in age from 18 to 25 years. Variations were noted within this range, with a few older persons falling slightly outside of it. The gender distribution among the student cohort favored male students, who had a larger representation than females. Notably, the sample includes a handful of transgenders, contributing to the group's diversity of gender identities. All 23 students in the sample were pursuing programming-related specialties. This emphasis demonstrates their unique interest in learning about creating, building, and maintaining software systems. The majority of the participants were from Lithuania, which corresponded to the location of Kaunas University of Technology.

The 49 Polish respondents were 1st-year students of IT in Business and IT & Econometrics attending the Introduction to Programming course at the Faculty of Economics, Finance, and Management of University of Szczecin. Almost all of them were 19 years old; 1/4 of them were female, and 3/4 male. Alike the students at Kaunas University of Technology, they have a passion for developing software systems, yet in contrast to them, they are more focused on enterprise information systems rather than software in general.

3.2. Pareto-Optimized Gamified Programming Task Selection Model

Designing a Pareto-optimized gamified programming task selection model (Figure 1) for adaptive personalized learning involves the creation of a multi-objective optimization model that seeks to balance multiple competing factors such as the learner's interests, abilities, task difficulty, novelty, and relevance to the curriculum. In this context, Pareto optimization refers to a state of allocation where it is impossible to make any one individual better off without making at least one individual worse off.

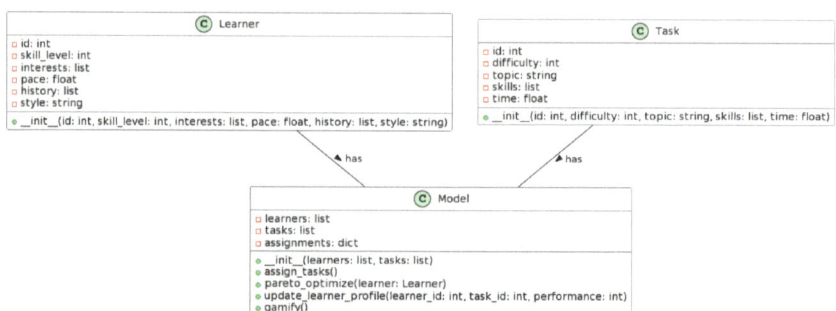

Figure 1. Pareto-optimized gamified programming task selection model.

The model has the following key components:

1. Programming task bank: A repository of programming tasks, each classified according to their difficulty level, related topic, required skills, estimated completion time, etc.
2. Learner profile: A dynamic profile for each learner capturing their programming skill level, areas of interest, learning pace, historical performance on tasks, preferred learning style, etc.

3. Gamification elements: Incorporation of game design elements such as points, badges, leaderboards, achievement tracking, feedback, progress bars, storyline, etc.

Formally, the model can be described as follows:

Let $L = \{L_1, L_2, \ldots, L_n\}$ be the set of learners.

Each learner L_i is represented as a tuple $(id_i, sl_i, in_i, p_i, h_i, st_i)$, where:

id_i is the id,

sl_i is the skill level,

in_i is the set of interests,

p_i is the learning pace,

h_i is the history of completed tasks,

st_i is the learning style.

Let $T = \{T_1, T_2, \ldots, T_m\}$ be the set of tasks.

Each task T_j is represented as a tuple $(id_j, d_j, t_j, s_j, time_j)$, where:

id_j is the id,

d_j is the difficulty level,

t_j is the topic,

s_j is the set of required skills,

$time_j$ is the estimated completion time.

Let $A : L \times T \to 2^T$ be the assignment function, where 2^T is the power set of T. This function assigns to each learner a set of tasks, i.e., $A(L_i) = \{T_{i1}, T_{i2}, \ldots\} \subseteq T$.

Let $O : L \times T \to 2^T$ be the Pareto optimization function, defined as:

$O(L_i, T) = \{T_j \in T \mid \text{there does not exist } T_k \in T \text{ such that } T_k \text{ is better than } T_j \text{ for } L_i\}$.

Let $U : L \times T \times \mathbb{R} \to L$ be the update function, defined as:

$U(L_i, T_j, performance) = L'_i$, where L'_i is the updated learner profile based on the performance on task T_j.

Here, L is the set of learners, each represented as a tuple of parameters. T is the set of tasks, each represented as a tuple of parameters. A is the assignment function that maps each learner to a set of tasks. O is the Pareto optimization function that assigns a learner a subset of tasks for which there are no better alternatives. U is the update function that updates a learner's profile based on their performance on a task.

The model starts by initializing the learners and tasks. Each learner is then assigned a set of tasks that are Pareto optimized for them. The Pareto optimization process involves finding the balance between different objectives (learner's interests, skill level, pace, etc., vs. the task's difficulty, skills required, topic relevance, etc.) to maximize the learning outcome. The learner's profile is updated after they complete a task based on their performance, and the task assignment process can be repeated as necessary. Gamification elements can be added to the model to increase learner engagement.

3.3. Implementation

Our goal was to fill a substantial knowledge gap on the potential benefits and inadequate use of Progressive Web Applications (PWAs) in the education sector, particularly for educational programming. The FGPE+ model provides programmers with a unique and engaging mobile user experience (see Figure 2). To create an effective and pleasant learning environment, it blends the ideas of Pareto optimization, gamification, and programming exercises. The PWA-based mobile-compatible solution has a clean and straightforward user experience that facilitates navigation and interaction. The interface is designed in a modern, minimalist style, with an emphasis on usability and clarity. When new users start the app, they are met with a full onboarding experience that introduces them to the features and capabilities of the FGPE+ model. Within the app, users are encouraged to create individualized profiles. They can choose their preferred programming language (Javascript, Python, Java, C#, Cpp), skill level, and areas of interest. This data allows the FGPE+ model to personalize exercise recommendations to the user's specific requirements and goals.

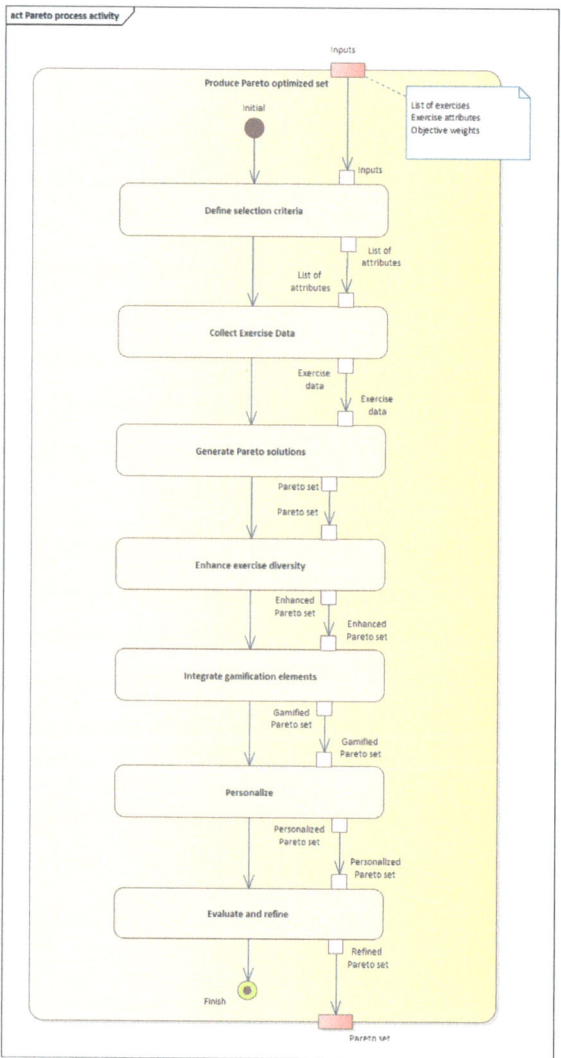

Figure 2. Implementation of the FGPE+ approach.

The app's fundamental feature relies around workout choices. The FGPE+ model curates a varied set of programming assignments using Pareto optimization methods. These exercises span a variety of programming languages, ideas, and levels of difficulty. Users can explore the exercise collection or let the computer recommend workouts based on their profile and learning objectives. The FGPE+ paradigm encourages users to participate in ongoing learning. It refreshes the exercise collection on a regular basis, providing new problems, upcoming programming languages, and trendy themes. As they move through their programming adventure, users may set objectives, measure their progress, and unlock advanced exercises. The UML class diagram represents the classes and their relationships involved in the algorithm for the Pareto-optimized gamified programming exercise selection (Figure 3).

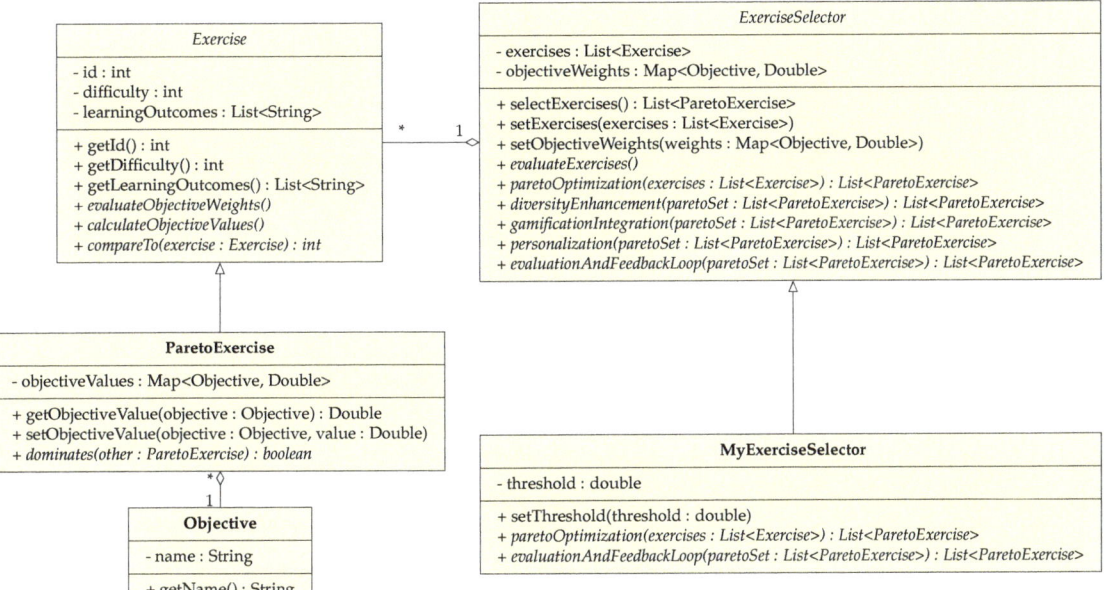

Figure 3. Pareto-optimized gamified programming exercise selection.

1. **The exercise** abstract class represents a programming exercise. It has attributes such as **id** (exercise identifier), **difficulty** (difficulty level of the exercise), and **learningOutcomes** (a list of learning outcomes associated with the exercise). It provides methods to access these attributes and defines three virtual methods: **evaluateObjectiveWeights**(to evaluate the objective weights of the exercise), **calculateObjectiveValues**() (to calculate the objective values of the exercise), and **compareTo**() (to compare two exercises based on their objective values).

2. **ParetoExercise**
class represents an exercise that includes objective values. It inherits from the **Exercise** class and has an additional attribute called **objectiveValues**, which is a map that stores objective values for the exercise. It provides methods to obtain and set objective values for specific objectives and defines the **dominates**() method to check whether it dominates another **ParetoExercise** based on their objective values.

3. **ExerciseSelector** abstract class serves as the base class for the exercise selection algorithm. It has attributes **exercises** (a list of exercises to select from) and **objectiveWeights** (a map that holds the weights of different objectives). It provides methods to select exercises, sets the exercises and objective weights, and defines six virtual methods that outline different steps of the algorithm: **evaluateExercises**() (to evaluate the exercises based on objectives), **paretoOptimization**() (to perform Pareto optimization on the exercises), **diversityEnhancement**() (to enhance the diversity of the exercise set), **gamificationIntegration**() (to integrate gamification elements into the exercises), **personalization**() (to personalize the exercise selection), and **evaluationAndFeedbackLoop**() (to evaluate and refine the exercise selection based on feedback).

4. **MyExerciseSelector**. This class represents a specific implementation of **ExerciseSelector**. It adds an additional attribute called the threshold (a threshold value for evaluation) and overrides the **paretoOptimization**() and **evaluationAndFeedbackLoop**() methods to provide custom implementation based on the defined threshold.
5. **Objective** class represents an objective to optimize in the exercise selection. It has an attribute **name** (the name of the objective) and provides a method to access the name.

The FGPE+ concept adds gamification aspects throughout the app to make the learning experience more interesting. For completing workouts, attaining milestones, and obtaining high scores, users gain points, badges, and virtual gifts. This gamified method encourages competitiveness, incentive, and ongoing progress. The app monitors and shows the progress and performance data of users. Users may check their completion rates, accuracy, and time required to complete each activity. The model offers customized reports and insights to assist users in identifying their own strengths, shortcomings, and opportunities for progress. The FGPE+ paradigm encourages user social engagement. Users may join groups, participate on discussion boards, and work together to solve puzzles. Users may also compare their performance to that of others, encouraging healthy rivalry and information exchange. Users obtain fast feedback on their workout answers, which helps them understand and improve from their mistakes. The software offers thorough explanations, code critiques, and advice to help users improve their programming skills. Users may also seek assistance from mentors or experienced programmers within the app.

Here is the pseudo code of the FGPE+ approach:

Inputs: List of exercises (ExerciseList); exercise attributes (e.g., learning outcomes, difficulty levels); objective weights (e.g., learning outcomes, engagement).

Outputs: Pareto optimal exercise set (ParetoSet).

Procedure: 1. DefineExerciseSelectionCriteria(); 2. CollectExerciseData(); 3. ParetoOptimization(); 4. DiversityEnhancement(); 5. GamificationIntegration(); 6. Personalization(); 7. EvaluationAndFeedbackLoop().

Procedure DefineExerciseSelectionCriteria(): Specify criteria for selecting exercises based on objectives and gamification elements; define attributes to consider, such as learning outcomes, difficulty levels, or programming concepts.

Procedure CollectExerciseData(): Gather information on exercises, including attributes and gamification elements; store exercise data in a suitable data structure.

Procedure ParetoOptimization(): Apply multi-objective optimization algorithm (e.g., NSGA-II, SPEA2) on the exercise data; generate a set of Pareto optimal solutions considering the defined objectives and weights.

Procedure DiversityEnhancement(): Enhance diversity in the selected exercise set; apply niching or crowding techniques to avoid redundancy and promote variety.

Procedure GamificationIntegration(): Integrate gamification elements into the selected exercises; consider elements such as points, badges, levels, leaderboards, or social interaction features.

Procedure Personalization(): Allow learners to personalize exercise selection based on preferences, prior knowledge, or skill levels; implement adaptive algorithms for dynamically adjusting exercise difficulty or sequence.

Procedure EvaluationAndFeedbackLoop(): Continuously evaluate the effectiveness of the exercise selection model; collect user feedback, learning outcomes, and engagement metrics; refine and improve the algorithm based on evaluation results.

Output ParetoSet: Return the set of exercises representing the Pareto optimal solutions.

4. Results

4.1. Evaluation of the PWA Version of the FGPE PLE Platform by Mobile Device Users

As the main purpose of redeveloping the FGPE PLE platform as a PWA was to make it more suitable for mobile device users, the first part of its evaluation was aimed at assessing to what extent we have succeeded. Although this could be evaluated with both self-report and behavioral measures [76], following the findings of [77] showing that the interpretation of the latter is not always straightforward and could be misleading, we decided to go with the former.

Apart from assessing the general students' attitude to using the FGPE PLE platform on a mobile device (Q1), we also strived to assess whether the learning place makes a difference in the mobile use of the platform (Q2 and Q3), as well as whether the mobile users still feel the need to use the PC version at all (Q4):

(Q1): How do you generally rate the mobile version of the FGPE PLE platform?
Answer range: 1 (bad)–5 (excellent).

(Q2): Do you think the mobile version of the FGPE PLE platform makes sense for students learning at home?
Answer range: 1 (bad)–5 (excellent).

(Q3): Do you think the mobile version of the FGPE PLE platform makes sense for students who study on the go to school/work?
Answer range: 1 (bad)–5 (excellent).

(Q4): Do you think it is possible to learn to write code only using the mobile version of the FGPE PLE platform—and without using the PC version at all?
Answer range: 1 (bad)–5 (excellent).

The results are summarized in Figure 4. The answers to Q1 demonstrate the overwhelmingly positive response to the mobile version of FGPE PLE in our study. The answers to Q2 and Q3 indicate the students see it as a convenient learning tool at home and during commuting to school/work (more so for the latter than the former). The answers to Q4 show that the students do not believe that programming can be learned only by using FGPE PLE on their mobile devices.

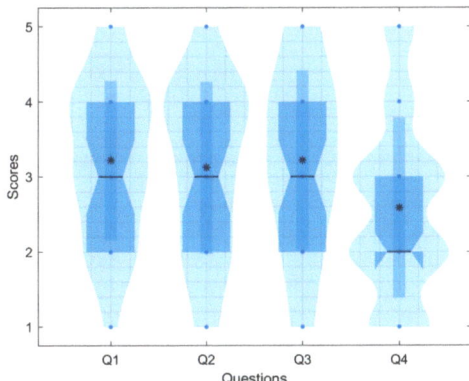

Figure 4. Results of questionnaire survey.

4.2. User Experience Analysis

In order to obtain a more detailed picture of how the users view the mobile version of FGPE PLE, the User Experience Questionnaire (UEQ) has been employed [78]. It is a well-established instrument used to evaluate the user experience of a product or service, designed to provide developers with a quick and straightforward way to assess the user experience of their product, be it a website, a software application, or any other kind of interactive system. It has been previously used for The Evaluation of User Experience on Learning Management Systems [79]. The UEQ measures six different scales:

- Attractiveness: covers the overall impression of the product, including whether it is pleasant or enjoyable to use.
- Perspicuity: measures how easy it is for users to understand how to use the product.
- Efficiency: evaluates the perception of how efficiently users can complete tasks using the product.

- Dependability: measures how reliable and predictable users find the product.
- Stimulation: evaluates how exciting and motivating the product is to use.
- Novelty: assesses whether the design of the product is creative and innovative and whether it meets users' expectations.

The questionnaire itself consists of 26 pairs of opposing adjectives (such as "complicated" vs. "easy"), and respondents rate their experience with the product on a 7-point Likert scale between these extremes (see Figure 5). The scores from these pairs of adjectives are then used to calculate scores on the six scales listed above. This provides a comprehensive and nuanced understanding of how users perceive the product or service.

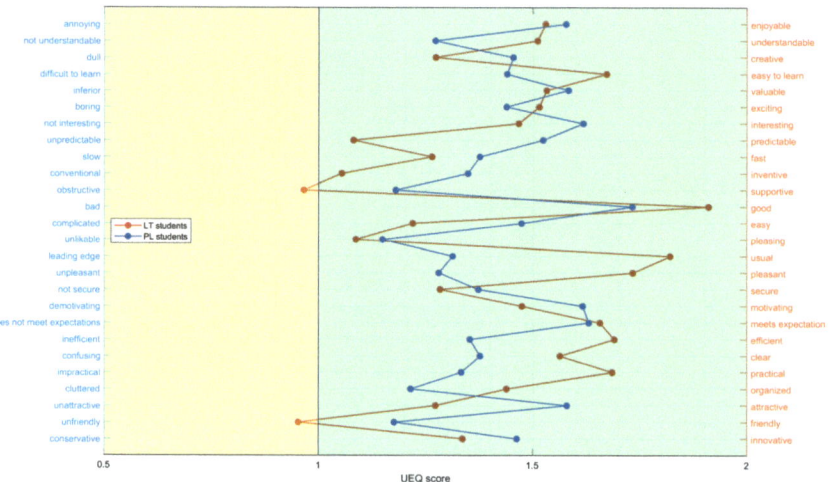

Figure 5. User Experience Questionnaire: responses from study participants.

Students from Poland and Lithuania provided answers to the User Experience Questionnaire (UEQ), reflecting how they used the FGPE+ model. Each input indicates a score for a particular aspect of the user experience, such as appeal, perspicacity, effectiveness, reliability, stimulation, and innovation (Figure 6). These replies come from many student groups; therefore, they represent the distinctive perspectives and experiences of these various cohorts.

The availability of data from two countries created an opportunity for a cross-country analysis. This analysis aimed at identifying discrepancies and/or parallels in the assessment that could be linked to differences in pedagogy and cultural backgrounds of the students taught in various countries. Here is a brief analysis of the results:

- Attractiveness: The Lithuanian group had a higher average score (mean = 1.4141, std = 0.6081) compared to the Polish group (mean = 0.6607, std = 0.5785). This indicates that the Lithuanian students found the learning environment more attractive and appealing than the Polish students.
- Perspicuity: The Lithuanian group also scored higher (mean = 1.4914, std = 0.7604) than the Polish group (mean = 0.5580, std = 0.7732), suggesting that the Lithuanian students found the learning environment more clear and understandable.
- Efficiency: Again, the Lithuanian group's score was higher (mean = 1.5193, std = 0.6714) than the Polish group (mean = 0.2727, std = 0.6650), indicating that the Lithuanian students found the learning environment more efficient for achieving their tasks.
- Dependability: The Lithuanian group had a higher mean score (mean = 1.2462, std = 0.8367) than the Polish group (mean = 0.7047, std = 0.7909), indicating they found the learning environment more reliable and dependable.

- Stimulation: The Lithuanian group scored slightly higher in this dimension (mean = 1.4972, std = 0.6408) than the Polish group (mean = 1.2548, std = 0.7053). This means that the Lithuanian students found the learning environment slightly more exciting and motivating.
- Novelty: Lastly, the Lithuanian group scored higher in terms of novelty (mean = 1.3701, std = 0.6361) compared to the Polish group (mean = 0.5738, std = 0.6177). This suggests that the Lithuanian students found the learning environment more innovative and creative.

By comparing the data, it is evident that Lithuanian students often score higher on the UEQ than Polish students. Such results indicates that, globally, Lithuanian students may have found the FGPE PLE mobile version easier to use. However, it is important to note that a thorough analysis is difficult without a complete understanding of the issues that correlate to the presented data. It may simply reflect differing cultural views, educational backgrounds, or degrees of knowledge with comparable systems rather than necessarily implying that the approach featuring mobile learning supported with FGPE is more frequently accepted in Lithuania. The lower results in the Polish student group may indicate an opportunity for the further development of the FGPE PLE in some areas to make it more flexible and advantageous for a wider range of users, but it could also be a reflection of various standards or expectations.

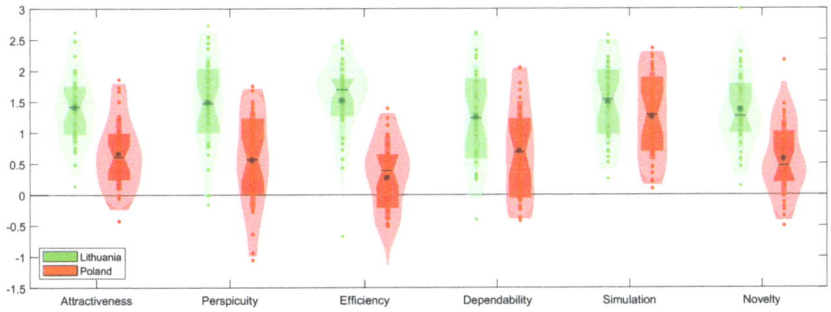

Figure 6. Cross-country comparison of aggregated User Experience Questionnaire scores.

4.3. Knowledge Evaluation Survey: FGPE Approach vs. Classic Moodle Course

Following the guidelines proposed in [80] regarding the evaluation of area-specific effects of gamification which suggest knowledge improvement as an indicator relevant for gamification applications aimed at supporting learning, we have used the opportunity that the Lithuanian group had a parallel group learning programming without the use of the FGPE toolset, to compare the two educational approaches.

A knowledge evaluation survey was conducted comparing two groups of learners: the first group used the PWA-based FGPE exercise selection model of the Python programming course, while the second one used the typical Moodle course format of Python programming (non-gamified, lecturer assigned, and ordered programming tasks). Table 1 demonstrates the perceived knowledge evaluation of groups using the FGPE+ model and the Moodle course, demonstrating the efficacy of the model. The 'N' column refers to the sample size for each group. 'M (SD)' represents the mean and standard deviation of the perceived knowledge scores. 't' and 'p' are the t-statistic and p-value, respectively, from a t-test comparing the game and Moodle course group scores. The FGPE+ course group had a higher average perceived knowledge score (M = 4.11, SD = 0.51) than the Moodle group (M = 3.67, SD = 0.56). The t-value indicates that the Moodle course group's score was higher, and the small p-value ($p < 0.05$) suggests this difference was statistically significant. This could imply that the FGPE+ model was effective in increasing the perceived knowledge.

Table 1. Perceived knowledge evaluation in groups using FGPE+ model vs. Moodle course.

	N	M (SD)	t	p
FGPE+ group	30	4.11 (0.51)	4.21	0.03
Moodle course group	35	3.67 (0.56)	3.63	0.04

4.4. Effectiveness of Using Sharable Content Object Reference Model

The Sharable Content Object Reference Model (SCORM) is a collection of standards and specifications for web-based educational content. It provides a standardized approach to creating and delivering online learning content, ensuring interoperability, accessibility, and reusability. SCORM has been widely adopted by e-learning providers as it ensures that learning content and Learning Management Systems (LMS) can work seamlessly together, regardless of the developer or platform. The results of evaluation according to SCORM criteria are presented in Figure 7). Data were collected using a descriptive survey following the practice established by [81], and assessment was conducted using their suggested quasi-experimental approach. ANCOVA results (FT = 3.76; FC = 8.11; $p = 0.04$) revealed a substantial difference in the impacts of SCORM-conformant e-content and conventional material on academic achievement.

Seventeen people (normalized) responded from the original groups of FPGE+ and Moodle users (see Table 1).

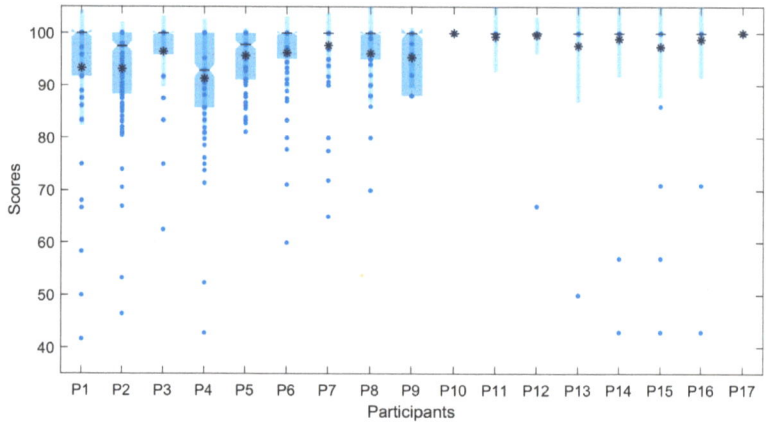

Figure 7. Results of SCORM questionnaire survey.

5. Discussion

We believe that PWAs have not yet gained full momentum in the education sector, and their use remains limited [82]. There is still somewhat of a scarcity of research and different uses of PWAs in education, demonstrating that the approach is not commonly recognized or used in the programming education arena [83]. Our Pareto-optimized approach enabled a personalized and adapted exercise selection to meet the specific needs and skill levels of individual learners, demonstrating that adaptive learning algorithms and techniques can dynamically adjust the difficulty, content, and feedback of gamified programming exercises based on learners' progress and performance. To address these issues, the Framework for Gamified Programming Education (FGPE) was developed. The FGPE framework includes requirement formulation, a collection of gamified activities, and the creation of supporting software.

We would like to bring up a few critical topics for reader consideration. The first step is to examine the long-term consequences. Longitudinal studies are needed to assess the long-term influence of gamified exercises on students' programming abilities and information retention. We investigate whether incorporating gamification into programming instruction results in better long-term learning outcomes than standard techniques [84]. Future study should look into how various gamification tactics and mechanics affect students' motivation, perseverance, and pleasure in programming tasks. Longer research would also aid in assessing social interaction and cooperation, since including social components such as competition, collaboration, and peer evaluation might improve students' learning experiences and outcomes in the long run.

5.1. Importance of FGPE+ for STEM Education

The FGPE+ model's potential extends far beyond the realm of programming education and has significant implications for the broader STEM (Science, Technology, Engineering, and Mathematics) education landscape. In an era characterized by rapid advancements in technology and the increasing relevance of digital literacy, effective STEM education is crucial in order to allow for the break-out of current rigid education schemes [85]. The FGPE+ model, with its innovative blend of personalization, gamification, and mobile learning, is a robust tool for enhancing STEM learning experiences.

One of the central tenets of the FGPE+ model is its ability to tailor learning experiences to individual learners. This approach aligns well with the diverse nature of STEM education, where learners often come with varying levels of background knowledge and abilities. The FGPE+ model can accommodate these differences, making learning more effective and enjoyable. The adaptive algorithm can be extended beyond programming tasks to include other STEM-related exercises, such as solving mathematical problems or designing engineering solutions.

The incorporation of gamification techniques in the FGPE+ model is a significant asset for STEM education [86]. Gamification elements such as badges, points, leaderboards, and levels can make complex STEM concepts more engaging and accessible, thereby fostering a positive attitude towards these subjects. These elements can also promote healthy competition and motivation, encouraging students to continually improve their understanding and mastery of STEM subjects.

With the prevalence of mobile devices, the mobile-compatible nature of the FGPE+ model offers vast potential for remote and flexible learning. This aspect can break down barriers to STEM education, allowing learners to access resources and engage with STEM concepts anytime, anywhere. Mobile learning also aligns with the digital habits of the current generation, thereby increasing its effectiveness and appeal.

The FGPE+ model's use of real-world tasks mirrors the application-based learning that is critical in STEM fields. By engaging with tasks that reflect real-world challenges, students can gain a deeper understanding of the practical applications of their learning, making the learning process more meaningful and relevant.

In conclusion, the FGPE+ model is a significant tool that can transform STEM education. By making learning more personalized, engaging, flexible, and application-focused, it aligns with the goals of modern STEM education, promoting increased participation and achievement in these crucial fields. Future research and development efforts should consider how the principles of FGPE+ can be effectively integrated and implemented across various STEM disciplines.

5.2. Limitations

Despite the potential contributions, this study has certain limitations that should be considered:

- This study used a subset of students from Poland and Lithuania, which may not be typical of the whole population. We believe that extending the study's size and altering the demographics of the participants would potentially increase generalizability.
- This research relied heavily on self-report measures, which are subjective and prone to bias. Incorporating objective measures, such as performance-based assessments or tracking system data, could provide a more comprehensive evaluation of the platform's effectiveness.

- The study aimed to assess the FGPE PLE platform in various educational settings and learning scenarios. The findings may not fully represent the intricacies and complexity of various educational settings. Future study might look at the platform's efficacy in other educational institutions, student backgrounds, and instructional environments.
- The PWA version of the FGPE PLE platform was assessed particularly for programming instruction in the research. The findings may not be applicable in other domains or topic areas. Replicating the study in additional educational fields would be advantageous in determining the platform's generalizability.
- The investigation focused on the initial user experience and perceived knowledge. Understanding the long-term impact of the FGPE PLE platform on learners' programming skill development and information retention would necessitate additional research outside the scope of this study.

5.3. Potential Lines of Research

The study on the evaluation of the PWA version of the FGPE PLE platform opened up several potential lines of research:

- We concentrated on evaluating the platform's initial user experience and perceived knowledge. More study might be conducted to investigate the long-term consequences of utilizing the FGPE PLE platform on mobile devices. Longitudinal studies might look into the long-term influence on learning outcomes, programming skill development, and knowledge retention.
- In terms of perceived knowledge, we compared the FGPE+ model to a traditional Moodle course. Future study might compare the efficacy of other educational methodologies, such as gamified platforms such as FGPE+ vs. traditional teaching methods. Comparative research might assist to uncover the advantages and disadvantages of each strategy and provide ideas into how to improve learning experiences.
- The platform was evaluated mostly using self-report measures in this study. Tracking user interactions, completion rates, and performance statistics, for example, might give a more objective evaluation of learners' progress and engagement with the platform if learning analytics approaches are used. Analyzing such data might aid in the discovery of trends, the identification of areas for development, and the implementation of individualized learning interventions.
- Further study might look into the efficacy of certain instructional tactics used inside the FGPE PLE platform. Investigating how various gamification aspects, adaptive learning algorithms, or social interaction features influence motivation, engagement, and learning results might provide useful insights for building and enhancing educational systems.
- Our investigation discovered some cross-national disparities in user experience perceptions. More thorough cross-cultural research might provide insight on how cultural backgrounds and educational environments impact FGPE PLE platform acceptability and efficacy. Understanding these cultural differences may help with the customization and localization of educational systems for varied learner groups.

6. Conclusions

This study explored the use of the FGPE+ model, a Pareto-optimized gamified programming exercise selection system, in a mobile learning context. The FGPE+ system, by integrating principles of Pareto optimization and gamification in a mobile-compatible platform, offered a unique, personalized, and engaging learning environment for programming students. The FGPE+ model's clean and user-friendly interface was another aspect that stood out, enabling learners to navigate and interact with the platform easily. The PWA-based system allowed learners to carry their learning environment with them, enhancing accessibility and convenience. The students appreciated the tailored exercise recommendations, adaptive difficulty level adjustments, and gamification elements that kept them motivated to learn continually. The study also shed light on the effectiveness of the model in catering to diverse learners, including those who preferred different programming languages and had varying levels of skills and interests.

The overwhelmingly positive response to the FGPE+ model in our study is a promising step towards transforming programming education, paving the way for an array of exciting future research opportunities. One immediate avenue for future research is to expand the scope of this investigation beyond the initial Polish and Lithuanian samples. Conducting comparative studies across different countries and cultures will provide a more comprehensive understanding of FGPE+'s cross-cultural efficacy and adaptability. This global approach can also reveal unique regional requirements or preferences, which can be integrated into the FGPE+ model to create a more universally effective learning platform. The current study focused on the learner's perspective. However, insights from educators who utilize FGPE+ could offer a different perspective, providing additional ways to improve and optimize the system. They could share valuable input on what works well in a classroom setting, potential areas of difficulty, or suggestions to improve learner engagement.

Additionally, the development of more sophisticated adaptive algorithms that leverage artificial intelligence (AI) and machine learning (ML) techniques could augment the FGPE+'s capabilities. These techniques could further personalize the learning experience, making it more responsive to individual learner's needs. For instance, the system could predict what a student might struggle with based on historical data and preemptively provide resources to mitigate these challenges [87]. Incorporating Virtual Reality (VR) or Augmented Reality (AR) could also enhance the FGPE+ model's immersive learning experiences. VR/AR could be used to simulate real-world programming scenarios, making abstract programming concepts more tangible and engaging for learners.

The FGPE+ system, although robust, currently supports only a limited set of programming languages. Extending its support to encompass a broader range of languages, including emerging ones, will make it more versatile and valuable to a wider audience. Finally, we observe potential in exploring the impact of different gamification elements on learning outcomes. While FGPE+ currently uses a set of gamification techniques, understanding which elements are most effective can help refine the system to maximize learner engagement and achievement.

In conclusion, the FGPE+ system is a promising approach to modernize programming education. By employing principles of gamification and Pareto optimization in a PWA platform, it provides an engaging, adaptive, and personalized learning experience. We believe that the positive results obtained highlight the system's potential for broader application. We anticipate that FGPE+ can be adapted to diverse educational contexts, playing a significant role in programming education across various age groups, skill levels, and cultural settings. However, more extensive and diverse studies are needed to validate these findings and refine the model to cater to a broader range of learners. By continuing to innovate and push boundaries, we believe that FGPE+ has the potential to revolutionize programming education. Through future research, the model could be further refined to make learning programming more accessible, engaging, and effective for all.

Author Contributions: Conceptualization, R.M. and J.S.; Data curation, R.D., R.Q., and J.C.P.; Formal analysis, R.M., R.D., J.S., R.Q., and J.C.P.; Funding acquisition, R.M.; Investigation, R.M., T.B., R.Q., and J.C.P.; Methodology, R.M. and J.S.; Project administration, R.M. and J.S.; Resources, T.B.; Software, T.B.; Supervision, R.M.; Validation, R.D.; Visualization, R.M. and R.D.; Writing—original draft, R.M.; Writing—review and editing, R.M., R.D., J.S., and J.C.P. All authors have read and agreed to the published version of the manuscript.

Funding: This research was funded by the Erasmus programme under the grant number 2020-1-PL01-KA226-HE-095786.

Institutional Review Board Statement: The study was conducted in accordance with the Declaration of Helsinki, and approved by the PT Institutional Review Board of Department of Information Technology in Management (protocol code 1547h 2021-11).

Informed Consent Statement: Informed consent was obtained from all subjects involved in the study.

Data Availability Statement: All data has been presented in main text.

Acknowledgments: We want to express our deep gratitude to all the great people involved in the FGPE+ project.

Conflicts of Interest: The authors declare no conflict of interest.

References

1. Mishra, L.; Gupta, T.; Shree, A. Online teaching-learning in higher education during lockdown period of COVID-19 pandemic. *Int. J. Educ. Res. Open* **2020**, *1*, 100012. [CrossRef]
2. Breiki, M.A.; Yahaya, W.A.J.W. Using Gamification to Promote Students' Engagement While Teaching Online During COVID-19. In *Teaching in the Post COVID-19 Era*; Springer International Publishing: Cham, Switzerland, 2021; pp. 443–453. [CrossRef]
3. Pedro, L.; Santos, C. Has Covid-19 emergency instruction killed the PLE? In Proceedings of the Ninth International Conference on Technological Ecosystems for Enhancing Multiculturality (TEEM'21), Barcelona, Spain, 26–29 October 2021; ACM: New York, NY, USA, 2021. [CrossRef]
4. Redondo, R.P.D.; Rodríguez, M.C.; Escobar, J.J.L.; Vilas, A.F. Integrating micro-learning content in traditional e-learning platforms. *Multimed. Tools Appl.* **2020**, *80*, 3121–3151. [CrossRef]
5. Jayalath, J.; Esichaikul, V. Gamification to Enhance Motivation and Engagement in Blended eLearning for Technical and Vocational Education and Training. *Technol. Knowl. Learn.* **2020**, *27*, 91–118. [CrossRef]
6. da Silva, J.P.; Silveira, I.F. A systematic review on open educational games for programming learning and teaching. *Int. J. Emerg. Technol. Learn.* **2020**, *15*, 156–172. [CrossRef]
7. Paiva, J.C.; Leal, J.P.; Queirós, R. Fostering programming practice through games. *Information* **2020**, *11*, 498. [CrossRef]
8. Maryono, D.; Budiyono, S.; Akhyar, M. Implementation of Gamification in Programming Learning: Literature Review. *Int. J. Inf. Educ. Technol.* **2022**, *12*, 1448–1457. [CrossRef]
9. Maskeliūnas, R.; Kulikajevas, A.; Blažauskas, T.; Damaševičius, R.; Swacha, J. An interactive serious mobile game for supporting the learning of programming in javascript in the context of eco-friendly city management. *Computers* **2020**, *9*, 102. [CrossRef]
10. Cuervo-Cely, K.D.; Restrepo-Calle, F.; Ramírez-Echeverry, J.J. Effect Of Gamification On The Motivation Of Computer Programming Students. *J. Inf. Technol. Educ. Res.* **2022**, *21*, 001–023. [CrossRef]
11. Chinchua, S.; Kantathanawat, T.; Tuntiwongwanich, S. Increasing Programming Self-Efficacy (PSE) Through a Problem-Based Gamification Digital Learning Ecosystem (DLE) Model. *J. High. Educ. Theory Pract.* **2022**, *22*, 131–143.
12. Aseriškis, D.; Damaševičius, R. Gamification Patterns for Gamification Applications. *Procedia Comput. Sci.* **2014**, *39*, 83–90. [CrossRef]
13. Panskyi, T.; Rowińska, Z. A Holistic Digital Game-Based Learning Approach to Out-of-School Primary Programming Education. *Inform. Educ.* **2021**, *20*, 1–22. [CrossRef]
14. Swacha, J.; Maskeliūnas, R.; Damaševičius, R.; Kulikajevas, A.; Blažauskas, T.; Muszyńska, K.; Miluniec, A.; Kowalska, M. Introducing sustainable development topics into computer science education: Design and evaluation of the eco jsity game. *Sustainability* **2021**, *13*, 4244. [CrossRef]
15. Damaševičius, R.; Maskeliūnas, R.; Blažauskas, T. Serious Games and Gamification in Healthcare: A Meta-Review. *Information* **2023**, *14*, 105.
16. Francillette, Y.; Boucher, E.; Bouchard, B.; Bouchard, K.; Gaboury, S. Serious games for people with mental disorders: State of the art of practices to maintain engagement and accessibility. *Entertain. Comput.* **2021**, *37*, 100396. [CrossRef]
17. Zhao, D.; Muntean, C.H.; Chis, A.E.; Rozinaj, G.; Muntean, G. Game-Based Learning: Enhancing Student Experience, Knowledge Gain, and Usability in Higher Education Programming Courses. *IEEE Trans. Educ.* **2022**, *65*, 502–513. [CrossRef]
18. Mohanarajah, S.; Sritharan, T. Shoot2learn: Fix-And-Play Educational Game For Learning Programming; Enhancing Student Engagement By Mixing Game Playing And Game Programming. *J. Inf. Technol. Educ. Res.* **2022**, *21*, 639–661. [CrossRef] [PubMed]
19. Xinogalos, S.; Satratzemi, M. The Use of Educational Games in Programming Assignments: SQL Island as a Case Study. *Appl. Sci.* **2022**, *12*, 6563. [CrossRef]
20. Barmpakas, A.; Xinogalos, S. Designing and Evaluating a Serious Game for Learning Artificial Intelligence Algorithms: SpAI War as a Case Study. *Appl. Sci.* **2023**, *13*, 5828. [CrossRef]
21. Costa, J.M. Using game concepts to improve programming learning: A multi-level meta-analysis. *Comput. Appl. Eng. Educ.* **2023**, *31*, 1098–1110. [CrossRef]
22. Soboleva, E.V.; Suvorova, T.N.; Grinshkun, A.V.; Bocharov, M.I. Applying Gamification in Learning the Basics of Algorithmization and Programming to Improve the Quality of Students' Educational Results. *Eur. J. Contemp. Educ.* **2021**, *10*, 987–1002.
23. Toda, A.M.; Valle, P.H.D.; Isotani, S. The Dark Side of Gamification: An Overview of Negative Effects of Gamification in Education. In *Communications in Computer and Information Science*; Springer International Publishing: Cham, Switzerland, 2018; pp. 143–156. [CrossRef]
24. Imran, H. An Empirical Investigation of the Different Levels of Gamification in an Introductory Programming Course. *J. Educ. Comput. Res.* **2022**, *61*, 847–874. [CrossRef]

25. Chatterjee, S.; Majumdar, D.; Misra, S.; Damaševičius, R. Adoption of mobile applications for teaching-learning process in rural girls' schools in India: An empirical study. *Educ. Inf. Technol.* **2020**, *25*, 4057–4076. [CrossRef]
26. Tuparov, G.; Keremedchiev, D.; Tuparova, D.; Stoyanova, M. Gamification and educational computer games in open source learning management systems as a part of assessment. In Proceedings of the 2018 17th International Conference on Information Technology Based Higher Education and Training (ITHET), Olhao, Portugal, 26–28 April 2018; pp. 1–5. [CrossRef]
27. Pérez-Berenguer, D.; García-Molina, J. A standard-based architecture to support learning interoperability: A practical experience in gamification. *Software Pract. Exp.* **2018**, *48*, 1238–1268. [CrossRef]
28. Calle-Archila, C.R.; Drews, O.M. Student-Based Gamification Framework for Online Courses. In *Communications in Computer and Information Science*; Springer International Publishing: Cham, Switzerland, 2017; pp. 401–414. [CrossRef]
29. Sheppard, D. Introduction to Progressive Web Apps. In *Beginning Progressive Web App Development*; Apress: New York, NY, USA, 2017; pp. 3–10. [CrossRef]
30. Hajian, M. PWA with Angular and Workbox. In *Progressive Web Apps with Angular*; Apress: New York, NY, USA, 2019; pp. 331–345. [CrossRef]
31. Devine, J.; Finney, J.; de Halleux, P.; Moskal, M.; Ball, T.; Hodges, S. MakeCode and CODAL: Intuitive and efficient embedded systems programming for education. *J. Syst. Archit.* **2019**, *98*, 468–483. [CrossRef]
32. Lee, J.; Kim, H.; Park, J.; Shin, I.; Son, S. Pride and Prejudice in Progressive Web Apps. In Proceedings of the 2018 ACM SIGSAC Conference on Computer and Communications Security, Toronto, ON, Canada, 15–19 October 2018; ACM: New York, NY, USA, 2018. [CrossRef]
33. FGPE PLE Environment. Available online: https://github.com/FGPE-Erasmus/fgpe-ple-v2 (accessed on 10 June 2023).
34. Sutadji, E.; Hidayat, W.N.; Patmanthara, S.; Sulton, S.; Jabari, N.A.M.; Irsyad, M. Measuring user experience on SIPEJAR as e-learning of Universitas Negeri Malang. *IOP Conf. Ser. Mater. Sci. Eng.* **2020**, *732*, 012116. [CrossRef]
35. Nah, F.F.H.; Zeng, Q.; Telaprolu, V.R.; Ayyappa, A.P.; Eschenbrenner, B. Gamification of Education: A Review of Literature. In *Lecture Notes in Computer Science*; Springer International Publishing: Cham, Switzerland, 2014; pp. 401–409. [CrossRef]
36. Barik, T.; Murphy-Hill, E.; Zimmermann, T. A perspective on blending programming environments and games: Beyond points, badges, and leaderboards. In Proceedings of the 2016 IEEE Symposium on Visual Languages and Human-Centric Computing (VL/HCC), Cambridge, UK, 4–8 September 2016; pp. 134–142. [CrossRef]
37. Prokhorov, A.V.; Lisovichenko, V.O.; Mazorchuk, M.S.; Kuzminska, O.H. Developing a 3D quest game for career guidance to estimate students' digital competences. *CEUR Workshop Proc.* **2020**, *2731*, 312–327. . [CrossRef]
38. Padilla-Zea, N.; Gutiérrez, F.L.; López-Arcos, J.R.; Abad-Arranz, A.; Paderewski, P. Modeling storytelling to be used in educational video games. *Comput. Hum. Behav.* **2014**, *31*, 461–474. [CrossRef]
39. Hadzigeorgiou, Y. Narrative Thinking and Storytelling in Science Education. In *Imaginative Science Education*; Springer International Publishing: Cham, Switzerland, 2016; pp. 83–119. [CrossRef]
40. Kusuma, G.P.; Wigati, E.K.; Utomo, Y.; Suryapranata, L.K.P. Analysis of Gamification Models in Education Using MDA Framework. *Procedia Comput. Sci.* **2018**, *135*, 385–392. [CrossRef]
41. Wu, M.; Liao, C.C.; Chen, Z.H.; Chan, T.W. Designing a Competitive Game for Promoting Students' Effort-Making Behavior by Virtual Pets. In Proceedings of the 2010 Third IEEE International Conference on Digital Game and Intelligent Toy Enhanced Learning, Kaohsiung, Taiwan, 12–16 April 2010; pp. 234–236. [CrossRef]
42. Chen, Z.H.; Liao, C.; Chien, T.C.; Chan, T.W. Animal companions: Fostering children's effort-making by nurturing virtual pets. *Br. J. Educ. Technol.* **2009**, *42*, 166–180. [CrossRef]
43. Slavin, R.E. Cooperative Learning: Applying Contact Theory in Desegregated Schools. *J. Soc. Issues* **1985**, *41*, 45–62. [CrossRef]
44. Zakaria, E.; Iksan, Z. Promoting Cooperative Learning in Science and Mathematics Education: A Malaysian Perspective. *EURASIA J. Math. Sci. Technol. Educ.* **2007**, *3*, 35–39. [CrossRef]
45. Correia, A.; Fonseca, B.; Paredes, H.; Martins, P.; Morgado, L. Computer-Simulated 3D Virtual Environments in Collaborative Learning and Training: Meta-Review, Refinement, and Roadmap. In *Progress in IS*; Springer International Publishing: Cham, Switzerland, 2016; pp. 403–440. [CrossRef]
46. Doumanis, I.; Economou, D.; Sim, G.R.; Porter, S. The impact of multimodal collaborative virtual environments on learning: A gamified online debate. *Comput. Educ.* **2019**, *130*, 121–138. [CrossRef]
47. Wanick, V.; Bui, H. Gamification in Management: A systematic review and research directions. *Int. J. Serious Games* **2019**, *6*, 57–74. [CrossRef]
48. Hooda, M.; Rana, C.; Dahiya, O.; Rizwan, A.; Hossain, M.S. Artificial Intelligence for Assessment and Feedback to Enhance Student Success in Higher Education. *Math. Probl. Eng.* **2022**, *2022*, 1–19. [CrossRef]
49. Maher, Y.; Moussa, S.M.; Khalifa, M.E. Learners on Focus: Visualizing Analytics Through an Integrated Model for Learning Analytics in Adaptive Gamified E-Learning. *IEEE Access* **2020**, *8*, 197597–197616. [CrossRef]
50. Dichev, C.; Dicheva, D. Gamifying education: What is known, what is believed and what remains uncertain: A critical review. *Int. J. Educ. Technol. High. Educ.* **2017**, *14*, 9. [CrossRef]
51. Skaalvik, E.M.; Skaalvik, S. Collective teacher culture and school goal structure: Associations with teacher self-efficacy and engagement. *Soc. Psychol. Educ.* **2023**, *26*, 945–969. [CrossRef]

52. Vasconcellos, D.; Parker, P.D.; Hilland, T.; Cinelli, R.; Owen, K.B.; Kapsal, N.; Lee, J.; Antczak, D.; Ntoumanis, N.; Ryan, R.M.; et al. Self-determination theory applied to physical education: A systematic review and meta-analysis. *J. Educ. Psychol.* **2020**, *112*, 1444–1469. [CrossRef]
53. dos Santos, W.O.; Bittencourt, I.I.; Isotani, S.; Dermeval, D.; Marques, L.B.; Silveira, I.F. Flow Theory to Promote Learning in Educational Systems: Is it Really Relevant? *Rev. Bras. Inform. Educ.* **2018**, *26*, 29. [CrossRef]
54. Schunk, D.H.; DiBenedetto, M.K. Motivation and social cognitive theory. *Contemp. Educ. Psychol.* **2020**, *60*, 101832. [CrossRef]
55. Torre, D.; Durning, S.J. Social cognitive theory: Thinking and learning in social settings. *Res. Med. Educ.* **2022**, 105–116. [CrossRef]
56. Gao, T.; Kuang, L. Cognitive Loading and Knowledge Hiding in Art Design Education: Cognitive Engagement as Mediator and Supervisor Support as Moderator. *Front. Psychol.* **2022**, *13*, 837374. [CrossRef] [PubMed]
57. Fleih, N.H.; Rushd, I. The theory of cognitive burden, its concept, importance, types, principles, strategies, in the educational learning process. *Ann. Fac. Arts* **2020**, *48*, 53–69. [CrossRef]
58. Armacost, R.; Pet-Armacost, J. Using mastery-based grading to facilitate learning. In Proceedings of the 33rd Annual Frontiers in Education, Westminster, CO, USA, 5–8 November 2003; FIE, 2003; Volume 1, pp. T3A–20. [CrossRef]
59. Bennani, S.; Maalel, A.; Ghezala, H.B. Adaptive gamification in E-learning: A literature review and future challenges. *Comput. Appl. Eng. Educ.* **2021**, *30*, 628–642. [CrossRef]
60. López, C.; Tucker, C. Toward Personalized Adaptive Gamification: A Machine Learning Model for Predicting Performance. *IEEE Trans. Games* **2020**, *12*, 155–168. [CrossRef]
61. Manzano-León, A.; Camacho-Lazarraga, P.; Guerrero, M.A.; Guerrero-Puerta, L.; Aguilar-Parra, J.M.; Trigueros, R.; Alias, A. Between Level Up and Game Over: A Systematic Literature Review of Gamification in Education. *Sustainability* **2021**, *13*, 2247. [CrossRef]
62. Rodrigues, L.; Toda, A.M.; Oliveira, W.; Palomino, P.T.; Avila-Santos, A.P.; Isotani, S. Gamification Works, but How and to Whom? In Proceedings of the 52nd ACM Technical Symposium on Computer Science Education, Virtual Event, 13–20 March 2021; ACM: New York, NY, USA, 2021. [CrossRef]
63. Hammerschall, U. A Gamification Framework for Long-Term Engagement in Education Based on Self Determination Theory and the Transtheoretical Model of Change. In Proceedings of the 2019 IEEE Global Engineering Education Conference (EDUCON), Dubai, United Arab Emirates, 8–11 April 2019; pp. 95–101. [CrossRef]
64. Huang, B.; Hew, K.F. Implementing a theory-driven gamification model in higher education flipped courses: Effects on out-of-class activity completion and quality of artifacts. *Comput. Educ.* **2018**, *125*, 254–272. [CrossRef]
65. Duggal, K.; Gupta, L.R.; Singh, P. Gamification and Machine Learning Inspired Approach for Classroom Engagement and Learning. *Math. Probl. Eng.* **2021**, *2021*, 1–18. [CrossRef]
66. Sánchez, D.O.; Trigueros, I.M.G. Gamification, social problems, and gender in the teaching of social sciences: Representations and discourse of trainee teachers. *PLoS ONE* **2019**, *14*, e0218869. [CrossRef]
67. Landers, R.N.; Armstrong, M.B.; Collmus, A.B. How to Use Game Elements to Enhance Learning. In *Serious Games and Edutainment Applications*; Springer International Publishing: Cham, Switzerland, 2017; pp. 457–483. [CrossRef]
68. Kalogiannakis, M.; Papadakis, S.; Zourmpakis, A.I. Gamification in Science Education. A Systematic Review of the Literature. *Educ. Sci.* **2021**, *11*, 22. [CrossRef]
69. Chou, Y.K. The Octalysis Framework for Gamification & Behavioral Design. 2013. Available online: https://yukaichou.com/gamification-examples/octalysis-complete-gamification-framework/ (accessed on 2 June 2023).
70. Werbach, K.; Hunter, D. *For the Win: How Game Thinking Can Revolutionize Your Business*; Wharton Digital Press: Philadelphia, PA, USA, 2012.
71. Kim, A.J. *Game Thinking: Innovate Smarter & Drive Deep Engagement with Design Techniques from Hit Games*; gamethinking.io: 2018.
72. Hunicke, R.; Leblanc, M.; Zubek, R. MDA: A Formal Approach to Game Design and Game Research. In Proceedings of the AAAI Workshop on Challenges in Game AI, San Jose, CA, USA, 25-29 July 2004; 2004; Volume 1.
73. Zichermann, G.; Cunningham, C. *Gamification by Design: Implementing Game Mechanics in Web and Mobile Apps*; O'Reilly Media, Inc.: Sebastopol, CA, USA, 2011.
74. Marczewski, A. The Intrinsic Motivation RAMP. 2014. Available online: https://www.gamified.uk/gamification-framework/the-intrinsic-motivation-ramp/ (accessed on 2 June 2023).
75. Keller, J.M. *Motivational Design for Learning and Performance: The ARCS Model Approach*; Springer: Berlin/Heidelberg, Germany, 2009. [CrossRef]
76. Hornbæk, K.; Law, E.L.C. Meta-Analysis of Correlations among Usability Measures. In Proceedings of the CHI '07 SIGCHI Conference on Human Factors in Computing Systems, San Jose, CA, USA, 28 April 2007–3 May 2007; pp. 617–626. [CrossRef]
77. O'Brien, H.L.; Lebow, M. Mixed-methods approach to measuring user experience in online news interactions. *J. Am. Soc. Inf. Sci. Technol.* **2013**, *64*, 1543–1556. [CrossRef]
78. Laugwitz, B.; Held, T.; Schrepp, M. Construction and Evaluation of a User Experience Questionnaire. In *Proceedings of the HCI and Usability for Education and Work*; Holzinger, A., Ed.; Springer: Berlin/Heidelberg, Germany, 2008; pp. 63–76.
79. Saleh, A.M.; Abuaddous, H.Y.; Alansari, I.S.; Enaizan, O. The Evaluation of User Experience on Learning Management Systems Using UEQ. *Int. J. Emerg. Technol. Learn. (iJET)* **2022**, *17*, 145–162. [CrossRef]
80. Swacha, J.; Queirós, R.; Paiva, J.C. GATUGU: Six Perspectives of Evaluation of Gamified Systems. *Information* **2023**, *14*, 136. [CrossRef]

81. Najafi, H.a. Shareable Content Object Reference Model: A model for the production of electronic content for better learning. *Bimon. Educ. Strateg. Med. Sci.* **2016**, *9*, 335–350.
82. Ng, J.Y.M.; Lim, T.W.; Tarib, N.; Ho, T.K. Development and validation of a progressive web application to educate partial denture wearers. *Health Inform. J.* **2022**, *28*, 146045822110695. [CrossRef] [PubMed]
83. Gómez-Sierra, C.J. Design and development of a PWA-Progressive Web Application, to consult the diary and programming of a technological event. *IOP Conf. Ser. Mater. Sci. Eng.* **2021**, *1154*, 012047. [CrossRef]
84. Case, D.M.; Steeve, C.; Woolery, M. Progressive Web Apps are a Game-Changer! Use Active Learning to Engage Students and Convert Any Website into a Mobile-Installable, Offline-Capable, Interactive App. In Proceedings of the 51st ACM Technical Symposium on Computer Science Education, Portland, OR, USA, 11–14 March 2020; ACM: New York, NY, USA, 2020. [CrossRef]
85. Sidekerskienė, T.; Damaševičius, R. Out-of-the-Box Learning: Digital Escape Rooms as a Metaphor for Breaking Down Barriers in STEM Education. *Sustainability* **2023**, *15*, 7393. [CrossRef]
86. Bonora, L.; Martelli, F.; Marchi, V.; Vagnoli, C. Gamification as educational strategy for STEM learning: DIGITgame project a collaborative experience between Italy and Turkey high schools around the Smartcity concept. In Proceedings of the IMSCI 2019-13th International Multi-Conference on Society, Cybernetics and Informatics, Proceedings, Orlando, FL, USA, 6–9 July 2019; Volume 2, pp. 122–127.
87. Paulauskas, L.; Paulauskas, A.; Blažauskas, T.; Damaševičius, R.; Maskeliūnas, R. Reconstruction of Industrial and Historical Heritage for Cultural Enrichment Using Virtual and Augmented Reality. *Technologies* **2023**, *11*, 36. [CrossRef]

Disclaimer/Publisher's Note: The statements, opinions and data contained in all publications are solely those of the individual author(s) and contributor(s) and not of MDPI and/or the editor(s). MDPI and/or the editor(s) disclaim responsibility for any injury to people or property resulting from any ideas, methods, instructions or products referred to in the content.

Review

Usage of Gamification Techniques in Software Engineering Education and Training: A Systematic Review

Vincenzo Di Nardo [†], Riccardo Fino [†], Marco Fiore [†], Giovanni Mignogna [†], Marina Mongiello [*,†] and Gaetano Simeone [†]

Department of Electrical and Information Engineering (DEI), Polytechnic University of Bari, 70125 Bari, Italy; v.dinardo@studenti.poliba.it (V.D.N.); r.fino@studenti.poliba.it (R.F.); marco.fiore@poliba.it (M.F.); g.mignogna@studenti.poliba.it (G.M.); g.simeone3@studenti.poliba.it (G.S.)
* Correspondence: marina.mongiello@poliba.it
[†] These authors contributed equally to this work.

Citation: Di Nardo, V.; Fino, R.; Fiore, M.; Mignogna, G.; Mongiello, M.; Simeone, G. Usage of Gamification Techniques in Software Engineering Education and Training: A Systematic Review. *Computers* **2024**, *13*, 196. https://doi.org/10.3390/computers13080196

Academic Editors: Carlos Vaz de Carvalho, Hariklia Tsalapatas and Ricardo Baptista

Received: 4 July 2024
Revised: 6 August 2024
Accepted: 8 August 2024
Published: 14 August 2024

Correction Statement: This article has been republished with a minor change. The change does not affect the scientific content of the article and further details are available within the backmatter of the website version of this article.

Copyright: © 2024 by the authors. Licensee MDPI, Basel, Switzerland. This article is an open access article distributed under the terms and conditions of the Creative Commons Attribution (CC BY) license (https://creativecommons.org/licenses/by/4.0/).

Abstract: Gamification, the integration of game design elements into non-game contexts, has gained prominence in the software engineering education and training realm. By incorporating elements such as points, badges, quests, and challenges, gamification aims to motivate and engage learners, potentially transforming traditional educational methods. This paper addresses the gap in systematic evaluations of gamification's effectiveness in software engineering education and training by conducting a comprehensive literature review of 68 primary studies. This review explores the advantages of gamification, including active learning, individualized pacing, and enhanced collaboration, as well as the psychological drawbacks such as increased stress and responsibility for students. Despite the promising results, this study highlights that gamification should be considered a supplementary tool rather than a replacement for traditional teaching methods. Our findings reveal significant interest in integrating gamification in educational settings, driven by the growing need for digital content to improve learning.

Keywords: gamification; software engineering; education; learning; literature review

1. Introduction

Gamification, the integration of game design elements into non-game contexts, has emerged as a significant approach in various domains, including education and training. It employs elements such as points, badges, quests, and challenges to motivate and engage users, thereby enhancing their experience and performance. Its application ranges from universities to industries.

In the realm of software engineering education and training (SEET), gamification has the potential to transform traditional learning methods by making them more interactive and engaging [1]. SEET encompasses the instructional methodologies, curricular designs, and practical experiences aimed at equipping students and professionals with the knowledge, skills, and competencies required in the field of software engineering. It includes both formal academic programs, such as university degrees and professional certifications, and informal learning opportunities, such as workshops, online courses, and bootcamps. The objective of SEET is to prepare individuals to effectively design, develop, test, and maintain software systems, ensuring they meet user needs and adhere to quality standards. SEET traditionally focuses on imparting technical skills and knowledge necessary for developing software systems. However, the conventional teaching methods often struggle to maintain student engagement and motivation [2]. By incorporating gamification, educators can create a more dynamic and stimulating learning environment, which can lead to improved learning outcomes. Gamification not only makes learning more enjoyable but also fosters a deeper understanding of complex concepts through active participation and immediate feedback [3].

In the context of professional training, gamification can play a crucial role in continuous learning and skill development. As the software industry evolves rapidly, professionals need to constantly update their skills and knowledge. Gamified training programs can offer a more compelling and effective way to achieve this, promoting sustained engagement and continuous professional development [4]. Moreover, gamification can increase the awareness of employees in critical scenarios [5] by letting them take action in less time to avoid dangerous situations [6].

Despite the promising potential of gamification, there is a need for a systematic evaluation of its application in the SEET topic. This involves examining both the result improvements and the technical challenges associated with implementing gamified systems. To address this gap, we conduct a systematic literature review to evaluate the maturity and impact of gamification in SEET. We propose six research questions aimed at exploring the effectiveness of gamification in enhancing learning and training outcomes in software engineering. This review is based on an extensive analysis of 68 primary studies, identified and filtered through the Preferred Reporting Items for Systematic Reviews and Meta-Analyses (PRISMA) methodology [7]. We assess the evolution of this topic over time, the application areas, and the results obtained after the implementation of gamification techniques in university courses. Additionally, we discuss future research directions and practical implications for educators and trainers in the software engineering domain.

This paper is structured as follows: Section 2 highlights the main points of the chosen methodology, Section 3 analyzes the background on gamification and SEET topics, Section 4 summarizes results, and Section 5 shows the current literature and the differences between other reviews and the one proposed in this work. Section 6 concludes the paper.

2. Research Methodology

Our research adheres to the guidelines for a Systematic Literature Review (SLR), as described in [8]. This study aims to explore the application of gamification in SEET, covering the literature published from 2015 to 2023, using the PRISMA methodology [7] and useful tools for paper gathering, such as Zotero https://www.zotero.org/ (accessed on 1 June 2024), and for tagging and data extraction, such as Python and Microsoft Excel.

2.1. PRISMA Methodology

This section outlines the systematic review methodology employed in this study, following the PRISMA guidelines. This approach ensures a rigorous and transparent review process, allowing for comprehensive identification, selection, and analysis of relevant studies.

The review is guided by some research questions (RQs) formulated to focus the scope of this study. These RQs are designed to capture the essence of the investigated topic and to guide the systematic review process.

The Population, Intervention, Comparison, Outcome (PICO) framework is used to refine the RQs and set clear criteria for study selection:

- Population: the group or individuals targeted by the intervention.
- Intervention: the specific intervention or exposure being investigated.
- Comparison: the control or comparison group, if applicable.
- Outcome: the outcomes or effects measured in this study.

A search strategy is developed to identify all relevant literature. The search string is constructed using keywords and phrases pertinent to the research question and is applied across multiple databases to ensure thorough coverage. Searches are conducted in major academic databases.

To ensure the selection of relevant and high-quality studies, specific inclusion and exclusion criteria are established. They are needed to ensure that filtered papers are relevant to the scope of the study.

The extracted data, based on defined RQs, are then synthesized to provide a comprehensive overview of the current state of research on the topic, identify trends, and highlight gaps in the literature.

The process for identifying relevant papers, based on the PRISMA guidelines, is illustrated in Figure 1, generated using the tool explained in [9].

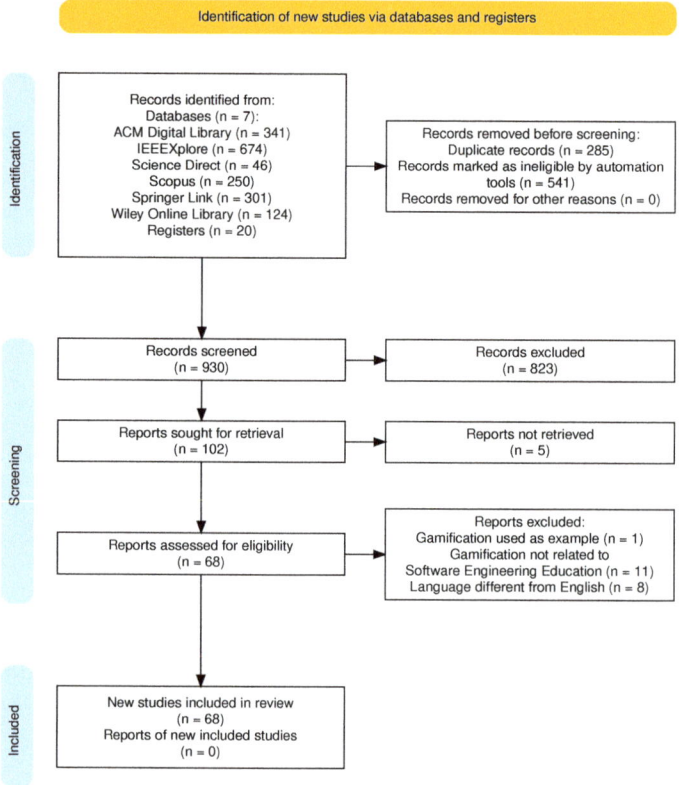

Figure 1. PRISMA search methodology.

2.2. Research Questions Definition

The objectives of this study are twofold: (a) to identify the current state of gamification in SEET, and (b) to provide a foundation for highlighting gaps and trends in this field, as well as suggesting future research directions. To achieve these goals, we formulate the following research questions (RQs):

- RQ1: What is the publication trend in the area of gamification applied to SEET?
  This question investigates the trend in publication quantity and the structure of publication venues, which are useful for understanding the progression of this topic.
- RQ2: In which areas of software engineering is gamification used?
  This question aims to identify the key areas of study and their contributions to the scientific community.
- RQ3: What are the analyzed application areas?
  This question explores the benefits of using gamification, considering its impact on learner engagement and performance.
- RQ4: What contribution does gamification offer when it is applied to SEET?
  This question examines the specific contribution and integration of gamification into educational practices.
- RQ5: On which continents is gamification mostly analyzed?
  This question aims to identify the continents that are most interested into gamification applied to SEET.

- RQ6: What are the advantages and disadvantages of gamification when applied to SEET?
  This question seeks to understand the the pros and cons to evaluate the success of gamification in educational settings.

2.3. Paper Selection

Relevant databases were systematically searched to ensure comprehensive coverage of the literature. A PICO approach was utilized to define the search keywords, grouped into two main categories:

- Population-related search terms: "Software Engineering education", "training".
- Intervention-related search terms: "gamification", "game-based learning".

The search string used was *"gamification AND (Software Engineering OR programming) AND (education OR training OR teaching OR learning)"*.

This search returned a total of 1756 results. The following databases were searched: ACM Digital Library (only Open Access content), IEEE Xplore, ScienceDirect, Scopus, SpringerLink, Wiley Online Library (only Open Access content), and Google Scholar (only Open Access content). This volume of results is considered appropriate for the scope of the review.

2.4. Inclusion and Exclusion Criteria

To enhance the reliability and relevance of the studies included in our review, specific inclusion and exclusion criteria were applied.

Inclusion criteria:

- Studies published between 2015 and 2023.
- Studies written in English.
- Studies published in peer-reviewed journals or conference proceedings.
- Studies focused on the application of gamification in software engineering education or training.
- Studies that present empirical evidence or substantial theoretical contributions.

Exclusion criteria:

- Studies for which the full text is not available (e.g., article not available online or DOI not found or not readable without subscriptions): 364 studies.
- Secondary or tertiary studies (e.g., reviews or surveys): 175 studies.
- Studies where gamification is not the main focus but is only mentioned: 284 studies.

By adhering to these criteria, we ensured that the selected studies were pertinent and of high quality, thereby providing a solid foundation for our systematic review.

A pool of 68 studies is included in the analysis after the application of the screening procedure. A replication package is available at https://github.com/Mackerkun/Usage-of-Gamification-techniques-in-Software-Engineering-Education-and-Training-A-Systematic-Review (accessed on 3 July 2024).

3. Background

This section introduces gamification by giving a common definition gathered from the analyzed papers. Moreover, it explores the main components of gamification and its role in SEET.

A brief definition of gamification can be given, based on different analyzed papers [10–14]: gamification is an approach characterized by the application of game design elements and principles in non-game contexts to enhance user engagement and motivation. Unlike traditional educational methods, which often rely on passive learning, gamification leverages the interactive and stimulating nature of games to create more dynamic learning experiences [15].

In the domain of SEET, gamification has shown significant potential. Traditional software engineering education typically involves theoretical learning and practical exercises designed to build technical skills. However, maintaining student motivation and engagement can be challenging with conventional teaching methods. By incorporating gamification, educators aim to make learning more interactive and enjoyable, thereby increasing student participation [1] and improving learning outcomes [16].

The core components of gamification include the integration of game mechanics, dynamics, and aesthetics [17].

- Game mechanics refer to the rules and feedback systems that drive gameplay, such as scoring, levels, and rewards.
- Game dynamics involve the emotions and behaviors induced by the mechanics, such as competition, collaboration, and achievement.
- Aesthetics pertain to the overall look and feel of the gamified experience, which can enhance its appeal and immersion.

Application of Gamification in Software Engineering Education

Figure 2 illustrates a typical architecture for gamified educational platforms. This architecture includes components such as the learning management system, game engine, and user interface [18]. The learning management system handles educational content and tracks student progress, while the game engine manages game mechanics and dynamics. The user interface presents the gamified experience to students, providing them with interactive and engaging learning activities, typically accessible as web apps [19], created using some frameworks (e.g., Angular, ReactJS).

Figure 2. A common architecture for gamified educational platforms.

The unique characteristics of gamification make it well-suited for SEET. Gamification can transform traditional learning environments by promoting active participation, immediate feedback, and a sense of progression. These features help address common challenges in education, such as student disengagement and the difficulty of maintaining sustained interest over time [20].

Gamification fosters an interactive learning environment where students can engage in problem-solving activities that mirror real-world software engineering tasks. This practical application of knowledge helps reinforce learning and develops critical thinking skills. For example, gamified platforms might simulate coding challenges or project management tasks, allowing students to apply theoretical concepts in a controlled, game-like setting.

Moreover, gamification can enhance collaborative learning. By incorporating team-based challenges and competitive elements, students are encouraged to work together, share knowledge, and develop essential soft skills such as communication, teamwork, and leadership [21]. This collaborative aspect is crucial in software engineering, where teamwork and communication are key to successful project completion.

Despite its benefits, the application of gamification in SEET also presents several challenges. Designing effective gamified systems requires a deep understanding of both game design and educational pedagogy. There is a need to balance game elements with educational content to ensure that learning objectives are met without compromising the fun and engagement aspects [22].

AC-contract is a systems design approach that uses cognitive psychology concepts, such as schemas, to ensure that systems remain adaptable and reliable during changes.

It inserts logical propositions into the source code, verified by a preprocessor, ensuring that the adaptable code meets the requirements even during changes. Applied to gamified educational platforms in software engineering (SEET), the use of methodologies such as AC-contract can ensure that these platforms are effective and reliable by adapting to various educational contexts and user interactions. This is crucial in dynamic environments where educational needs and interactions can change rapidly. In the field of gamification, the adoption of AC-contract principles allows gamified platforms to maintain effectiveness and reliability, addressing challenges such as active learning, personalization of study rhythms, and collaboration while reducing stress and responsibilities for students. This underscores the importance of adaptable and reliable systems and suggests that gamification should complement, not replace, traditional methods [23].

Moreover, tools like PrOnto, an ontology-driven business process mining tool, demonstrate the importance of identifying and modeling processes within organizations to improve their effectiveness and competitiveness. PrOnto's approach of utilizing business ontologies to classify and abstract business processes can be analogously applied to educational settings. In gamified educational environments, understanding and modeling the learning processes can enhance the personalization and effectiveness of gamified interventions. By leveraging ontologies to dynamically exploit knowledge at runtime, similar to PrOnto, educational systems can better adapt to the needs and contexts of learners. This dynamic adaptation is crucial in gamified platforms where the engagement and motivation of learners are influenced by how well the system can personalize the experience based on real-time data analysis and contextual understanding [24].

To address the proposed challenges and explore the potential of gamification in SEET, a systematic evaluation of existing studies is necessary. This involves assessing the design, implementation, and outcomes of gamified educational tools and identifying best practices and areas for improvement.

4. Results and Discussion

The following section analyzes results of the performed review. Each subsection answers one of the six proposed research questions.

4.1. RQ1: What Is the Publication Trend in the Area of Gamification Applied to SEET?

To analyze the temporal evolution of research on gamification in SEET, we examined the publication trends over the past years. Figure 3 illustrates the number of publications per year from 2015 to 2023.

The analysis reveals several notable trends. Starting in 2015, there were a modest number of publications, increasing over the next few years. A significant rise is observed in 2018, with the number of publications peaking at 11. This increase indicates growing interest and recognition of the potential benefits of gamification in SEET during this period.

Interestingly, the number of publications remained stable in 2019, again with six papers, before experiencing another rise in 2020 with eight publications. The most substantial growth occurred in 2021, with a peak of 16 publications. This surge can be attributed to the COVID-19 pandemic, which necessitated remote learning solutions and the adoption of innovative teaching methodologies, including gamification, to engage students in virtual environments. The continued interest in gamification indicates that it remains a relevant and important area of research in SEET, together with the higher number of journal publications in recent years.

Figure 4 presents the distribution of publication types, differentiating between conference papers and journal articles.

Figure 3. Publication trends over time for gamification in SEET.

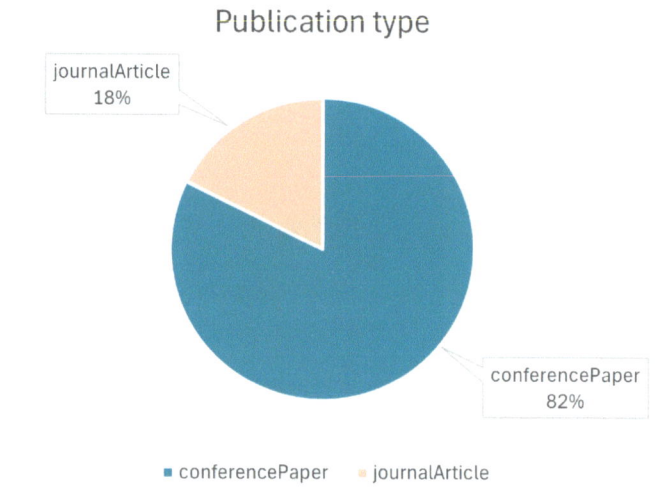

Figure 4. Distribution of publication types for gamification in SEET.

The analysis reveals that the majority of the research output on gamification in SEET has been in the form of conference papers. This is indicative of the dynamic and rapidly evolving nature of the field, where researchers prefer the relatively faster dissemination route offered by conferences to share their latest findings and innovations. Approximately 80% of the total publications were conference papers, reflecting the community's emphasis on quick dissemination and discussion of new ideas.

In contrast, journal articles, which typically undergo a more rigorous and lengthy peer-review process, constituted around 20% of the publications. This lower percentage suggests that while there is a substantial amount of exploratory and preliminary research being conducted, fewer studies have reached the level of maturity required for journal publication. The presence of journal articles, however, highlights that some research in this domain has achieved significant depth and rigor, contributing to a more formal and comprehensive understanding of gamification in SEET.

4.2. RQ2: In Which Areas of Software Engineering Is Gamification Used?

Figure 5 illustrates the sectors where gamification has been applied in SEET.

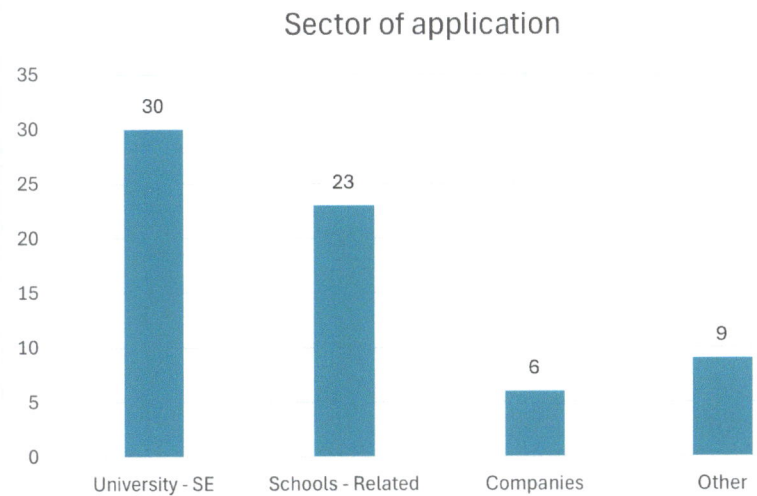

Figure 5. Sectors of application for gamification in SEET.

The analysis indicates that the predominant application area for gamification in SEET is within university-level software engineering programs, which accounts for 30 studies, about 44% of the analyzed studies. This reflects a significant focus on integrating gamified approaches to enhance the learning experiences of students in higher education, particularly in courses related to software engineering, as shown in papers [11,14,25].

Schools and related educational institutions represent the second most common sector, with 23 studies, 34% of the analyzed papers. This includes primary, secondary, and other non-university educational contexts where gamification is used to make learning more engaging and effective.

Companies are also exploring the use of gamification (9% of the papers). In the corporate sector, gamification is employed to improve employee training, professional development, and motivation, as shown in [26]. The lower number of studies in this sector could be due to the proprietary nature of corporate training programs, which might not be as widely documented in academic literature.

The "Other" category, encompassing nine studies, includes various applications that do not fit neatly into the previously mentioned sectors. This involves informal learning environments, online courses, or interdisciplinary studies where gamification is applied.

4.3. RQ3: What Are the Analyzed Application Areas?

To explore how gamification is used within SEET, we first understand the search type of selected papers, then we analyze the various application areas where gamified approaches are implemented.

To evaluate the depth of analysis on gamification in SEET, we categorized the 68 primary studies into five distinct groups based on their focus: Proposal, Analysis, Implementation/Tool, Validation, and Other. The distribution of studies across these categories provides insight into the current state and focus areas of research in this field. Results are shown in Figure 6.

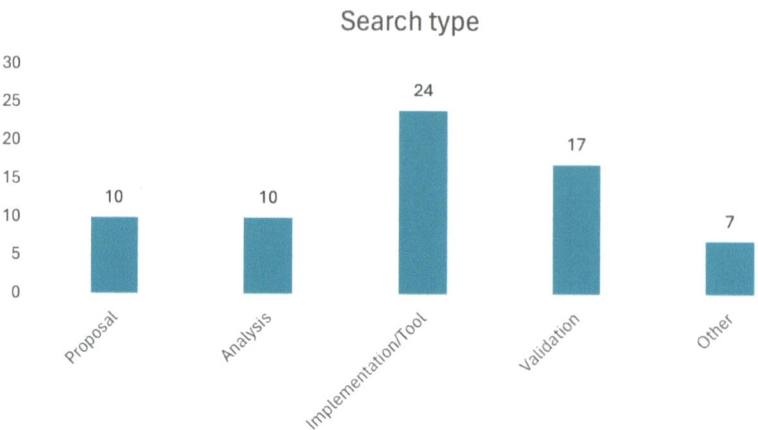

Figure 6. Distribution by search type.

- Proposal: This category includes studies that introduce new concepts, frameworks, or methodologies for applying gamification in SEET. We identified 10 papers that primarily focus on theoretical foundations and suggest innovative approaches to integrating gamification into educational contexts.
- Analysis: In this category, 10 studies provide detailed examinations of existing gamification techniques and their impacts on learning outcomes.
- Implementation/Tool: The largest category, with 24 studies, focuses on the practical aspects of implementing gamification. These papers describe the development and deployment of specific tools, platforms, or software that incorporate gamification elements into SEET. They often include case studies or reports on pilot projects.
- Validation: Comprising 17 studies, this category includes empirical research that evaluates the effectiveness of gamification through experiments, surveys, or longitudinal studies. These papers provide evidence-based insights into how gamification influences student engagement, motivation, and learning outcomes.
- Other: The remaining seven studies cover various other aspects of gamification that do not fit neatly into the above categories. This includes research on the broader impacts of gamification, such as its effects on educational policy, its role in lifelong learning, and interdisciplinary applications.

The categorization reveals a balanced approach to exploring gamification in SEET, with significant emphasis on practical implementation and empirical validation, and while a substantial number of studies propose and analyze gamification frameworks, the majority focus on real-world applications and their validation, highlighting a strong interest in understanding the practical benefits and challenges of gamification in education.

Figure 7 illustrates the distribution of the application areas. The analysis reveals that the primary use of gamification in SEET is for teaching support, which constitutes 56% of the studies. This significant proportion indicates that gamification is predominantly applied to enhance educational experiences, making learning more engaging and interactive for students. Examples of this include the incorporation of game elements into lectures, assignments, and assessments to motivate students and improve their learning outcomes. Work improvement represents 19% of the applications. This area focuses on using gamification to enhance productivity, collaboration, and efficiency in software engineering practices. For instance, gamified tools and platforms are used to improve team dynamics, project management, and individual performance within software development teams. Approach analysis accounts for 18% of the studies. This area involves evaluating and analyzing the effectiveness of different gamified approaches and methodologies in SEET. Research in this domain aims to understand the impact of gamification on learning outcomes, student engagement, and overall educational quality. Serious games constitute 7% of the applications. These are games designed for purposes beyond mere entertainment, specifically to educate and train individuals in software engineering concepts and practices. Serious games provide immersive and interactive learning environments that simulate real-world software engineering challenges. For instance, a serious game might involve a scenario where players must collaboratively debug a software application or manage a software project with constraints and deadlines [27]. These games are categorized based on their educational objectives and the inclusion of realistic software engineering tasks. An example is "SimSE" [28], a game that simulates software engineering processes, allowing students to experience the impact of their decisions on the project's outcome. Another example is "CodeSpells" [29], where players write code to cast spells, learning programming logic and problem-solving skills in a magical context. Serious games enhance understanding and retention of complex software engineering principles.

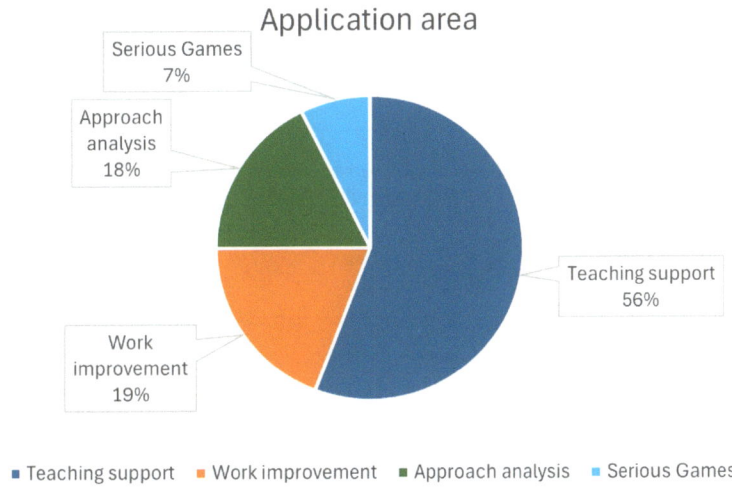

Figure 7. Application areas of gamification in SEET.

4.4. RQ4: What Contribution Does Gamification Offer When It Is Applied to SEET?

To answer this RQ, we summarize the main contribution of each paper in Table 1. This table provides an overview of the key findings from each study, highlighting the specific ways in which gamification has been applied to the education topic. By examining these contributions, we can identify common themes, benefits, and challenges associated with gamification in this context, offering insights into its effectiveness and areas for further research.

Table 1. Main finding for each analyzed paper.

Paper	Contribution
Ortega-Arranz et al. [13]	This paper discusses the use of gamification in MOOCs for a large number of students, using automatic tools to assign rewards (flipped-classroom tickets, quiz benefits, extra learning content) for a course on Spanish history from the 11th to the 16th century.

Table 1. *Cont.*

Paper	Contribution
Arif et al. [30]	This paper covers the use of gamification for web programming in high schools, specifically for learning HTML, CSS, and JavaScript. The gamification involves a web app with rewards such as avatars, lives, and time constraints.
Rahim et al. [31]	This paper focuses on using gamification to learn linear algebra, with storytelling as the main gamified element, along with avatars (king or queen), levels representing different topics, and scoring.
Hajarian and Diaz [32]	This paper describes creating an application with gamification techniques, emphasizing a reward-based system with customizable items, score saving, and leaderboards.
Iquira et al. [33]	This paper presents a mobile application using gamification to understand software engineering, particularly extreme programming (XP), with points and level progression as elements. Positive results were achieved in testing.
Robledo-Rella et al. [34]	This paper describes a mobile and web application for learning discrete math, physics, and chemistry through gamification, using quizzes, points, customizations, and avatars, with positive feedback from students.
Gomes Fernandes Matsubara and Lima Corrêa Da Silva [35]	This paper mentions using a gamified platform to learn software engineering, utilizing missions, experience points (XPs), and level progression.
Rodrigues et al. [36]	This paper surveys software engineering professors to determine if GBL and gamification improve learning outcomes, with positive results and gamified elements such as quizzes, points, levels, and badges.
Quinde et al. [37]	This paper uses gamification in a penitentiary for digital literacy, with tutorials guiding inmates through basic literacy and computing concepts.
John and Fertig [38]	This paper uses Moodle for gamification in agile and scrum model learning, with points, badges, anonymous leaderboards, quizzes, and storytelling, though the latter was less engaging over time.
Ivanova et al. [39]	This paper utilizes various existing platforms with gamification for software engineering learning, including Kahoot and "Who Wants to Be a Millionaire?" for theoretical testing, and platforms for project role division and coding.
Gasca-Hurtado et al. [40]	This paper describes creating a tool for educators to develop educational content with gamification elements, tested on a software engineering course using Happy Faces for points and Kahoot for web 2.0.
Carreño-León et al. [41]	This paper applies gamification to basic programming learning using playing cards with commands, forming groups to solve assigned algorithms, with different difficulty levels.
Sherif et al. [42]	This paper describes a platform (CoverBot) using gamification to teach code debugging, with levels, graphics, and sounds to enhance user experience.
Norsanto and Rosmansyah [43]	This paper applies gamification to civil service training with a custom application using missions, points, ranking systems, levels, and badges.

Table 1. *Cont.*

Paper	Contribution
Call et al. [44]	This paper uses gamification for understanding algorithms and data structures in C++, with Moodle incorporating points and leaderboards to motivate faster assignment completion, and a Q&A forum for extra points.
Trinidad et al. [45]	This paper analyzes a multi-context, narrative-based platform (GoRace) for educational and workplace use, with storytelling, challenges, rewards, penalties, rankings, and a shop for advantageous items.
Prasetya et al. [46]	This paper uses a tower defense game for learning formal languages, where users defend a processor from bugs while creating abstract syntax trees for assigned commands.
Bucchiarone et al. [47]	This paper discusses gamification in programming and modeling (UML diagrams) using PolyGlot and PapyGame platforms with points, XP, levels, coins, and rewards.
Lema Moreta et al. [48]	This paper applies gamification to a risk management course with a web app using points, levels, and leaderboards for competition, with positive results.
Ouhbi and Pombo [49]	This review surveys instructors, identifying gaps in SEE teaching and proposing solutions like SWEBOK guidance, Mentimeter, and Flipped Classroom.
Villagra et al. [50]	This paper provides gamification implementation examples like Flipped Classroom, recorded short lessons, and group projects.
Moser et al. [51]	This paper uses gamification for university project development, suggesting characteristics like negative points for wrong code and positive points for solving software quality issues.
Rattadilok et al. [52]	This paper presents "iGaME", a bot for teaching machine learning algorithms in classrooms using gamification.
Bucchiarone et al. [53]	This paper uses gamification in "Minecraft" to teach Scrum development methods to electrical engineering students.
Ebert et al. [54]	This paper describes applying gamification in Vector to develop software applications, enhancing user engagement and learning.
Maxim et al. [55]	This paper describes teachers using gamification principles like realistic stories for students to immerse in software creation tasks.
Jiménez-Hernández et al. [56]	This paper presents the serious game "Tree Legend" for studying trees/graphs.
Nagaria et al. [57]	This paper describes MOOC platforms like Moodle using the "CodeRunner" plugin for coding questions and "Pacman" for pathfinding algorithms.
Margalit [58]	This paper describes "Capture the Flag" for understanding AI, machine learning, and microprocessor decoding.
Stol et al. [59]	This paper discusses gamification in software engineering training to expand knowledge of new development technologies, with younger SWE more receptive than seniors. Stackoverflow's gamification with badges and reputation is also mentioned.
Fulcini and Torchiano [25]	This paper proposes using ChatGPT to find strategies for implementing gamification in software engineering Education.
Đambić et al. [60]	This paper presents an experiment in a Croatian university during COVID-19, using a mobile app for short lessons and gamified elements like leaderboards, points, and rewards.

Table 1. Cont.

Paper	Contribution
Mi et al. [61]	This paper discusses GamiCRS, a web application using PBL (Points-Badges-Levels) for coding skill improvement and student motivation, tested in a Hong Kong university with positive feedback.
Monteiro et al. [14]	This paper presents MEEGA+, a framework for evaluating educational games in software engineering using the GQIM approach, evaluated by three researchers in five phases.
Takbiri et al. [11]	This paper discusses gamification's impact on students and teachers in software engineering, education, and psychology, highlighting improvements in individual skills and teamwork.
Molins-Ruano et al. [12]	This paper discusses e-valUAM, an adaptive gamified system tested in a Madrid university using the MUD model to enhance engagement.
Tsunoda and Yumoto [10]	This paper compares the PRBL (points-ranking-badges-levels) gamification method with traditional teaching, highlighting its advantages and disadvantages.
Skalka et al. [62]	This paper discusses Microlearning, an action-oriented approach with short lessons, combined with interactive gamification elements using the Octalysis Framework.
Silvis-Cividjian [63]	This paper discusses a course for medical, aerospace, and IT equipment testers using gamification to address various teaching challenges and enhance realism.
Makarova et al. [64]	This paper highlights the advantages and disadvantages of e-learning, showing how gamification can improve teaching and training with role-playing, exercise games, and simulation games.
de Paula Porto et al. [65]	This paper characterizes how gamification has been applied in software engineering, identifying benefits and challenges.
Vlahu-Gjorgievska et al. [66]	This paper discusses the inclusion of computational thinking in curricula and the need for an educational approach involving various stakeholders.
Chan et al. [67]	This paper examines a course on professional software development and the integration of gamification to enhance learning outcomes.
Figueiredo and García-Peñalvo [68]	This paper highlights the motivational power of games and explores gamification's potential to increase student engagement in programming courses.
Pratama et al. [69]	This paper describes the development and impact of Rimigs, a gamification system aimed at improving student engagement and learning outcomes.
Naik and Jenkins [70]	This paper reviews the role of agile methodologies in software development education and how gamification can enhance collaborative learning.
Swacha and Szydłowska [71]	This paper evaluates the effectiveness of gamification in computer programming education through various case studies and learning outcomes.

Table 1. *Cont.*

Paper	Contribution
Sousa-Vieira et al. [72]	This paper analyzes the impact of social learning and gamification on higher education, focusing on activity levels and learning results.
Ren and Barrett [26]	This paper explores the importance of communication in software management and how gamification can improve team interactions and project outcomes.
Monteiro et al. [73]	This paper presents the recurring theme of gamification in software engineering education literature and its influence on student engagement.
Jusas et al. [74]	This paper assesses the potential of gamification to enhance student engagement, drive learning, and support sustainable educational practices.
Maher et al. [75]	This paper introduces the Personalized Adaptive Gamified E-learning (PAGE) model, which extends MOOCs with enhanced learning analytics and visualization to support learner intervention. The results indicate a positive potential for learning adaptation and the necessity of focusing on gamification.
Bachtiar et al. [76]	This paper develops an e-learning system named Code Mania (CoMa) that integrates gamification elements like leaderboards and badges to increase student engagement in a Java Programming course. The system performs well as specified, demonstrating the potential of gamification in enhancing e-learning environments.
Laskowski [77]	This paper investigates the applicability of gamification across different higher education courses through an experiment involving computer science students. The study shows the comparative results of gamified and non-gamified groups, indicating the impact of gamification on student performance.
Fuchs and Wolff [78]	This paper presents an online learning platform with gamification elements designed for software engineering education. It combines formative assessment with gamification to enhance learning experiences, providing detailed examples and system design.
Bucchiarone et al. [79]	This paper reports on the outcomes of the 6th International Workshop on Games and Software Engineering, highlighting the growing complexity and need for theoretical frameworks in gamification. The workshop covered perspectives on software projects, testing, and design, with insights from keynotes and panel discussions.
Bucchiarone et al. [80]	This paper presents POLYGLOT, a gamified programming environment targeting programming languages education and text-based modeling languages like SysML v2. The approach allows for the creation of heterogeneous gamification scenarios, enhancing the learning experience.
Poecze and Tjoa [81]	This paper explores the relevance of publication bias tests in meta-analytical approaches to gamification in higher education. It discusses the challenges in conducting meta-analyses due to heterogeneity and compares methods for correcting publication bias.

Table 1. Cont.

Paper	Contribution
Cabezas [82]	This paper introduces a continuous improvement cycle for teaching scenarios in engineering, combining gamification theory and ABET criteria. The proposed cycle is applied in a computer programming course, showing a positive impact on student engagement and learning outcomes.
Bucchiarone et al. [83]	This paper discusses the convergence of game engineering, software engineering, and user experience to create solutions blending game strengths with real-world applications. It highlights the potential benefits of gamification and serious games in various domains such as education and healthcare.
Ristov et al. [84]	This paper presents a gamification approach in a hardware-based course on microprocessors and microcontrollers for computer science students. The approach improved course grades and motivated students to enroll in other hardware courses, demonstrating the positive impact of gamification on student interest and performance.
Bernik et al. [85]	This paper presents empirical research on the use of gamification in online programming courses. A gamified e-course was designed, and its impact on student engagement and use of learning materials was examined, showing potential benefits of gamification in e-learning.
Schäfer [86]	This paper reports on a gamification approach using Minecraft to train students in Scrum, an agile project management method. The study compares two teaching periods, highlighting findings and lessons learned from using game-based learning to teach Scrum principles.
Petrov et al. [87]	This paper analyzes gamification software for promoting minority languages. It provides an overview of current educational software and assesses the need for new gamification solutions to support regional and minority languages.
Tsalikidis and Pavlidis [88]	This paper presents jLegends, an online multiplayer platform game designed to teach programming with JavaScript. The game employs a role-playing approach to enhance learning through game mechanics, demonstrating the effectiveness of game-based learning in programming education.

4.5. RQ5: In Which Continents Is Gamification Mostly Analyzed?

An analysis on the geographical distribution of relevant papers reveals that Europe is the leading region, with 41% of the studies, as shown in Figure 8. This significant proportion indicates a robust interest and investment in gamification strategies within European educational institutions and research communities.

The prominence of European research in this field suggests that many universities and educational bodies in Europe are actively exploring and implementing gamification to enhance learning outcomes. This focus could be driven by several factors, including the strong support for educational innovation in European countries, the availability of funding for educational research, and the collaborative networks among European researchers. The "hybrid" field includes papers where authors have affiliations from two or more continents, to avoid inconsistencies in the data extraction.

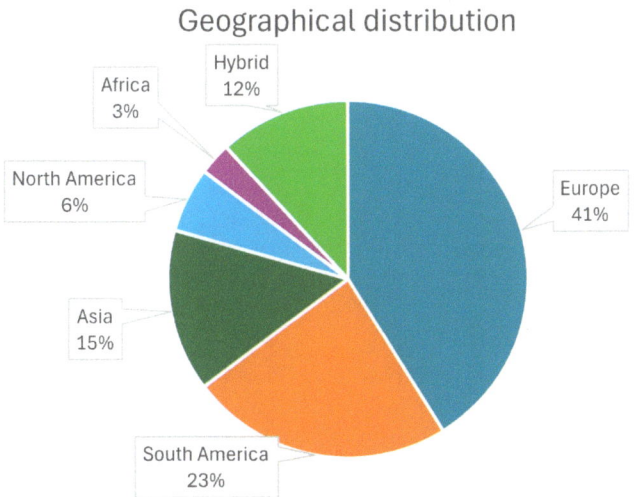

Figure 8. Geographical distribution of studies on gamification in SEET.

4.6. RQ6: What Are the Advantages and Disadvantages of Gamification When Applied to SEET?

The application of gamification as an educational tool has been the subject of extensive analysis, revealing both its advantages and disadvantages. We examine the potential benefits and drawbacks of this technique below.

Among the advantages, gamification has been shown to significantly enhance student participation [89] and interest [68] compared to traditional teaching methods. For instance, in the study conducted at the University of Applied Sciences Würzburg-Schweinfurt [38], 40 students were surveyed, with 27 respondents. Initially, 35% of students reported moderate motivation levels before engaging with the gamified course. Post-intervention, 40% of students indicated increased motivation due to the gamification content, although nearly 20% did not find it motivating at all. This finding aligns with the broader research on gamification, highlighting that motivation varies among individuals and is influenced by different types of gamified elements [35].

Furthermore, the economic growth associated with gamification is notable. As reported in [45], the gamification market is projected to expand from USD 9.1 billion in 2020 to USD 30.7 billion by 2025, with a growth rate of 27.4% per year. This underscores the increasing interest and investment in gamification as a promising educational approach. Gamification also provides intrinsic benefits for student motivation. By offering rewards and real-time feedback, students can visualize their progress and achievements, thereby enhancing their learning experience.

However, the application of gamification is not without its challenges. Some limitations of gamification in SEET are summarized below [36]:

- Difficulty in measuring performance improvements;
- Increased workload for educators [75];
- Lack of digital platforms to implement gamified techniques [36];
- Challenges in engaging all students;
- Difficulty for some students to understand the gamification method;
- Lack of appreciation for the method by some students;
- Difficulty for students in gaining human feedback, for platforms in which gamification is used alongside Artificial Intelligence [25];
- Insufficient knowledge of gamification approaches;

- Limited time and interest: students can sometimes become annoyed or disengaged by gamification elements if they find them distracting [90];
- Scarcity of materials and resources [75];
- Ensuring proper use of gamification by students.

The analysis indicates that gamification, like any pedagogical method, presents both strengths and weaknesses. The primary challenge lies in the nascent stage of this technique, which precludes definitive conclusions about its long-term efficacy [51]. Nevertheless, short- to medium-term studies and experiences in educational settings provide valuable insights into its potential benefits and limitations [91]. Furthermore, the novelty of gamification means that large-scale, comprehensive studies are currently lacking, making it difficult to obtain a fully reliable overview of all its advantages and disadvantages. Continued research and practical implementation are essential to fully understand the impact of gamification on education.

5. Comparison with Other Review Papers

Different reviews on the topic of gamification applied to SEET were examined. The main ones are proposed below, together with analyzed databases and main findings.

Dal Sasso et al. [92] propose a critical overview of gamification and its application in supporting software engineering tasks, starting from the IEEEXplore database. They highlight how to implement gamified approaches and propose a method to evaluate gamification systems. However, the presented work is not a systematic review, but a first approach in evaluating the literature.

Pedreira et al. [93] carry out a systematic mapping study to analyze papers between 2011 and mid-2014. They use Scopus, Science Direct, IEEEXplore, ACM Digital Library, and Springer databases to gather papers and find 29 primary contributions: main results show that research in this field is still preliminary, and most of the considered papers focus on software development and not on proving empirical evidence of pros and cons of gamification.

Barreto et al. [94] carry a mixed-approach literature review, composed of both an ad hoc review in which they manually select relevant papers and a systematic review following software engineering guidelines based on ACM Digital Library, IEEEXplore, and Science Direct databases. They conclude that "researchers in the field tend towards a strict view of gamification, the practical results of gamification are unclear and polemic, and this research area has still much to improve".

Ngandu et al. [95] analyze 15 papers, conducting a literature review on IEEEXplore and Science Direct to understand the impact of gamification and its key elements to student interest in software engineering. Their main findings regard points and leaderboards, considered as the main elements used in this topic.

Chamorro-Atalaya et al. [96] study the impact of COVID-19 pandemics in the education shift and the opportunities carried by gamification in online teaching. Analyzed papers are gathered from Taylor & Francis, IEEEXplore, and Scopus. Software engineering and Computer Science result to be the main topics in which gamification is applied, and the application of gamification effectively generates motivation, commitment, and permanent participation of students.

Monteiro et al. [97] analyze papers coming from IEEEXplore and ACM Digital Library that report procedures for the evaluation of gamification. However, only three of the 64 studies actually propose evaluation models for gamification. The main finding is that "the evaluation of gamification requires a mix of subjective and objective inputs, and qualitative and quantitative data analysis approaches".

Analysis of related works shows the need for developing an updated literature review, to analyze not only teaching effectiveness, but also the role of gamification in software engineering-based jobs. Our review also makes an analysis on how the effectiveness of gamification in SEET is measured. The comparison between the different reviews is summarized in Table 2. The columns in the table include the reference for the reviewed work, key terms used in this study to highlight its focus areas, whether this study followed a systematic review or mapping methodology, the primary setting or context where gamification was applied, whether this study involved or considered industrial applications or partnerships or their involvement in the SEET topic, and a summary of the practical findings regarding the effectiveness or impact of gamification as reported by the study.

Table 2. Comparison between related works and our review.

Reference	Keywords	Systematic Approach	Main Application Field	Industrial Applications	Findings
Dal Sasso et al. [92]	Games, Software Engineering, Context, Psychology, Computer Bugs, Collaboration, Systematics	No	Universities	No	No findings
Pedreira et al. [93]	Gamification, Software Engineering, Systematic Mapping	Yes	Universities	No	No findings
Barreto and França [94]	Motivation, Engagement, Gamification, Software Engineering	Mixed	Universities	No	Unclear practical results
Ngandu et al. [95]	Gamification, Software Engineering, Student Interest, Game Elements, Engagement, Motivation, Participation	Yes	Student engagement	No	No findings
Chamorro-Atalaya et al. [96]	Gamification, Engineering Education, Design, Success Factors, Motivation	Yes	Universities after COVID-19 pandemics	No	No findings
Barbosa Monteiro et al. [97]	Gamification, Systematic Mapping, Evaluation, Software Engineering, Education	Yes	Universities	Yes	Yes
Our proposal	Gamification, Software Engineering, Education, Learning, Literature Review	Yes	Universities and companies	Yes	Yes

6. Conclusions

The use of gamification in SEET is a contentious and highly debated topic among experts in teaching, while disagreements exist regarding the integration of gaming aspects into educational systems, numerous proposed projects and studies have demonstrated positive outcomes that support the efficacy of gamification.

The advantages of using game environments for teaching are numerous. Gamification allows students to actively construct their understanding of topics, learn at their own pace individually or collectively in spontaneous groups, and proceed on different paths at varying speeds according to their interests and abilities. It also promotes collaboration and encourages just-in-time learning, as opposed to general training.

However, this teaching method also has some psychological drawbacks, particularly concerning its potential negative effects on students. Students, with gamification, are free to fail and free to experiment, and while these freedoms aim to provide students with the ideal tools to build their own experiences, they can also lead to an overclock of responsibility. This can result in a stressful and unmanageable situation for some students.

Our review, based on the analysis of 68 papers, reveals a significant interest in using gamification in universities and secondary schools. Despite its recent development and the promising studies showcasing its effectiveness, there remains a tendency to rely on traditional learning methods, often overlooking the potential of gamification. Nonetheless, the growing need to integrate digital content to enhance learning has spurred an increased interest in this alternative method. It is crucial to recognize that not all students or users are motivated by gamification, and while the majority may find it effective, there will always be a percentage of students who prefer traditional teaching methods due to differing attitudes and learning times. Therefore, it is unrealistic to expect a single tool to be universally effective.

In conclusion, gamification in education should be viewed as a valuable support tool rather than a total replacement for traditional teaching methods. As highlighted in many of the analyzed papers, gamification is often used alongside classic, proven teaching methods and sometimes as an alternative support. Its potential to enhance learning experiences is significant, but it must be integrated thoughtfully to complement and not completely replace traditional educational approaches.

Author Contributions: Conceptualization, M.F.; methodology, M.F.; software, V.D.N., R.F., M.F., G.M. and G.S.; validation, V.D.N., R.F., G.M. and G.S.; formal analysis, M.F.; investigation, M.F.; resources, V.D.N., R.F., G.M. and G.S.; data curation, V.D.N., R.F., G.M. and G.S.; writing—original draft preparation, V.D.N., R.F., G.M. and G.S.; writing—review and editing, M.F. and M.M.; visualization, V.D.N.; supervision, M.F. and M.M.; project administration, M.M.; funding acquisition, M.M. All authors have read and agreed to the published version of the manuscript.

Funding: This research received no external funding

Conflicts of Interest: The authors declare no conflicts of interest.

Abbreviations

The following abbreviations are used in this manuscript:

SEET	Software Engineering Education and Training
PRISMA	Preferred Reporting Items for Systematic Reviews and Meta-Analyses
PICO	Population, Intervention, Comparison, Outcome

References

1. Malhotra, R.; Massoudi, M.; Jindal, R. An Innovative Approach: Coupling Project-Based Learning and Game-Based Learning Approach in Teaching Software Engineering Course. In Proceedings of the 2020 IEEE International Conference on Technology, Engineering, Management for Societal Impact Using Marketing, Entrepreneurship and Talent (TEMSMET), Bengaluru, India, 10 December 2020; pp. 1–5. [CrossRef]
2. Kim, M.K.; Kim, S.M. Dynamic learner engagement in a wiki-enhanced writing course. *J. Comput. High. Educ.* **2020**, *32*, 582–606. [CrossRef]
3. Ng, D.T.K.; Xinyu, C.; Leung, J.K.L.; Chu, S.K.W. Fostering students' AI literacy development through educational games: AI knowledge, affective and cognitive engagement. *J. Comput. Assist. Learn.* 2024; *online version of record before inclusion in an issue*. [CrossRef]
4. Magioli Sereno, M.; Ang, H.B. The impact of gamification on training, work engagement, and job satisfaction in banking. *Int. J. Train. Dev.* **2024**, *28*, 362–384. [CrossRef]
5. Mongiello, M.; Nocera, F.; Parchitelli, A.; Patrono, L.; Rametta, P.; Riccardi, L.; Sergi, I. A smart iot-aware system for crisis scenario management. *J. Commun. Softw. Syst.* **2018**, *14*, 91–98. [CrossRef]
6. Cavalera, G.; Rosito, R.C.; Lacasa, V.; Mongiello, M.; Nocera, F.; Patrono, L.; Sergi, I. An innovative smart system based on IoT technologies for fire and danger situations. In Proceedings of the 2019 4th International Conference on Smart and Sustainable Technologies (SpliTech), Split, Croatia, 18–21 June 2019; pp. 1–6. [CrossRef]
7. Moher, D.; Liberati, A.; Tetzlaff, J.; Altman, D.G. Preferred reporting items for systematic reviews and meta-analyses: The PRISMA statement. *Int. J. Surg.* **2010**, *8*, 336–341. [CrossRef] [PubMed]
8. Petersen, K.; Feldt, R.; Mujtaba, S.; Mattsson, M. Systematic mapping studies in software engineering. In Proceedings of the 12th International Conference on Evaluation and Assessment in Software Engineering (EASE), Bari, Italy, 26–27 June 2008; BCS Learning & Development: Swindon, UK, 2008.
9. Haddaway, N.R.; Page, M.J.; Pritchard, C.C.; McGuinness, L.A. PRISMA2020: An R package and Shiny app for producing PRISMA 2020-compliant flow diagrams, with interactivity for optimised digital transparency and Open Synthesis. *Campbell Syst. Rev.* **2022**, *18*, e1230. [CrossRef] [PubMed]
10. Tsunoda, M.; Yumoto, H. Applying Gamification and Posing to Software Development. In Proceedings of the 2018 25th Asia-Pacific Software Engineering Conference (APSEC), Nara, Japan, 4–7 December 2018; pp. 638–642. [CrossRef]

11. Takbiri, Y.; Amini, A.; Bastanfard, A. A Structured Gamification Approach for Improving Children's Performance in Online Learning Platforms. In Proceedings of the 2019 5th Iranian Conference on Signal Processing and Intelligent Systems (ICSPIS), Shahrood, Iran, 18–19 December 2019; pp. 1–6. [CrossRef]
12. Molins-Ruano, P.; Jurado, F.; Rodríguez, P.; Atrio, S.; Sacha, G.M. An Approach to Gamify an Adaptive Questionnaire Environment. In Proceedings of the 2016 IEEE Global Engineering Education Conference (EDUCON), Abu Dhabi, United Arab Emirates, 10–13 April 2016; pp. 1129–1133. [CrossRef]
13. Ortega-Arranz, A.; Kalz, M.; Martínez-Monés, A. Creating Engaging Experiences in MOOCs through In-Course Redeemable Rewards. In Proceedings of the 2018 IEEE Global Engineering Education Conference (EDUCON), Canary Islands, Spain, 17–20 April 2018; pp. 1875–1882. [CrossRef]
14. Monteiro, R.H.B.; Oliveira, S.R.B.; De Almeida Souza, M.R. A Standard Framework for Gamification Evaluation in Education and Training of Software Engineering: An Evaluation from a Proof of Concept. In Proceedings of the 2021 IEEE Frontiers in Education Conference (FIE), Lincoln, NE, USA, 13–16 October 2021; pp. 1–7. [CrossRef]
15. Nistor, G.C.; Iacob, A. The advantages of gamification and game-based learning and their benefits in the development of education. In Proceedings of the International Scientific Conference eLearning and Software for Education, Bucharest, Romania, 19–20 April 2018; "Carol I" National Defence University: București, Romania, 2018; Volume 1, pp. 308–312.
16. Fiore, M.; Gattullo, M.; Mongiello, M. First Steps in Constructing an AI-Powered Digital Twin Teacher: Harnessing Large Language Models in a Metaverse Classroom. In Proceedings of the 2024 IEEE Conference on Virtual Reality and 3D User Interfaces Abstracts and Workshops (VRW), Orlando, FL, USA, 16–21 March 2024; pp. 939–940. [CrossRef]
17. Limantara, N.; Gaol, F.L.; Prabowo, H. Mechanics, dynamics, and aesthetics Framework on gamification at university. In Proceedings of the 2020 International Conference on Informatics, Multimedia, Cyber and Information System (ICIMCIS), Jakarta, Indonesia, 19–20 November 2020; pp. 34–39.
18. Strmecki, D.; Bernik, A.; Radosevic, D. Gamification in E-Learning: Introducing Gamified Design Elements into E-Learning Systems. *J. Comput. Sci.* 2015, *11*, 1108–1117. [CrossRef]
19. Zichermann, G.; Cunningham, C. *Gamification by Design: Implementing Game Mechanics in Web and Mobile Apps*; O'Reilly Media, Inc.: Sebastopol, CA, USA, 2011.
20. Alsawaier, R.S. The effect of gamification on motivation and engagement. *Int. J. Inf. Learn. Technol.* 2018, *35*, 56–79. [CrossRef]
21. Apandi, A.M. Gamification meets mobile learning: Soft-skills enhancement. In *Research Anthology on Developments in Gamification and Game-Based Learning*; IGI Global: Hershey, PA, USA, 2022; pp. 1280–1299.
22. Kabilan, M.K.; Annamalai, N.; Chuah, K.M. Practices, purposes and challenges in integrating gamification using technology: A mixed-methods study on university academics. *Educ. Inf. Technol.* 2023, *28*, 14249–14281. [CrossRef] [PubMed]
23. Mongiello, M.; Pelliccione, P.; Sciancalepore, M. AC-Contract: Run-Time Verification of Context-Aware Applications. In Proceedings of the 2015 IEEE/ACM 10th International Symposium on Software Engineering for Adaptive and Self-Managing Systems, Florence, Italy, 18–19 May 2015; Volume 15, pp. 24–34. [CrossRef]
24. Bistarelli, S.; Di Noia, T.; Mongiello, M.; Nocera, F. Pronto: An ontology driven business process mining tool. *Procedia Comput. Sci.* 2017, *112*, 306–315. [CrossRef]
25. Fulcini, T.; Torchiano, M. Is ChatGPT Capable of Crafting Gamification Strategies for Software Engineering Tasks? In Proceedings of the 2nd International Workshop on Gamification in Software Development, Verification, and Validation, San Francisco, CA, USA, 4 December 2023; pp. 22–28. [CrossRef]
26. Ren, W.; Barrett, S. An Empirical Investigation on the Benefits of Gamification in Communication within University Development Teams. *Comput. Appl. Eng. Educ.* 2023, *31*, 1808–1822. [CrossRef]
27. Miljanovic, M.A.; Bradbury, J.S. Robobug: A serious game for learning debugging techniques. In Proceedings of the 2017 ACM Conference on International Computing Education Research, Tacoma, WA, USA, 18–20 August 2017; pp. 93–100.
28. Navarro, E.O.; van der Hoek, A. SIMSE: An Interactive Simulation Game for Software Engineering Education. In Proceedings of the CATE, Kauai, HI, USA, 16–18 August 2004; Volume 1, pp. 12–17.
29. Esper, S.; Foster, S.R.; Griswold, W.G. CodeSpells: Embodying the metaphor of wizardry for programming. In Proceedings of the 18th ACM Conference on Innovation and Technology in Computer Science Education, Canterbury, UK, 1–3 July 2013; pp. 249–254.
30. Arif, R.F.; Rosyid, H.A.; Pujianto, U. Design and Implementation of Interactive Coding with Gamification for Web Programming Subject for Vocational High School Students. In Proceedings of the 2019 International Conference on Electrical, Electronics and Information Engineering (ICEEIE), Bali, Indonesia, 3–4 October 2019; Volume 6, pp. 177–182. [CrossRef]
31. Rahim, R.H.A.; Tanalol, S.H.; Ismail, R.; Baharum, A.; Rahim, E.A.; Noor, N.A.M. Development of Gamification Linear Algebra Application Using Storytelling. In Proceedings of the 2019 International Conference on Information and Communication Technology Convergence (ICTC), Jeju Island, Republic of Korea, 16–18 October 2019; pp. 133–137. [CrossRef]
32. Hajarian, M.; Diaz, P. Effective Gamification: A Guideline for Gamification Workshop of WEEF-GEDC 2021 Madrid Conference. In Proceedings of the 2021 World Engineering Education Forum/Global Engineering Deans Council (WEEF/GEDC), Madrid, Spain, 15–18 November 2021; pp. 506–510. [CrossRef]
33. Iquira, D.; Galarza, M.; Sharhorodska, O. Enhancing Software Engineering Courses with a Mobile Gamified Platform: Results of a Mixed Approach. In Proceedings of the 2021 XVI Latin American Conference on Learning Technologies (LACLO), Arequipa, Peru, 19–21 October 2021; pp. 534–537. [CrossRef]

34. Robledo-Rella, V.; de Lourdes Quezada Batalla, M.; Ramírez-de-Arellano, J.M.; Acosta, R.D.S. Gam-Mate: Gamification Applied to an Undergrad Discrete Math Course. In Proceedings of the 2022 10th International Conference on Information and Education Technology (ICIET), Matsue, Japan, 9–11 April 2022; pp. 135–139. [CrossRef]
35. Gomes Fernandes Matsubara, P.; Lima Corrêa Da Silva, C. Game Elements in a Software Engineering Study Group: A Case Study. In Proceedings of the 2017 IEEE/ACM 39th International Conference on Software Engineering: Software Engineering Education and Training Track (ICSE-SEET), Buenos Aires, Argentina, 20–28 May 2017; pp. 160–169. [CrossRef]
36. Rodrigues, P.; Souza, M.; Figueiredo, E. Games and Gamification in Software Engineering Education: A Survey with Educators. In Proceedings of the 2018 IEEE Frontiers in Education Conference (FIE), San Jose, CA, USA, 3–6 October 2018; pp. 1–9. [CrossRef]
37. Quinde, C.P.; Paredes, R.I.; Maldonado, S.A.; Guerrero, J.S.; Toro, M.F.V. Gamification as a Didactic Strategy in a Digital Literacy: Case Study for Incacerated Individuals. In Proceedings of the 2018 IEEE Global Engineering Education Conference (EDUCON), Canary Islands, Spain, 17–20 April 2018; pp. 1314–1319. [CrossRef]
38. John, I.; Fertig, T. Gamification for Software Engineering Students—An Experience Report. In Proceedings of the 2022 IEEE Global Engineering Education Conference (EDUCON), Tunis, Tunisia, 28–31 March 2022; pp. 1942–1947. [CrossRef]
39. Ivanova, G.; Kozov, V.; Zlatarov, P. Gamification in Software Engineering Education. In Proceedings of the 2019 42nd International Convention on Information and Communication Technology, Electronics and Microelectronics (MIPRO), Opatija, Croatia, 20–24 May 2019; pp. 1445–1450. [CrossRef]
40. Gasca-Hurtado, G.P.; Gómez-Álvarez, M.C.; Hincapié, J.A.; Zepeda, V.V. Gamification of an Educational Environment in Software Engineering: Case Study for Digital Accessibility of People with Disabilities. *IEEE Rev. Iberoam. Tecnol. Aprendiz.* 2021, 16, 382–392. [CrossRef]
41. Carreño-León, M.; Sandoval-Bringas, A.; Álvarez-Rodríguez, F.; Camacho-González, Y. Gamification Technique for Teaching Programming. In Proceedings of the 2018 IEEE Global Engineering Education Conference (EDUCON), Canary Islands, Spain, 17–20 April 2018; pp. 2009–2014. [CrossRef]
42. Sherif, E.; Liu, A.; Nguyen, B.; Lerner, S.; Griswold, W.G. Gamification to Aid the Learning of Test Coverage Concepts. In Proceedings of the 2020 IEEE 32nd Conference on Software Engineering Education and Training (CSEE&T), Munich, Germany, 9–12 November 2020; pp. 1–5. [CrossRef]
43. Norsanto, D.; Rosmansyah, Y. Gamified Mobile Micro-Learning Framework: A Case Study of Civil Service Management Learning. In Proceedings of the 2018 International Conference on Information and Communications Technology (ICOIACT), Yogyakarta, Indonesia, 6–7 March 2018; pp. 146–151. [CrossRef]
44. Call, M.W.; Fox, E.; Sprint, G. Gamifying Software Engineering Tools to Motivate Computer Science Students to Start and Finish Programming Assignments Earlier. *IEEE Trans. Educ.* 2021, 64, 423–431. [CrossRef]
45. Trinidad, M.; Calderón, A.; Ruiz, M. GoRace: A Multi-Context and Narrative-Based Gamification Suite to Overcome Gamification Technological Challenges. *IEEE Access* 2021, 9, 65882–65905. [CrossRef]
46. Prasetya, W.; Leek, C.; Melkonian, O.; ten Tusscher, J.; van Bergen, J.; Everink, J.; van der Klis, T.; Meijerink, R.; Oosenbrug, R.; Oostveen, J.; et al. Having Fun in Learning Formal Specifications. In Proceedings of the 2019 IEEE/ACM 41st International Conference on Software Engineering: Software Engineering Education and Training (ICSE-SEET), Montreal, QC, Canada, 25–31 May 2019; pp. 192–196. [CrossRef]
47. Bucchiarone, A.; Cicchetti, A.; Bassanelli, S.; Marconi, A. How to Merge Gamification Efforts for Programming and Modelling: A Tool Implementation Perspective. In Proceedings of the 2021 ACM/IEEE International Conference on Model Driven Engineering Languages and Systems Companion (MODELS-C), Fukuoka, Japan, 10–15 October 2021; pp. 721–726. [CrossRef]
48. Lema Moreta, L.; Gamboa, A.C.; Palacios, M.G. Implementing a Gamified Application for a Risk Management Course. In Proceedings of the 2016 IEEE Ecuador Technical Chapters Meeting (ETCM), Guayaquil, Ecuador, 12–14 October 2016; pp. 1–6. [CrossRef]
49. Ouhbi, S.; Pombo, N. Software Engineering Education: Challenges and Perspectives. In Proceedings of the 2020 IEEE Global Engineering Education Conference (EDUCON), Porto, Portugal, 27–30 April 2020; pp. 202–209. [CrossRef]
50. Villagra, S.; De Benedetti, G.; Bruno, T.; Fernández, L.; Outeda, N. Teaching Software Engineering: An Active Learning Experience. In Proceedings of the 2020 IEEE Congreso Bienal de Argentina (ARGENCON), Resistencia, Argentina, 1–4 December 2020; pp. 1–6. [CrossRef]
51. Moser, G.; Vallon, R.; Bernhart, M.; Grechenig, T. Teaching Software Quality Assurance with Gamification and Continuous Feedback Techniques. In Proceedings of the 2021 IEEE Global Engineering Education Conference (EDUCON), Vienna, Austria, 21–23 April 2021; pp. 505–509. [CrossRef]
52. Rattadilok, P.; Roadknight, C.; Li, L. Teaching Students About Machine Learning Through a Gamified Approach. In Proceedings of the 2018 IEEE International Conference on Teaching, Assessment, and Learning for Engineering (TALE), Wollongong, NSW, Australia, 4–7 December 2018; pp. 1011–1015. [CrossRef]
53. Bucchiarone, A.; Cicchetti, A.; Loria, E.; Marconi, A. Towards a Framework to Assist Iterative and Adaptive Design in Gameful Systems. In Proceedings of the 2021 36th IEEE/ACM International Conference on Automated Software Engineering Workshops (ASEW), Melbourne, Australia, 15–19 November 2021; pp. 78–84. [CrossRef]
54. Ebert, C.; Vizcaino, A.; Grande, R. Unlock the Business Value of Gamification. *IEEE Softw.* 2022, 39, 15–22. [CrossRef]
55. Maxim, B.R.; Brunvand, S.; Decker, A. Use of Role-Play and Gamification in a Software Project Course. In Proceedings of the 2017 IEEE Frontiers in Education Conference (FIE), Indianapolis, IN, USA, 18–21 October 2017; pp. 1–5. [CrossRef]

56. Jiménez-Hernández, E.M.; Jiménez-Murillo, J.A.; Segura-Castruita, M.A.; González-Leal, I. Using a Serious Video Game to Support the Learning of Tree Traversals. In Proceedings of the 2021 9th International Conference in Software Engineering Research and Innovation (CONISOFT), San Diego, CA, USA, 25–29 October 2021; pp. 238–244. [CrossRef]
57. Nagaria, B.; Evans, B.C.; Mann, A.; Arzoky, M. Using an Instant Visual and Text Based Feedback Tool to Teach Path Finding Algorithms: A Concept. In Proceedings of the 2021 Third International Workshop on Software Engineering Education for the Next Generation (SEENG), Virtual, 24 May 2021; pp. 11–15. [CrossRef]
58. Margalit, O. Using Computer Programming Competition for Cyber Education. In Proceedings of the 2016 IEEE International Conference on Software Science, Technology and Engineering (SWSTE), Beer Sheva, Israel, 23–24 June 2016; pp. 104–107. [CrossRef]
59. Stol, K.J.; Schaarschmidt, M.; Goldblit, S. Gamification in Software Engineering: The Mediating Role of Developer Engagement and Job Satisfaction. *Empir. Softw. Eng.* **2022**, *27*, 35. [CrossRef] [PubMed]
60. Đambić, G.; Keščec, T.; Kučak, D. A Blended Learning with Gamification Approach for Teaching Programming Courses in Higher Education. In Proceedings of the 2021 44th International Convention on Information, Communication and Electronic Technology (MIPRO), Opatija, Croatia, 27 September–1 October 2021; pp. 843–847. [CrossRef]
61. Mi, Q.; Keung, J.; Mei, X.; Xiao, Y.; Chan, W.K. A Gamification Technique for Motivating Students to Learn Code Readability in Software Engineering. In Proceedings of the 2018 International Symposium on Educational Technology (ISET), Osaka, Japan, 31 July–2 August 2018; pp. 250–254. [CrossRef]
62. Skalka, J.; Drlík, M.; Obonya, J.; Cápay, M. Architecture Proposal for Micro-Learning Application for Learning and Teaching Programming Courses. In Proceedings of the 2020 IEEE Global Engineering Education Conference (EDUCON), Porto, Portugal, 27–30 April 2020; pp. 980–987. [CrossRef]
63. Silvis-Cividjian, N. Awesome Bug Manifesto: Teaching an Engaging and Inspiring Course on Software Testing (Position Paper). In Proceedings of the 2021 Third International Workshop on Software Engineering Education for the Next Generation (SEENG), Madrid, Spain, 24 May 2021; pp. 16–20. [CrossRef]
64. Makarova, I.; Pashkevich, A.; Shubenkova, K. Blended Learning Technologies in the Automotive Industry Specialists' Training. In Proceedings of the 2018 32nd International Conference on Advanced Information Networking and Applications Workshops (WAINA), Krakow, Poland, 16–18 May 2018; pp. 319–324. [CrossRef]
65. de Paula Porto, D.; de Jesus, G.M.; Ferrari, F.C.; Fabbri, S.C.P.F. Initiatives and Challenges of Using Gamification in Software Engineering: A Systematic Mapping. *J. Syst. Softw.* **2021**, *173*, 110870. [CrossRef]
66. Vlahu-Gjorgievska, E.; Videnovik, M.; Trajkovik, V. Computational Thinking and Coding Subject in Primary Schools: Methodological Approach Based on Alternative Cooperative and Individual Learning Cycles. In Proceedings of the 2018 IEEE International Conference on Teaching, Assessment, and Learning for Engineering (TALE), Wollongong, NSW, Australia, 4–7 December 2018; pp. 77–83. [CrossRef]
67. Chan, Y.C.; Min Gan, C.; Lim, C.Y.; Hwa Tan, T.; Cao, Q.; Seow, C.K. Learning CS Subjects of Professional Software Development and Team Projects. In Proceedings of the 2022 IEEE International Conference on Teaching, Assessment and Learning for Engineering (TALE), Hung Hom, Hong Kong, 4–7 December 2022; pp. 71–77. [CrossRef]
68. Figueiredo, J.; García-Peñalvo, F.J. Increasing Student Motivation in Computer Programming with Gamification. In Proceedings of the 2020 IEEE Global Engineering Education Conference (EDUCON), Porto, Portugal, 27–30 April 2020; pp. 997–1000. [CrossRef]
69. Pratama, F.A.; Silitonga, R.M.; Jou, Y.T. Rimigs: The Impact of Gamification on Students' Motivation and Performance in Programming Class. *Indones. J. Electr. Eng. Comput. Sci.* **2021**, *24*, 1789–1795. [CrossRef]
70. Naik, N.; Jenkins, P. Relax, It'sa Game: Utilising Gamification in Learning Agile Scrum Software Development. In Proceedings of the 2019 IEEE Conference on Games (CoG), London, UK, 20–23 August 2019; pp. 1–4.
71. Swacha, J.; Szydłowska, J. Does Gamification Make a Difference in Programming Education? Evaluating FGPE-Supported Learning Outcomes. *Educ. Sci.* **2023**, *13*, 984. [CrossRef]
72. Sousa-Vieira, M.E.; López-Ardao, J.C.; Fernández-Veiga, M.; Rodríguez-Rubio, R.F. Study of the Impact of Social Learning and Gamification Methodologies on Learning Results in Higher Education. *Comput. Appl. Eng. Educ.* **2023**, *31*, 131–153. [CrossRef]
73. Monteiro, R.; Souza, M.; Oliveira, S.; Soares, E. The Adoption of a Framework to Support the Evaluation of Gamification Strategies in Software Engineering Education. In Proceedings of the 14th International Conference on Computer Supported Education, Online, 22–24 April 2022; pp. 450–457. [CrossRef]
74. Jusas, V.; Barisas, D.; Jančiukas, M. Game Elements towards More Sustainable Learning in Object-Oriented Programming Course. *Sustainability* **2022**, *14*, 2325. [CrossRef]
75. Maher, Y.; Moussa, S.M.; Khalifa, M.E. Learners on Focus: Visualizing Analytics through an Integrated Model for Learning Analytics in Adaptive Gamified E-Learning. *IEEE Access* **2020**, *8*, 197597–197616. [CrossRef]
76. Bachtiar, F.A.; Pradana, F.; Priyambadha, B.; Bastari, D.I. CoMa: Development of Gamification-based E-learning. In Proceedings of the 2018 10th International Conference on Information Technology and Electrical Engineering (ICITEE), Bali, Indonesia, 24–26 July 2018; pp. 1–6. [CrossRef]
77. Laskowski, M. Implementing Gamification Techniques into University Study Path - A Case Study. In Proceedings of the 2015 IEEE Global Engineering Education Conference (EDUCON), Tallinn, Estonia, 18–20 March 2015; pp. 582–586. [CrossRef]

78. Fuchs, M.; Wolff, C. Improving Programming Education through Gameful, Formative Feedback. In Proceedings of the 2016 IEEE Global Engineering Education Conference (EDUCON), Abu Dhabi, United Arab Emirates, 10–13 April 2016; pp. 860–867. [CrossRef]
79. Bucchiarone, A.; Cooper, K.M.L.; Lin, D.; Melcer, E.F.; Sung, K. Games and Software Engineering: Engineering Fun, Inspiration, and Motivation. *ACM SIGSOFT Softw. Eng. Notes* **2023**, *48*, 85–89. [CrossRef]
80. Bucchiarone, A.; Martorella, T.; Colombo, D.; Cicchetti, A.; Marconi, A. POLYGLOT for Gamified Education: Mixing Modelling and Programming Exercises. In Proceedings of the 2021 ACM/IEEE International Conference on Model Driven Engineering Languages and Systems Companion (MODELS-C), Fukuoka, Japan, 10–15 October 2021; pp. 605–609. [CrossRef]
81. Poecze, F.; Tjoa, A.M. Meta-Analytical Considerations for Gamification in Higher Education: Existing Approaches and Future Research Agenda. In Proceedings of the 2020 4th International Conference on Informatics and Computational Sciences (ICICoS), Semarang, Indonesia, 10–11 November 2020; pp. 1–6. [CrossRef]
82. Cabezas, I. On Combining Gamification Theory and ABET Criteria for Teaching and Learning Engineering. In Proceedings of the 2015 IEEE Frontiers in Education Conference (FIE), Washington, DC, USA, 21–24 October 2015; pp. 1–9. [CrossRef]
83. Bucchiarone, A.; Cooper, K.M.L.; Lin, D.; Smith, A.; Wanick, V. Fostering Collaboration and Advancing Research in Software Engineering and Game Development for Serious Contexts. *ACM SIGSOFT Softw. Eng. Notes* **2023**, *48*, 46–50. [CrossRef]
84. Ristov, S.; Ackovska, N.; Kirandziska, V. Positive Experience of the Project Gamification in the Microprocessors and Microcontrollers Course. In Proceedings of the 2015 IEEE Global Engineering Education Conference (EDUCON), Tallinn, Estonia, 18–20 March 2015; pp. 511–517. [CrossRef]
85. Bernik, A.; Radošević, D.; Bubaš, G. Introducing Gamification into E-Learning University Courses. In Proceedings of the 2017 40th International Convention on Information and Communication Technology, Electronics and Microelectronics (MIPRO), Opatija, Croatia, 22–26 May 2017; pp. 711–716. [CrossRef]
86. Schäfer, U. Training Scrum with Gamification: Lessons Learned after Two Teaching Periods. In Proceedings of the 2017 IEEE Global Engineering Education Conference (EDUCON), Athens, Greece, 25–28 April 2017; pp. 754–761. [CrossRef]
87. Petrov, E.; Mustafina, J.; Alloghani, M. Overview on Modern Serious Games for Regional and Minority Languages Promotion. In Proceedings of the 2017 10th International Conference on Developments in eSystems Engineering (DeSE), Paris, France, 14–16 June 2017; pp. 120–123. [CrossRef]
88. Tsalikidis, K.; Pavlidis, G. jLegends: Online Game to Train Programming Skills. In Proceedings of the 2016 7th International Conference on Information, Intelligence, Systems & Applications (IISA), Chalkidiki, Greece, 13–15 July 2016; pp. 1–6. [CrossRef]
89. Gamarra, M.; Dominguez, A.; Velazquez, J.; Páez, H. A Gamification Strategy in Engineering Education—A Case Study on Motivation and Engagement. *Comput. Appl. Eng. Educ.* **2022**, *30*, 472–482. [CrossRef]
90. Kadar, R.; Wahab, N.A.; Othman, J.; Shamsuddin, M.; Mahlan, S.B. A study of difficulties in teaching and learning programming: A systematic literature review. *Int. J. Acad. Res. Progress. Educ. Dev.* **2021**, *10*, 591–605. [CrossRef] [PubMed]
91. Fiore, M.; Mongiello, M. Using Peer Assessment Leveraging Large Language Models in Software Engineering Education. *Int. J. Softw. Eng. Knowl. Eng.* **2024**, *34*. [CrossRef]
92. Dal Sasso, T.; Mocci, A.; Lanza, M.; Mastrodicasa, E. How to gamify software engineering. In Proceedings of the 2017 IEEE 24th International Conference on Software Analysis, Evolution and Reengineering (SANER), Klagenfurt, Austria, 20–24 February 2017; pp. 261–271. [CrossRef]
93. Pedreira, O.; García, F.; Brisaboa, N.; Piattini, M. Gamification in software engineering—A systematic mapping. *Inf. Softw. Technol.* **2015**, *57*, 157–168. [CrossRef]
94. Barreto, C.F.; França, C. Gamification in software engineering: A literature review. In Proceedings of the 2021 IEEE/ACM 13th International Workshop on Cooperative and Human Aspects of Software Engineering (CHASE), Madrid, Spain, 20–21 May 2021; pp. 105–108.
95. Ngandu, M.R.; Risinamhodzi, D.; Dzvapatsva, G.P.; Matobobo, C. Capturing student interest in software engineering through gamification: A systematic literature review. *Discov. Educ.* **2023**, *2*, 47. [CrossRef]
96. Chamorro-Atalaya, O.; Morales-Romero, G.; Trinidad-Loli, N.; Caycho-Salas, B.; Guía-Altamirano, T.; Auqui-Ramos, E.; Rocca-Carvajal, Y.; Arones, M.; Arévalo-Tuesta, J.A.; Gonzales-Huaytahuilca, R. Gamification in engineering education during COVID-19: A systematic review on design considerations and success factors in its implementation. *Int. J. Learn. Teach. Educ. Res.* **2023**, *22*, 301–327. [CrossRef]
97. Barbosa Monteiro, R.H.; de Almeida Souza, M.R.; Bezerra Oliveira, S.R.; dos Santos Portela, C.; de Cristo Lobato, C.E. The Diversity of Gamification Evaluation in the Software Engineering Education and Industry: Trends, Comparisons and Gaps. In Proceedings of the 2021 IEEE/ACM 43rd International Conference on Software Engineering: Software Engineering Education and Training (ICSE-SEET), Virtual Event, 25–28 May 2021; pp. 154–164. [CrossRef]

Disclaimer/Publisher's Note: The statements, opinions and data contained in all publications are solely those of the individual author(s) and contributor(s) and not of MDPI and/or the editor(s). MDPI and/or the editor(s) disclaim responsibility for any injury to people or property resulting from any ideas, methods, instructions or products referred to in the content.

Article

Adaptive Gamification in Science Education: An Analysis of the Impact of Implementation and Adapted Game Elements on Students' Motivation

Alkinoos-Ioannis Zourmpakis [1], Michail Kalogiannakis [2] and Stamatios Papadakis [1,*]

1 Department of Preschool Education, Faculty of Education, University of Crete, 74100 Crete, Greece
2 Department of Special Education, University of Thessaly, 38221 Volos, Greece; mkalogian@uth.gr
* Correspondence: stpapadakis@uoc.gr

Abstract: In recent years, gamification has captured the attention of researchers and educators, particularly in science education, where students often express negative emotions. Gamification methods aim to motivate learners to participate in learning by incorporating intrinsic and extrinsic motivational factors. However, the effectiveness of gamification has yielded varying outcomes, prompting researchers to explore adaptive gamification as an alternative approach. Nevertheless, there needs to be more research on adaptive gamification approaches, particularly concerning motivation, which is the primary objective of gamification. In this study, we developed and tested an adaptive gamification environment based on specific motivational and psychological frameworks. This environment incorporated adaptive criteria, learning strategies, gaming elements, and all crucial aspects of science education for six classes of third-grade students in primary school. We employed a quantitative approach to gain insights into the motivational impact on students and their perception of the adaptive gamification application. We aimed to understand how each game element experienced by students influenced their motivation. Based on our findings, students were more motivated to learn science when using an adaptive gamification environment. Additionally, the adaptation process was largely successful, as students generally liked the game elements integrated into their lessons, indicating the effectiveness of the multidimensional framework employed in enhancing students' experiences and engagement.

Keywords: adaptive gamification; science education; adapted game elements; students' motivation

Citation: Zourmpakis, A.-I.; Kalogiannakis, M.; Papadakis, S. Adaptive Gamification in Science Education: An Analysis of the Impact of Implementation and Adapted Game Elements on Students' Motivation. *Computers* **2023**, *12*, 143. https://doi.org/10.3390/computers12070143

Academic Editors: Carlos Vaz de Carvalho, Hariklia Tsalapatas and Ricardo Baptista

Received: 21 June 2023
Revised: 10 July 2023
Accepted: 15 July 2023
Published: 18 July 2023

Copyright: © 2023 by the authors. Licensee MDPI, Basel, Switzerland. This article is an open access article distributed under the terms and conditions of the Creative Commons Attribution (CC BY) license (https://creativecommons.org/licenses/by/4.0/).

1. Introduction

Educators have consistently prioritized students' active participation in the classroom as a fundamental aspect [1]. With the dynamic influence and continuous progress of technology, it is essential to develop innovative learning environments that cater to the requirements and interests of contemporary learners, thereby fostering an engaging and inspiring educational experience [2,3]. The attainment of high levels of student engagement and motivation is of great importance, as research has demonstrated its positive impact on academic achievement [4,5]. In recent years, there has been a notable increase in the utilization of digital games across various domains, including academia [6]. This trend has sparked the interest of researchers and practitioners, leading to the emergence of a novel approach known as gamification [7]. The COVID-19 pandemic has further highlighted an ongoing challenge wherein numerous students struggle to effectively regulate their motivation, particularly within digital learning settings [8,9].

Promoting science education is crucial for the progress of our society and the development of individuals who possess scientific literacy, enabling them to comprehend and appreciate the intricacies of the world [10]. Science education nurtures essential learning skills and fosters attitudes that emphasize the significance of evidence-based decision-making while nurturing social and environmental consciousness. These benefits extend to individuals regardless of their future involvement in the fields of science and technology [10,11]. Students at all educational levels have consistently encountered challenges when grasping scientific concepts [12,13]. The complexity associated with comprehending and understanding science-related concepts often leads to negative emotions, unfavourable experiences, and diminished motivation for learning among students [14,15]. However, other factors can influence science learning, such as teacher self-efficacy and motivation [16], geographical differences [16], gender differences [15], and school setting [15].

The increasing popularity of games in our society has sparked significant interest among educators and instructional developers in a concept known as gamification. Although the term "gamification" was initially introduced in 2008, it was not until 2010 that it gained broader acceptance [7]. Since then, its popularity has continued to grow steadily and remains a central concept today [17]. Gamification in education refers to "incorporating game mechanics, aesthetics, and the cognitive and behavioural aspects associated with games into non-game-related educational content" [18]. This approach aims to engage and motivate students, address challenging situations, and enhance the learning experience through digital materials.

Extensive research has solidified the understanding that gamification holds significant potential to influence and drive desired changes in behaviour [17,19]. Despite being applied for nearly a decade, the existing literature on gamification reveals varied outcomes concerning its effectiveness in enhancing learning and motivation [20]. These mixed results suggest that the commonly employed "one size fits all" approach to gamification, which assumes similar reactions from all users towards gamification elements, may be insufficient [21,22]. The lack of adaptation of game elements and the absence of an appropriate didactic approach tailored to the individual needs of each learner, combined with the frequent presentation of repetitive game elements, can contribute to higher levels of abandonment over time [23]. Furthermore, the absence of a well-defined and carefully planned design can also result in adverse outcomes [19,24,25]. It is essential to consider these parameters when designing a gamified app [26] and implementing it in a classroom setting [27].

Previous studies have powerfully shown that for gamification to be effective, it should be adapted to align with users' expectations and individual preferences [3,28,29]. Indeed, adapting game elements to cater to individual preferences can be challenging, considering the diverse range of learners and their varying motivations for learning. However, it is essential to note that adaptive gamification, which involves tailoring game elements based on individual user actions, preferences, and characteristics [30], is still a relatively new concept. There are currently only a limited number of approaches described in the literature, and even fewer specifically designed to be content-specific and not generic [29,31].

The development of the adaptive gamification environment to teach scientific concepts related to the water cycle was based on a framework encompassing adaptive criteria, learning strategies, gaming elements, and all vital aspects of the learning process related to specific science education [31]. Our main objectives are to understand primary education students' motivation and engagement when utilizing an adaptive gamified application specified for science-related content and the motivational impact game elements had on students. More specifically, our research questions are:

- What was the motivational impact of this adaptive gamification environment on the students regarding science education?
- How did the adaptive game elements motivate students?

2. Literature Review

2.1. Adaptable Gamification and Science Education

Overall, the interaction and correlation between students and science education in schools is often described as problematic. It is believed that there is a decline in motivation, attitudes, and interest, mainly as students grow up. Similar thoughts are derived when considering gender differences favouring boys. The motivation to learn science is seen as an influential factor in the development of scientific literacy among individuals [32]. Motivation is crucial in science education and acquiring scientific knowledge and skills [32]. Empirical studies have already established a connection between motivation to engage in science and academic performance [33,34]. Even though motivation towards science can impact the learning process [35], it is essential to note that various factors, like personal interests, personality traits, and cognitive style, can contribute to individual motivation [34,36].

In recent years, gamification has gained considerable attention as a concept that has been shown to enhance student engagement and motivation to learn, particularly in various fields and science education [17,20]. This aspect has become even more crucial as learners no longer seem as engaged with traditional teaching approaches as they once were [15]. The concept of gamification utilizes game elements and mechanics, known for their ability to motivate and engage players over extended periods, and applies them to nongame contexts. Its main goal is to replicate the same level of motivation and engagement for other purposes beyond gaming [37]. In addition, technological developments have facilitated the expansion of gamification into digital environments. This includes using applications or platforms that leverage digital devices such as computers, tablets, or smartphones [38]. As a result, science gamification applications have seen a noticeable increase [17]. Unlike other educational games, the primary objective of these applications is not solely focused on learning, although learning is an indirect outcome. The main goal is to modify learner behaviour or attitude within a specific context [39] (p. 759).

Nevertheless, it is essential to note that motivation varies among individuals, and different people can be motivated by various elements in specific ways. Consequently, in a gamified environment, interactions with the game can impact individuals differently based on their unique motivational factors [26]. Additionally, it is worth mentioning that gamification has its challenges, as several studies have raised concerns regarding its impact on learning outcomes and motivation [40,41], including science education [42,43]. As such, it has become apparent that gamification's "one size fits all" approach is not enough [22,30]. The nonadaptation of game elements and the appropriate didactic approach to the needs of each trainee individually, as well as the frequent presentation of similar and repetitive game elements, can increase abandonment levels in the long run [23]. Users do not necessarily have the same style when playing and achieving their tasks and goals, and they are usually motivated by specific game elements and mechanics [44]. Consequently, researchers have focused on adapting gamified environments to meet students' characteristics [45,46].

As per the Horizon Report 2021 [47], adaptive learning uses technologies that track students' progress and modify instructional approaches by leveraging data and various information. Adaptive learning technologies "dynamically adapt the level or type of course content based on an individual's abilities or skill acquisition. This process involves automated interventions and interventions from instructors, all aimed at accelerating the learner's performance" [48]. Adaptive gamification enhances learner participation by adapting and integrating different game elements and mechanics according to the user's characteristics [30]. It is designed to incorporate specific elements responsive to the learner. Including all elements relevant to different types of learners carries the risk of creating an excessive user interface overload [49]. However, adaptive gamification is still in its early stages [31].

Based on the available literature, adaptive gamification has yielded various outcomes. While there is evidence supporting the positive motivational impact of adaptive gamification, such as increased engagement [50], motivation [23,51], and retention [49], these effects may not apply to all player types [52,53]. Furthermore, there are instances where students using both adapted and nonadapted game elements have reported similar flow experiences [54]. In terms of learning performance, students have demonstrated higher task completion [22] and course completion rates [23], as well as improved learning outcomes [51]. However, studies investigating student motivation and learning outcomes still need to be expanded and cannot provide a clear overview. Additionally, according to

Hallifax et al. [55], conflicting results raise concerns regarding user modelling and selecting relevant game elements.

As Seaborn and Fels [56] have highlighted, the challenges the adaptive gamification approach faces are the interplay between gamification elements, dynamics, and user characteristics. They noted the need for an ideal gamification system to integrate game elements seamlessly. Therefore, the main challenge lies in developing a feasible design and implementation of adaptive gamification that effectively tackles these concerns. This entails utilizing customised game elements to match the students' characteristics and needs.

2.2. Gaming Elements

Game mechanics are a crucial component of gamification applications, and their significance becomes even more significant in adaptive gamification. According to Kapp [18], game application elements encompass various aspects, including challenges, badges, points, storytelling, etc. These mechanics are employed to enhance engagement and motivation within the gamified experience. Given their influence on students' behaviour, engagement, and motivation, it is crucial to recognize, select, and apply game elements within gamification. Carefully considering and implementing these elements can significantly impact the overall effectiveness and success of the gamified experience [18,57]. Designing a successful gamification system is complex and contains several inherent difficulties. It is important to note that incorporating numerous game elements simultaneously only sometimes guarantees a practical gamification experience [58]. Indeed, within the literature, different terminologies are used to define what could be the same game element. This variation arises because some works employ definitions at different levels of abstraction. For example, "progression" and "level" may refer to the same game element, depending on how the gamification system is structured or conceptualized. This highlights the need for clarity and consensus in defining and categorizing game elements to ensure effective communication and understanding in gamification research and practice [59].

Furthermore, despite the growing utilization of gamification in education [60], the assessment of the impact of the various game elements in education needs to be supported by factual, empirical findings. More rigorous empirical studies and research in this area are necessary to understand better how specific game elements impact learning outcomes and student engagement. Following the call for a deeper understanding of incentives in such contexts [61], it is strongly suggested [62] that additional studies explore the degree to which game mechanics influence the overall motivations of participants in idea contests. This highlights how specific game mechanics impact individuals' motivations and engagement in collaborative innovation processes such as idea contests.

2.3. Framework

The framework approach employed in the adaptive gamification environment aligns with the suggestions made by Zourmpakis et al. [31]. This framework revolves around two key factors. The first factor is the player model, which categorizes students' preferences for playing modes and game elements into six categories based on the Hexad model [2,63]. The categories are [64]:

- Achievers: They are primarily driven by a sense of competence. They enjoy engaging in new experiences and taking on challenges to demonstrate their abilities and accomplishments.
- Players: They are primarily motivated by external rewards. The rewards the system provides highly influence them, significantly impacting their behaviour, even if unrelated to their main progress or objectives. The reward system is crucial in motivating and shaping their engagement within the gamification environment.
- Philanthropists: They are primarily motivated by a sense of purpose. They derive satisfaction from helping others and are willing to offer assistance without expecting anything in return. Their motivation is driven by the desire to contribute and positively impact others rather than seeking personal rewards or gains.

- Disruptors: They are primarily motivated by change. Disruptors tend to push the system's boundaries, either in a negative manner, such as by spoiling the game for others, or in a positive manner, such as by identifying flaws and working towards improving the system. They desire to challenge and disrupt the status quo, seeking ways to bring change and innovation within the gamified environment.
- Socializers: They are primarily motivated by the need for relatedness. Socializers are intrinsically motivated by interactions with other players and establishing relationships with them (social relatedness). They find fulfilment and enjoyment in engaging with others, fostering social connections, and building community within the gamification context. Interpersonal interactions and social engagement are central to motivating and satisfying their gaming experience.
- Free spirits: They are mainly motivated by autonomy and self-expression. They have a strong desire to be in control of their actions and decisions, preferring to explore the system independently rather than being tightly regulated or controlled. They value the freedom to express themselves and engage with the gamified environment in ways that align with their preferences and interests. Autonomy and the opportunity for self-expression are critical drivers of motivation for free spirits.

The second factor in the proposed framework is learning strategies. Learning strategies significantly shape the learning process's goals, objectives, paths, and stages. However, due to the potential burden of switching between learning strategies and the nature of science education, the framework focuses on only two preferred learning strategies. This approach aims to reduce the workload associated with adaptation and familiarize students with both learning strategies if adaptation is required. The chosen learning strategies also share common aspects, facilitating the transition between them if necessary. The proposed approach encompasses two adaptation processes: the adaptation of game elements and the adaptation of the learning process. Three main points are considered in adapting the game elements: user feedback, profiling, and adaptation. The system continuously updates the player's profile throughout the course. This is achieved through in-app dialogues designed to gather the user's feedback and opinions regarding the game elements. At the end of the lesson, the user is asked questions based on their updated profile. These questions help to customize the game elements and allow the user to select a game element from the second and third-player categories, based on their ranking, to be included.

The basic architecture of the proposed adaptive gamification is as follows:

1. The user completes a questionnaire using the Hexad model to create the initial player profile.
2. The system selects and applies game elements to the environment based on the player's profile.
3. The player's profile is updated throughout the course through in-app dialogues to gather feedback and preferences.
4. At the end of the lesson, the user is asked questions based on their updated profile to customize the game elements.
5. The user selects a game element from the second and third player categories to be included according to the ranking.
6. The system adjusts the gamified environment based on the selected game element and updates the player's profile accordingly.

This basic architecture allows for the personalized adaptation of game elements based on the user's preferences and feedback, enhancing the overall gamification experience. The framework utilized in the adaptive gamification environment aligns with and builds upon the work conducted by Zourmpakis et al. [31] (Figure 1). For a comprehensive understanding of the adaptive gamification framework, including additional details and explanations, we recommend referring to the research conducted by Zourmpakis et al. [31].

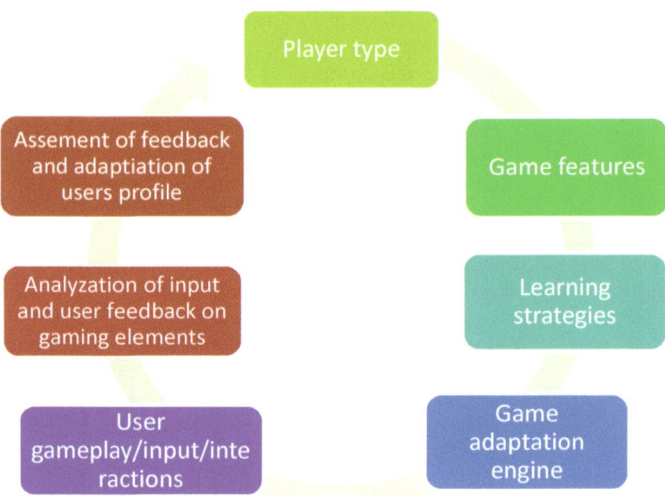

Figure 1. The architecture of the adaptive gamification environment from the proposal of Zourmpakis et al. [31].

Based on the provided guidelines and methodology, an adaptive gamification environment called "Water Cycle" was developed using the Unity3D game engine. This gamification application focuses on teaching phenomena related to the water cycle, explicitly melting, freezing, boiling, and evaporation.

The "Water Cycle" application is designed as an open-world simulation environment incorporating avatars. After the user completes the questionnaire and creates a profile, they are presented with a text that describes the game's story based on their profile and the chosen learning strategy.

In the case of exploratory learning, the user assumes the role of a friend of a policewoman, assisting her in investigating various natural phenomena. On the other hand, if problem-solving learning is selected, the user takes on the role of a police officer's assistant and is assigned tasks to solve specific cases related to the water cycle.

Figure 2 illustrates the user's freedom to navigate and explore the application. Users can interact with nonplayer characters, engage in conversations, collect materials, conduct experiments, and observe real-life phenomena and sounds. This immersive environment provides users a rich and interactive experience, allowing them to actively participate in and learn about the water cycle through various activities and interactions.

Figure 2. Water Cycle in-game environment.

3. Methodology

This quantitative study was conducted in primary schools of Heraklion, Crete, Greece. This research was carried out with the main purpose of defining the motivational impact of primary students on the use of adaptive gamification environments in science education based on their views. The students' views regarding the game elements they used were also examined.

Initially, it is essential to acknowledge that the examination of water cycle phenomena was selected due to its wide range of practical applications in daily life, as we believed it would captivate students' interest. Additionally, the water cycle concept and its associated ideas have been incorporated into the recently developed Comprehensive Curriculum in Greece, with teacher training already underway. In this study, we utilized a semi-experimental design with a convenience sample. It occurred in 6 3rd-grade classes in 3 different schools in Heraklion. Students in all classes were taught similarly, using the adaptive gamification application, which ran exclusively on computers. The students that collectively took part were 80. Each class followed four lessons. The concepts taught were coagulation, melting, evaporation, and boiling. The research was implemented in 2 phases. In the first phase, the four teaching interventions were implemented. In the second phase, the students completed the questionnaire. Each class's teachers conducted the lessons. The elementary teachers were trained beforehand using a theoretical framework centred on integrating technology into the educational process. The teacher training was therefore designed based on the TPASK (technological pedagogical science knowledge) model [65] as it enables the analysis and development of a multifaceted phenomenon such as technology integration while helping to formulate the kind of knowledge that teachers need in order to integrate technology while teaching science concepts into the actual classroom setting. During the research process, we systematically followed all national and international rules for ethics and ethics in research [66], and permission was obtained from the Ethics Committee of the PTPE of the University of Crete.

We initiated the process by considering the primary research inquiries when developing the questionnaire. We then proceeded by consulting the pertinent literature in the field, explicitly referring to Rajendran et al. [67], Halim et al. [68], and Melkersson and Lundin [69] for guidance, insights and designing the survey questionnaire. Furthermore, specific questions were appropriately revised for technological advancements and research on their content or context. Taggart et al. [70] suggested involving experts in developing a questionnaire can enhance its content validity. In creating the questionnaire for this particular study, we employed an iterative approach, which involved the research team in generating the items and sought input from educational technology experts at the University of Crete to review the questionnaire items. The development process encompassed multiple cycles of iteration until the final version was reached.

The Likert scale was used as a measurement for data collection. All categories of questions were scored from 1 to 5 (strongly disagree, disagree, neither agree nor disagree, agree, strongly agree). In total, the questionnaire provided included two categories of questions. In the first category, there were questions about students' motivation and views on learning science using the adaptive gamification application and their motivation regarding game mechanisms and elements. Regarding the second category, there were questions about all the elements and mechanisms included in the application. However, before answering about a mechanism, students were asked to select whether they had encountered or used that particular element. Students answered only about game elements they had encountered and used in the app. Each item had the same six appropriately tailored questions. The questionnaire included nine questions for Category 1 and 66 questions (11 items) for Category 2.

It should be noted that in two out of the six classes, there were more students than there were computers, by a few. In order to avoid splitting the classes in half, some students were grouped the first time based on their profiling and, more accurately, their dominant type. This was performed to ensure that the grouped students would have, as much as possible, similar characteristics relative to each other. Though this ensured that most students had similar characteristics, not all students that were put together liked teamwork necessarily. As such, though the grouped students had roughly similar characteristics based on their primary player type, some had to cooperate even though they would not have wanted to. However, students in the free spirits category were not grouped, as this would have substantially affected their experience. The results were analyzed using the SPSS statistical package.

4. Results

The first category had nine items, graded on a Likert-type scale of 1 to 5, with the intent to examine students' motivation and views about learning science using the adaptive gamification application. The information collected was organized into Table 1, as presented below, and was subjected to descriptive analysis.

Table 1. Students' motivation and views about learning science using the adaptive gamification application.

	Questions	Absolutely Disagree (%)	Disagree (%)	Neither Agree nor Disagree (%)	Agree (%)	Strongly Agree (%)	Mean Average	Std Deviation
1.	I like to learn about natural phenomena using applications like the one with the water cycle application	0	0	7.5	32.5	60.0	4.53	0.636
2.	I am interested in learning about natural phenomena in school.	0	3.8	7.5	37.5	51.3	4.36	0.783
3.	I prefer to spend more time learning about natural phenomena from other subjects.	5	2.5	22.5	28.8	41.3	3.99	1.097
4.	I can learn natural phenomena using applications such as the Water Cycle.	2.5	2.5	6.3	36.3	52.5	4.34	0.899
5.	I feel more confident learning about natural phenomena in a fun way.	1.3	2.5	13.8	30.0	52.5	4.30	0.892
6.	The classroom lesson on natural phenomena is boring.	26.3	11.3	17.5	12.5	32.5	3.14	1.613
7.	Learning about natural phenomena such as the water cycle increases my interest in learning.	0	5.0	6.3	31.3	57.5	4.41	0.822
8.	Using apps like the one with the water cycle makes me want to learn more about natural phenomena and be good at it.	7.5	1.3	6.3	30.0	55.0	4.24	1.139
9.	Using applications like the Water Cycle to learn about natural phenomena makes me feel less nervous in class.	10.0	2.5	21.3	26.3	40.0	3.84	1.267

Based on Table 1, students showed their substantially high-levelled fondness for learning science while using the adaptive gamification application (M= 4.53), with 92.5% agreeing or strongly agreeing (Q1). Though students seem to be highly interested in learning science concepts in the school setting (Q2) (M = 4.36) and even showed a preference for science education and would rather spend more time in it than other subjects (M = 3.99), nearly half of them (45%, agree or strongly agree) consider the traditional teaching that occurs in the classroom regarding science education to be dull (Q6). Furthermore, this impacted their level of confidence in learning, as students expressed feeling more assured in understanding natural phenomena through an enjoyable approach (M = 4.30) (Q5). A significant 52.5% of students strongly agreed with this notion. However, in contrast to traditional classroom teaching methods, students reported a significant increase in their interest in learning through the adaptive gamification environment (mean = 4.41) (Q7). Only a tiny percentage (5%) expressed disagreement with this statement.

Additionally, students displayed higher motivation and perceived the application as a valuable learning tool for studying science (mean = 4.34) (Q4). Moreover, 85.5% of students agreed or strongly agreed that the application fostered a greater desire to learn and excel (Q8). Finally, using the application positively affected students' nervousness (mean = 3.84), with nearly two-thirds of students agreeing or strongly agreeing that they felt more relaxed and less nervous while engaging in the learning process (Q9).

The second category aimed to assess the impact of game elements and mechanisms students encountered in the adaptive gamification environment on their motivation. This category comprised 11 subcategories, each representing a distinct game element. Within each subcategory, six questions were modified to convey the same meaning. At the start of each subcategory, the initial question was to establish whether the student had encountered that specific element in their playthroughs. If the student had not encountered it, they would mark "NO" and move on to the following subcategory, skipping the current one. If the student answered affirmatively, they would respond to the following five questions within that subcategory. Like the first category, the descriptive analysis examined all the questions.

Table 2 presents the first two game elements: badges and currency. Among the 80 students, it was found that 60 students had encountered badges in at least one of their playthroughs, while 61 students had experienced the currency mechanism at some point during their learning process. In the case of the badges element, students admitted to being fond of the application since it contained badges (M = 4.23), with almost half of them strongly agreeing with the notion. Furthermore, the participants expressed their strong fondness for the badges (M = 4.55), with no one expressing disagreement or strong disagreement. They also demonstrated significant effort in actively seeking to obtain badges while using the app (M = 4.33). The majority of students were successful in acquiring the desired badges (M = 4.03). However, almost one out of three students (36.9%) agreed or strongly agreed that they experienced nervousness while attempting to collect badges.

Table 2. Badges and currency.

Subcategory	Questions	Participants Answered	Absolutely Disagree (%)	Disagree (%)	Neither Agree nor Disagree (%)	Agree (%)	Strongly Agree (%)	Mean Average	Std Deviation
Students' motivation for badges	I liked the app because it had badges.	60	5	0	10	36.7	48.3	4.23	0.998
	I was actively trying to get badges.	60	0	3.3	11.7	33.3	51.7	4.33	0.816
	I collected the badges I wanted to collect.	60	3.3	10	15	23.3	48.3	4.03	1.164
	I loved collecting badges.	60	0	0	11.7	21.7	66.7	4.55	0.699
	I got nervous trying to collect badges (reversed question).	60	46.7	6.7	10	18.3	18.3	2.55	1.641
Students' motivation for currency	I liked the app because it had currency/money.	61	6.6	9.8	19.7	26.2	37.7	3.79	1.24
	I was actively trying to earn money.	61	4.9	3.3	8.2	29.5	54.1	4.25	1.075
	I earned the money I wanted to raise.	61	13.1	11.5	24.6	19.7	31.1	3.44	1.385
	I liked to earn money.	61	0	1.6	19.7	24.6	54.1	4.31	0.847
	I was stressed trying to save money (reversed question).	61	36.1	19.7	9.8	18	16.4	2.59	1.532

Regarding the in-app currency, the students were highly interested in earning money (M = 4.31) and actively pursued it within the game (M = 4.25). However, their fondness for the application decreased slightly due to the currency mechanism (M = 3.79). This decline could be attributed to their inability to earn the desired amount of money, as only about half admitted to achieving this goal (50.8% agreed or strongly agreed). Additionally, over 34.4% of participants reported feeling stressed while trying to acquire money (agreed or strongly agreed).

As indicated in Table 3, the following game elements include the storytelling/cases element and points. The data from Table 3 reveals that most students experienced these elements, with 72 students reporting the point element, while only 3 out of 80 individuals did not report encountering the storytelling/cases element. The cases/storytelling game element appears to be highly appreciated, as nearly 9 out of 10 students (88.3%) agreed or strongly agreed that they enjoyed solving cases (M = 4.47) and liked the application more because of the inclusion of this specific game element (89.6%) (M = 4.47). Furthermore, a similarly high percentage of participants expressed their active and enthusiastic engagement with this element (89.6% agree or strongly agree) (M = 4.42) and reported success in following the story and solving the cases (90.9% agree or absolutely agree) (M = 4.51). However, almost 3 out of 10 individuals (29.9% agree or absolutely agree) admitted feeling nervous while attempting to solve the cases.

Table 3. Cases/storytelling and points.

Subcategory	Questions	Participants Answered	Absolutely Disagree (%)	Disagree (%)	Neither Agree nor Disagree (%)	Agree (%)	Strongly Agree (%)	Mean Average	Std Deviation
Students' motivation for the cases/storytelling	I liked the game because it had cases.	77	1.3	1.3	7.8	28.6	61	4.47	0.804
	I was actively trying to solve the cases.	77	2.6	1.3	6.5	31.2	58.4	4.42	0.879
	I investigated and solved the cases to the best of my ability.	77	3.9	0	5.2	23.4	67.5	4.51	0.912
	I liked solving cases.	77	1.3	1.3	9.1	26	62.3	4.47	0.821
	I used to get nervous when solving cases (reversed questions).	77	40.3	13	16.9	13	16.9	2.53	1.535
Students' motivation for points	I liked the app because it had points.	72	4.2	0	13.9	41.7	40.3	4.14	0.954
	I was actively trying to score points.	72	4.2	2.8	1.4	33.3	58.3	4.39	0.972
	I collected the points I wanted to collect.	72	5.6	9.7	16.7	22.2	45.8	3.93	1.237
	I liked to score points.	72	5.6	1.4	4.2	26.4	62.5	4.39	1.042
	I used to get nervous when I was trying to score points (reversed question).	72	41.7	20.8	8.3	11.1	18.1	2.43	1.555

Regarding the points element, the majority of participants expressed their fondness for it (M = 4.39), and nearly 8 out of 10 individuals (80%, agree or absolutely agree) appreciated the adaptive gamification environment more because it included this specific game element (M = 4.14). Most students were highly focused and made significant efforts to accumulate points (M = 4.39). However, not all students successfully attained the desired points, as approximately 2 out of 3 (67%) agreed or strongly agreed that they could collect the points they wanted (M = 3.93). Furthermore, almost 1 out of 3 participants (33%) reported feeling nervous (agree or strongly agree) while trying to earn points.

Table 4 presents the motivation levels related to the gift and levels of game elements. According to Table 4, a small number of students (24 participants) encountered or noticed the gift game element, whereas most students (76 out of 80) experienced the level mechanism. As indicated in Table 4, a few students (24 participants) encountered or noticed the gift game element. However, despite the limited exposure to this element, the students showed a strong affinity towards it (M = 4.50) and liked the application because of its inclusion (M = 4.54). Surprisingly, none of these students disagreed with the gift game element (disagree or absolutely disagree). Most students actively engaged with the gift game element and tried to acquire gifts (M = 4.21). Two out of three students (66.7% agree or absolutely agree) reported successfully collecting a significant number of the gifts they desired (M = 4.00). However, similar to previous findings, almost three out of ten individuals (29.2%) experienced significant stress (agreed or absolutely agreed) while engaging with the gift game element. The levels element was also well-received, with 84.2% of participants liking this specific game element (M = 4.39).

Table 4. Gifts and levels.

Subcategory	Questions	Participants Answered	Absolutely Disagree (%)	Disagree (%)	Neither Agree nor Disagree (%)	Agree (%)	Strongly Agree (%)	Mean Average	Std Deviation
Students' motivation for gifts	I liked the app because it had gifts.	24	0	0	4.2	37.5	58.3	4.54	0.588
	I was actively trying to get gifts.	24	8.3	0	8.3	29.2	54.2	4.21	1.179
	I collected the gifts I wanted to collect.	24	8.3	0	25	16.7	50	4	1.251
	I like collecting gifts.	24	0	0	12.5	25	62.5	4.5	0.722
	I used to get nervous when collecting gifts (reversed question).	24	41.7	12.5	16.7	16.7	12.5	2.46	1.503
Students' motivation for levels	I liked the app because it had levels.	76	1.3	0	17.1	36.8	44.7	4.24	0.831
	I was actively trying to climb levels.	76	2.6	5.3	10.5	19.7	61.8	4.33	1.038
	I raised to the level I wanted to.	76	9.2	9.2	18.4	22.4	40.8	3.76	1.325
	I liked going up levels.	76	1.3	1.3	13.2	25	59.2	4.39	0.865
	I used to get nervous when trying to go up levels (reversed question).	76	47.4	21.1	9.2	3.9	18.4	2.25	1.533

Additionally, 80.5% of students were fond of the application because it included the element of the level (M = 4.24) (agreed or absolutely agreed). Furthermore, a significant majority of students (81.5%) actively attempted to reach higher levels during their playthrough (M = 4.33), with nearly two out of three students (66.7%) successfully achieving this goal (agreed or absolutely agreed) (M = 3.76). The stress levels associated with levelling up were relatively low, as only about one out of five students (22.3%) absolutely agreed or agreed that they felt nervous while trying to progress to higher levels.

Based on the data in Table 5, 61 students came across the promotion element, while 57 students cooperated with other peers and essentially used the cooperation element. The promotion element was highly regarded, as most students expressed their fondness for this element (M = 4.43) and approved of the adaptive gamification environment because of its inclusion (M = 4.30). Additionally, nearly three out of four individuals (77.1% agree or absolutely agree)) actively made an effort to get promoted during their gameplay (M = 4.13), and approximately seven out of ten students (72.1% agree or absolutely agree) successfully achieved their desired promotion (M = 4.00). However, it is worth noting that a fair number of students (31.1% agree or absolutely agree) reported feeling stressed concerning the promotion element. The cooperation element garnered significant appreciation, as many individuals expressed their fondness for working with other students (86% agree or absolutely agree) (M = 4.26).

Similarly, the same percentage of students liked the application specifically because it included the cooperation mechanism (86% agree or absolutely agree) (M = 4.36). However, it is worth noting that the negative responses for both questions were around 10%, one of the highest among all the other game elements. Furthermore, a substantial majority of students (80.7%) agreed or absolutely agreed that they actively tried to cooperate with their peers (M = 4.09), and the same percentage believed that they were successful in achieving cooperation based on their perspective (M = 4.25). Unfortunately, the stress levels reported by the students concerning the cooperation element were relatively high, with 42.1% agreeing or strongly agreeing that they felt stressed. This could be understood considering that a few students were not keen on working together, leading to higher stress levels than other game elements.

Table 5. Promotion and cooperation.

Subcategory	Questions	Participants Answered	Absolutely Disagree (%)	Disagree (%)	Neither Agree nor Disagree (%)	Agree (%)	Strongly Agree (%)	Mean Average	Std Deviation
Students' motivation for promotion	I liked the app because I could get promoted.	61	1.6	0	14.8	34.4	49.2	4.3	0.843
	I was actively trying to get promoted.	61	3.3	6.6	13.1	27.9	49.2	4.13	1.087
	I got the promotions I wanted in the app.	61	6.6	6.6	14.8	24.6	47.5	4	1.225
	I liked getting promoted.	61	1.6	1.6	6.6	32.8	57.4	4.43	0.826
	I used to get nervous when trying to get promoted (reversed question).	61	42.6	16.4	9.8	9.8	21.3	2.51	1.619
Students' motivation for cooperation	I liked the app because I could collaborate with/help other students.	57	5.3	5.3	3.5	22.8	63.2	4.33	1.123
	I actively tried to cooperate with/help other students.	57	7	5.3	7	33.3	47.4	4.09	1.184
	I cooperated/helped the other students I wanted to help.	57	1.8	7	10.5	26.3	54.4	4.25	1.023
	I liked working with/helping other students.	57	5.3	3.5	5.3	31.6	54.4	4.26	1.078
	I was stressed when trying to cooperate/help other students (reversed question).	57	38.6	14	5.3	14	28.1	2.79	1.719

The following two game elements, as highlighted in Table 6, include challenges and customization. According to the methodology employed in the adaptive gamification environment, all profiles created by the dominant player type and the two higher player types lead to the inclusion of the challenges in all possible profiles. As depicted in Table 6, all participants noticed and experienced the element of the challenge, while only 25 individuals encountered the customization element. Students demonstrated a strong affinity for the element of the challenge, with 88.8% agreeing or agreeing that they enjoyed facing and overcoming challenges (M = 4.39). Furthermore, 85.1% of students agreed or absolutely agreed that including challenges in the application made them appreciate it more (M = 4.24). Most students actively worked to overcome the challenges they encountered (M = 4.31), and most expressed satisfaction in successfully overcoming the challenges they set out to do (M = 4.03). However, like other game elements, nearly 3 out of 10 students reported feeling nervous while attempting to resolve the challenges.

Regarding the customization element, students displayed a significant fondness for the ability to change their appearance (M = 4.08). However, they could have appreciated the adaptive gamification environment more due to the inclusion of this game mechanism (M = 3.81). A total of 15% of the students stated that they did not like or dislike the application because it contained this element, the highest among all other elements. This response is understandable, considering the data presented in Table 6. Although students actively tried to change their appearance (M = 4.04), only a few could do so to the extent they desired, as more than half of the students stated that they could not achieve their desired customization (57.7% absolutely disagree or disagree). Fortunately, the stress levels associated with the customization element were similar to some of the other game elements, with 23.1% of students reporting significant stress levels (agree or absolutely agree).

Table 7 illustrates the last game element included in the application, the roles game element. According to the data, almost all participants (except for 2) noticed and experienced the roles game element. This element received highly positive feedback (M = 4.47), with 91.1% of participants (agree or absolutely agree) expressing their liking for having a role in the application. Additionally, an almost similar percentage (93.6%, agree or absolutely agree) expressed fondness for the application due to the inclusion of this element (M = 4.53). Students showed great interest and actively engaged in "playing" their roles (M = 4.45), with approximately 8 out of 10 students (80.7%, agree or absolutely agreeing) succeeding in improving their role during gameplay (M = 4.22). However, nearly 1 out of 4 individuals reported feeling nervous to a significant degree (agree or absolutely agree) while participating in their assigned roles.

Table 6. Challenges and customization.

Subcategory	Questions	Participants Answered	Absolutely Disagree (%)	Disagree (%)	Neither Agree nor Disagree (%)	Agree (%)	Strongly Agree (%)	Mean Average	Std Deviation
Student motivation for the challenges	I liked the app because it had challenges.	80	2.5	2.5	10	38.8	46.3	4.24	0.917
	I was actively trying to overcome challenges.	80	3.8	1.3	11.3	27.5	56.3	4.31	0.988
	I overcame the challenges I wanted to overcome.	80	6.3	7.5	13.8	22.5	50	4.03	1.232
	I like to overcome challenges.	80	2.5	5	3.8	28.8	60	4.39	0.961
	I used to get nervous when overcoming challenges (reversed question).	80	51.3	11.3	8.8	13.8	15	2.3	1.562
Student motivation for customization	I liked the game because of the change in appearance.	26	11.5	3.8	19.2	23.1	42.3	3.81	1.357
	I was actively trying to change my appearance.	26	0	11.5	15.4	30.8	42.3	4.04	1.038
	I managed to change my appearance as many times as I wanted.	26	42.3	15.4	19.2	7.7	15.4	2.38	1.499
	I like to change my appearance.	26	0	4	32	16	48	4.08	0.997
	I used to get anxious when I tried to change my appearance (reversed question).	26	50	7.7	19.2	15.4	7.7	2.23	1.423

Table 7. Roles.

Subcategory	Questions	Participants Answered	Absolutely Disagree (%)	Disagree (%)	Neither Agree nor Disagree (%)	Agree (%)	Strongly Agree (%)	Mean Average	Std Deviation
Student motivation for the roles	I liked the game because I had a role.	78	1.3	5.1	5.1	32.1	61.5	4.53	0.716
	I was actively trying to "play" my role.	78	0	2.6	9	29.5	59	4.45	0.767
	I wanted to and succeeded in improving my role within the application.	78	9	2.6	7.7	19.2	61.5	4.22	1.255
	I liked having a role within the app.	78	0	2.6	6.4	32.1	59	4.47	0.734
	I was nervous when I had a role within the app (reversed question).	78	48.7	10.3	12.8	10.3	17.9	2.38	1.589

5. Discussion

The current research study provides valuable insights into the field of adaptive gamification, particularly in the context of science education. While limited research exists on gamification in science education [17], and even fewer studies focus specifically on students' motivation using adaptive gamification applications, most existing studies generally compare gamification and adaptive gamification in a general manner [71]. This study acknowledges that students' motivational aspects can be influenced by their activities and the specific domain or content [59]. Therefore, drawing generalized conclusions about the affordances of adaptive gamification becomes challenging. The objective of this study is to offer an understanding of the motivational impact of a domain-specific adaptive gamification framework on students in the field of science education, as well as how the gaming elements they encountered influenced them.

In terms of motivational aspects, the findings of this study indicate that participants displayed a higher level of motivation towards learning science when using an adaptive gamification environment. Students strongly preferred the application as a means of learning science, with no students disagreeing or strongly disagreeing. They also reported an increased interest in learning through this approach, with only a tiny percentage (5%) expressing disagreement. These results demonstrate that students felt motivated and highly interested in learning science when incorporating the adaptive gamification environment. Additionally, a significant portion of students found traditional classroom lessons boring. However, the data suggests that it is not the content that discourages students from learning science, as they demonstrated a high interest in learning about natural phenomena and even preferred science over other subjects. Students associated their confidence with the "fun" aspect of learning, indicating that a more engaging and enjoyable learning approach can significantly impact their level of engagement and confidence. This is evident because students believed they could learn science concepts through the application and were willing to enhance their understanding of natural phenomena through its use. This finding aligns with previous studies indicating that gamification environments, which offer an enjoyable approach to learning, can enhance student engagement and foster a greater willingness to engage with similar applications in the future [72]. Furthermore, many students reported feeling less nervous during the learning process when using the adaptive gamification environment, which increased their attentiveness in the classroom and their readiness to improve their performance in science lessons.

In addition, we examined the impact of game elements on students' motivation. It is important to note that the game elements and mechanisms were not implemented simultaneously in this adaptive gamification application. Our framework employed a multidimensional adaptive design, unlike most previous studies that followed a single-dimensional personalization approach [71]. This design allowed for runtime adaptation, resulting in a more comprehensive personalization [31]. Based on our findings, it became evident that at least some students experienced adaptation in their profiles, leading to variations in the game elements they encountered. This was reflected in the number of elements reported by students and the possible combinations they experienced. These insights highlight the dynamic and personalized nature of the adaptive gamification approach used in this study [31].

Based on these findings, the adaptation process was largely successful, as students generally liked the game elements integrated into their lessons. However, the meagre negative ratings regarding the likeness and appreciation of the application suggest that either the adaptation was applied to a small number of individuals or it was implemented on a limited scale, such as switching between closely related user types (e.g., player and achiever) or adjusting the secondary and tertiary dominant user types. Nevertheless, regardless of the specifics, the multidimensional framework, which considered more than just a single user type, effectively enhanced students' experiences and engagement. A similar study also supports this conclusion [71].

Additionally, it was observed that badges, storytelling/cases, gifts, and roles received the most positive feedback from students and contributed to their appreciation of the application. This suggests that students' motivation is greatly influenced by their immersion in the storytelling aspect, the role they assume within the application, and the rewards that symbolize their status and self-improvement. The connection between immersion in storytelling/narrative elements and its positive impact on student engagement aligns with findings from other studies [73]. Similarly, the positive relationship between badges and student motivation and engagement has been documented in previous research [19]. These findings further support that these game elements effectively enhance students' motivation and engagement in the learning process.

What is more, the satisfaction of individuals with using game elements did not necessarily impact their level of active engagement or fondness for the application. Lower scores in satisfaction did not lead to a corresponding decrease in fondness of the application. However, user satisfaction was generally lower than active engagement and fondness, except for cases/storytelling, roles, and cooperation. This suggests that students were primarily satisfied by the immersive aspects of the application, which is consistent with the findings of Aldemir et al. [73], highlighting the importance of immersion for the storytelling element to affect motivation and engagement. The application has room for improvement concerning other game elements, particularly in customization, levels, and currency, which received lower satisfaction scores.

Regarding cooperation, it is not easy to draw confident conclusions about student satisfaction since it was also used by some students who may have yet to be inclined to use it. Students were less active in the cooperation element, but their satisfaction was higher. This indicates that even though some students did not put in their maximum effort to cooperate with others, possibly because they did not have a strong desire for it, the results were better than expected, resulting in higher satisfaction. We cannot make assumptions about the specific user types and their interactions with most of the game elements, as we need to have information about the profile of each user and the ability for user types to change during gameplay. However, all user profiles consistently utilized the challenges game element [31]. It was found that challenges received high approval rates among all user types, promoting high levels of active engagement and satisfaction [73]. This suggests that the intrinsic motivation for accomplishment has a noticeable effect on the achiever type [64] and all user types. However, it is impossible to determine to what extent each user type is affected by this game element.

Moreover, games are commonly associated with stress reduction for individuals [74]. Based on our findings, most game elements did not induce high-stress levels, as the percentage of individuals who agreed or mostly agreed that a game element or mechanism made them anxious was generally below 30%. However, this differed for the cooperation, badges, and currency elements. The higher level of anxiety reported for the cooperation element can be understood since some students were required to cooperate even if they did not want to. The previous literature has shown that badges can generate a certain level of anxiety [69], which aligns with our results. However, the same does not apply to the currency element. The anxiety associated with this element is likely a result of students being highly active and striving to earn money during their playthrough but ultimately failing to do so.

Consequently, not being able to accumulate enough money to make in-game purchases and acquire items they had set their minds on could contribute to their anxiety. Furthermore, the element of the challenge involved the use of timers, as difficulty and assistance were linked to time. However, students did not report higher stress levels than most other game elements, contrasting with findings from other research studies [75].

6. Limitations

The present study has certain limitations. Adaptive gamification was introduced to enhance the effectiveness of one-size-fits-all gamification. While we examined the motivational outcomes of students in the adapted gamification environment, we did not compare it to a one-size-fits-all gamification approach. As a result, this study cannot provide definitive results on whether this objective was achieved. Additionally, our findings are limited to the specific game elements incorporated in our adaptive gamification approach. We did not include every game element mentioned in gamification literature, such as leaderboards and team chats. Different outcomes may be observed by integrating these or other game elements.

Moreover, in this study, we did not consider other essential user characteristics regarding gamification, such as gender differences. In addition, each learning course was carried out by the teacher of that class. Though all teachers had been trained to use the adaptive gamification environment before teaching in the school setting, it is still possible to have affected students' motivation from the learning experience. Furthermore, knowing students' profiles would provide more insight into the accuracy of the Hexad typology, the frequency of profile changes, whether changes occurred collectively or individually, and the frequency of adjustments, particularly to the dominant profile. Additionally, it would offer a more detailed and precise understanding of how each game element affects each profile [3,76]. Finally, this study needs to be more extensive concerning our results' context dependency and generalizability. Long-term studies could also ensure that the motivational outcomes are not due to the novelty effect. The framework used was specifically designed for the science education learning environment, and results may differ in other domains, as Hallifax et al. [3] suggested.

7. Conclusions

This article showcased the results of a quantitative semi-experimental investigation into the impact of adaptive gamification on learner motivation and perceptions in science education. The study was conducted in four primary schools located in Heraklion, Crete, Greece. The data collected and analyzed involved 80 students from six distinct classes who had completed four custom-designed learning modules focused on water cycle science concepts. These modules incorporated a specialized adaptive gamification framework tailored to the subject matter.

Our findings offer new perspectives and enhance the utilization of adaptive gamification. We discovered valuable insights into how to customize gamification, particularly by demonstrating that the Hexad user typology is one of the most relevant approaches for identifying user preferences regarding game elements. However, in this implementation, the adaptation of game elements and learning strategies was not solely based on a single user type. We recognized that motivations can be fluid and may change during gameplay. Therefore, the foundation for adaptation was the learner's profile rather than a static adaptation that classifies individuals solely based on their user profiles after an initial selection, as proposed in the literature [2,55].

Nonetheless, since the appropriate game element is tailored to the specific learner profile, it still promotes self-determination. The results obtained by examining each game element emphasize that they have varying effects on learners' motivation, as shown by previous research [75]. It is crucial to exercise caution when suggesting game elements to learners, as these elements may have conflicting impacts depending on their profiles [75].

According to Hallifax et al. [3], user motivation in tailored gamification is influenced by two significant factors: implementing a specific motivational strategy and selecting the user typology or profile. Both factors are considered in this case, with the results being extremely positive. However, it is essential to note that teachers also play a crucial role. Though the teachers in this study had volunteered and were trained in utilizing this application in class, the inability to effectively utilize digital content from a pedagogical standpoint has been shown to impact students negatively, such as increased discouragement and reduced learning motivation [77,78]. Thus, considering that the influence of teachers' views and attitudes is essential when introducing new teaching methodologies into the learning process, it is crucial to maximize a learning tool's positive benefits on students' motivation to learn [27,79].

Moreover, future research is recommended to examine personalization's effects in user studies conducted over an extended period. This would enable investigating whether the enhanced user experience translates into improved performance over time. Alternatively, offering users the option to decide whether they want to use the system regularly would provide further insights into potential effects on user behaviour, and this aspect could be explored in future studies as well.

Additional studies that address our research's limitations and explore the effectiveness of similar adaptive gamification applications in science education would be beneficial. These studies could validate and generalize the results while providing valuable insights for future design modifications to enhance the learning experience. Additionally, it is crucial to investigate learning achievements. Measuring both learning outcomes and motivational aspects is equally important to comprehensively assess adaptive gamification's effectiveness in education. By examining the impact on learning outcomes, such as knowledge retention and skill acquisition, alongside motivational factors, we can better understand the effectiveness and potential benefits of implementing adaptive gamification in the learning process [72,80]. Consequently, properly developing applications for science teaching concepts is crucial for motivating students to actively participate and engage with the material [17,81]. Moreover, the development of similar content-specific adaptive gamification environments holds the potential to benefit other subject areas beyond science education.

Author Contributions: Conceptualization, A.-I.Z., M.K. and S.P.; methodology, A.-I.Z., M.K. and S.P.; software, A.-I.Z.; validation, A.-I.Z., M.K. and S.P.; formal analysis, A.-I.Z., M.K. and S.P.; investigation, A.-I.Z., M.K. and S.P.; resources, A.-I.Z., M.K. and S.P.; data curation, A.-I.Z., M.K. and S.P.; writing—original draft preparation, A.-I.Z., M.K. and S.P.; writing—review and editing, A.-I.Z., M.K. and S.P.; visualization, A.-I.Z., M.K. and S.P.; supervision, A.-I.Z., M.K. and S.P.; project administration, A.-I.Z., M.K. and S.P. All co-authors contributed to data collection and/or analysis of project results. All authors have read and agreed to the published version of the manuscript.

Funding: This research received no external funding.

Data Availability Statement: The data presented in this study are available on request from the corresponding author.

Conflicts of Interest: The authors declare no conflict of interest.

References

1. Amado, C.M.; Roleda, L.S. Game Element Preferences and Engagement of Different Hexad Player Types in a Gamified Physics Course. In *Proceedings of the 2020 11th International Conference on E-Education, E-Business, E-Management, and E-Learning, Osaka, Japan, 10–12 January 2020*; Association for Computing Machinery: New York, NY, USA, 2020; pp. 261–267.
2. Lavoué, É.; Monterrat, B.; Desmarais, M.; George, S. Adaptive Gamification for Learning Environments. *IEEE Trans. Learn. Technol.* **2019**, *12*, 16–28. [CrossRef]
3. Hallifax, S.; Serna, A.; Marty, J.C.; Lavoué, É. Adaptive Gamification in Education: A Literature Review of Current Trends and Developments. In *Proceedings of the Transforming Learning with Meaningful Technologies: 14th European Conference on Technology Enhanced Learning, Delft, The Netherlands, 16–19 September 2019*; Springer: Cham, Switzerland, 2019; Volume 11722, pp. 294–307.
4. Gunuc, S. The Relationships between Student Engagement and Their Academic Achievement. *Int. J. New Trends Educ. Their Implic.* **2014**, *5*, 199–214.
5. Wara, E.; Aloka, J.O.; Benson, C.O. Relationship between Emotional Engagement and Academic Achievement among Kenyan Secondary School Students. *Acad. J. Interdiscip. Stud.* **2018**, *7*, 107–118. [CrossRef]
6. Damaševičius, R.; Maskeli, R.; Blažauskas, T. Serious Games and Gamification in Healthcare: A Meta-Review. *Information* **2023**, *14*, 105. [CrossRef]
7. Rodrigues, L.F.; Oliveira, A.; Rodrigues, H. Main Gamification Concepts: A Systematic Mapping Study. *Heliyon* **2019**, *5*, e01993. [CrossRef]
8. Sakkir, G.; Dollah, S.; Ahmad, J. E-Learning in COVID-19 Situation: Students' Perception. *EduLine J. Educ. Learn. Innov.* **2021**, *1*, 9–15. [CrossRef]
9. Tan, C. The Impact of COVID-19 on Student Motivation, Community of Inquiry and Learning Performance. *Asian Educ. Dev. Stud.* **2021**, *10*, 308–321. [CrossRef]
10. Council, N.R. *A Framework for K-12 Science Education: Practices, Crosscutting Concepts, and Core Ideas*; The National Academies Press: Washington, DC, USA, 2012.
11. Obe, W.H. *The Teaching of Science in Primary Schools*, 7th ed.; Routledge: Oxford, UK, 2018.
12. Bilal, E.; Erol, M. Hypothesis-Experiment-Instruction (HEI) Method for Investigation and Elimination of Misconceptions on Friction. *Balk. Phys. Lett.* **2010**, *18*, 269–276.
13. Alesandrini, A.T.; Heron, P.R. Types of Explanations Students Use to Explain Answers to Conceptual Physics Questions. In *Proceedings of the Physics Education Research Conference, Provo, UT, USA, 24–29 July 2019*; American Association of Physics Teachers: College Park, MD, USA, 2019; pp. 21–25.
14. Brígido, M.; Borrachero, A.B.; Bermejo, M.L.; Mellado, V. Prospective Primary Teachers' Self-Efficacy and Emotions in Science Teaching. *Eur. J. Teach. Educ.* **2013**, *36*, 200–217. [CrossRef]
15. Fortus, D.; Vedder-Weiss, D. Measuring Students' Continuing Motivation for Science Learning. *J. Res. Sci. Teach.* **2014**, *51*, 497–522. [CrossRef]
16. Taştan, S.B.; Davoudi, S.M.M.; Masalimova, A.R.; Bersanov, A.S.; Kurbanov, R.A.; Boiarchuk, A.V.; Pavlushin, A.A. The Impacts of Teacher's Efficacy and Motivation on Student's Academic Achievement in Science Education among Secondary and High School Students. *Eurasia J. Math. Sci. Technol. Educ.* **2018**, *14*, 2353–2366. [CrossRef]
17. Kalogiannakis, M.; Papadakis, S.; Zourmpakis, A.-I. Gamification in Science Education. A Systematic Review of the Literature. *Educ. Sci.* **2021**, *11*, 22. [CrossRef]

18. Kapp, K.M. The Gamification of Learning and Instruction: Game-Based Methods and Strategies for Training and Education. *Int. J. Gaming Comput. Mediat. Simul.* **2012**, *4*, 81–83. [CrossRef]
19. Bai, S.; Hew, K.F.; Huang, B. Does Gamification Improve Student Learning Outcome? Evidence from a Meta-Analysis and Synthesis of Qualitative Data in Educational Contexts. *Educ. Res. Rev.* **2020**, *30*, 100322. [CrossRef]
20. Zainuddin, Z.; Chu, S.K.W.; Shujahat, M.; Perera, C.J. The Impact of Gamification on Learning and Instruction: A Systematic Review of Empirical Evidence. *Educ. Res. Rev.* **2020**, *30*, 100326. [CrossRef]
21. Klock, A.C.T.; da Cunha, L.F.; de Carvalho, M.F.; Rosa, B.E.; Anton, A.J.; Gasparini, I. Gamification in E-Learning Systems: A Conceptual Model to Engage Students and Its Application in an Adaptive e-Learning System. In *Proceedings of the Learning and Collaboration Technologies: Second International Conference, Los Angeles, CA, USA, 2–7 August 2015*; Springer: Cham, Switzerland, 2015; Volume 9192, pp. 595–607.
22. Jagušt, T.; Botički, I.; So, H.J. Examining Competitive, Collaborative and Adaptive Gamification in Young Learners' Math Learning. *Comput. Educ.* **2018**, *125*, 444–457. [CrossRef]
23. Hassan, M.A.; Habiba, U.; Majeed, F.; Shoaib, M. Adaptive Gamification in E-Learning Based on Students' Learning Styles. *Interact. Learn. Environ.* **2019**, *29*, 545–565. [CrossRef]
24. Toda, A.M.; Valle, P.H.D.; Isotani, S. The Dark Side of Gamification: An Overview of Negative Effects of Gamification in Education. In *Proceedings of the Communications in Computer and Information Science, Moscow, Russia, 17–21 September 2018*; Springer: Cham, Switzerland, 2018; Volume 832, pp. 143–156.
25. Koivisto, J.; Hamari, J. The Rise of Motivational Information Systems: A Review of Gamification Research. *Int. J. Inf. Manag.* **2019**, *45*, 191–210. [CrossRef]
26. Botte, B.; Bakkes, S.; Veltkamp, R. Motivation in Gamification: Constructing a Correlation Between Gamification Achievements and Self-Determination Theory. In *Proceedings of the International Conference on Games and Learning Alliance, Laval, France, 9–10 December 2020*; Springer Science and Business Media Deutschland GmbH: Berlin, Germany, 2020; Volume 12517, pp. 157–166.
27. Zourmpakis, A.I.; Papadakis, S.; Kalogiannakis, M. Education of Preschool and Elementary Teachers on the Use of Adaptive Gamification in Science Education. *Int. J. Technol. Enhanc. Learn.* **2022**, *14*, 1–16. [CrossRef]
28. Lopez, C.; Tucker, C. Toward Personalized Adaptive Gamification: A Machine Learning Model for Predicting Performance. *IEEE Trans. Games* **2020**, *12*, 155–168. [CrossRef]
29. Monterrat, B.; Lavoué, É.; George, S. Adaptation of Gaming Features for Motivating Learners. *Simul. Gaming* **2017**, *48*, 625–656. [CrossRef]
30. Codish, D.; Ravid, G. Personality Based Gamification–Educational Gamification for Extroverts and Introverts. In *Proceedings of the 9th CHAIS Conference for the Study of Innovation and Learning Technologies: Learning in the Technological Era, Ra'anana, Israel, 11–12 February 2014*; The Open University of Israel: Ra'anana, Israel, 2015; pp. 36–44.
31. Zourmpakis, A.-I.; Kalogiannakis, M.; Papadakis, S. A Review of the Literature for Designing and Developing a Framework for Adaptive Gamification in Physics Education. In *The International Handbook of Physics Education Research: Teaching Physics*; AIP Publisher: New York, NY, USA, 2023; pp. 1–26. [CrossRef]
32. Glynn, S.M.; Brickman, P.; Armstrong, N.; Taasoobshirazi, G. Science Motivation Questionnaire II: Validation with Science Majors and Nonscience Majors. *J. Res. Sci. Teach.* **2011**, *48*, 1159–1176. [CrossRef]
33. Velayutham, S.; Aldridge, J.; Fraser, B. Development and Validation of an Instrument to Measure Students' Motivation and Self-Regulation in Science Learning. *Int. J. Sci. Educ.* **2011**, *33*, 2159–2179. [CrossRef]
34. Schumm, M.F.; Bogner, F.X. The Impact of Science Motivation on Cognitive Achievement within a 3-Lesson Unit about Renewable Energies. *Stud. Educ. Eval.* **2016**, *50*, 14–21. [CrossRef]
35. Schönfelder, M.L.; Bogner, F.X. Between Science Education and Environmental Education: How Science Motivation Relates to Environmental Values. *Sustainability* **2020**, *12*, 1968. [CrossRef]
36. Zeyer, A. Motivation to Learn Science and Cognitive Style. *Eurasia J. Math. Sci. Technol. Educ.* **2010**, *6*, 121–128. [CrossRef]
37. Deterding, S. Gamification: Designing for Motivation. *Interactions* **2012**, *19*, 14–17. [CrossRef]
38. Klock, A.C.T.; Ogawa, A.N.; Gasparini, I.; Pimenta, M.S. Does Gamification Matter? A Systematic Mapping about the Evaluation of Gamification in Educational Environments. In *Proceedings of the ACM Symposium on Applied Computing, Pau, France, 9 April 2018*; Association for Computing Machinery: New York, NY, USA, 2018; Volume 7, pp. 2006–2012.
39. Landers, R.N. Developing a Theory of Gamified Learning. *Simul. Gaming* **2014**, *45*, 752–768. [CrossRef]
40. De-Marcos, L.; Domínguez, A.; Saenz-De-Navarrete, J.; Pagés, C. An Empirical Study Comparing Gamification and Social Networking on E-Learning. *Comput. Educ.* **2014**, *75*, 82–91. [CrossRef]
41. Rachels, J.R.; Rockinson-Szapkiw, A.J. The Effects of a Mobile Gamification App on Elementary Students' Spanish Achievement and Self-Efficacy. *Comput. Assist. Lang. Learn.* **2018**, *31*, 72–89. [CrossRef]
42. Rose, J.A.; O'Meara, J.M.; Gerhardt, T.C.; Williams, M. Gamification: Using Elements of Video Games to Improve Engagement in an Undergraduate Physics Class. *Phys. Educ.* **2016**, *51*, 055007. [CrossRef]
43. Henukh, A.; Guntara, Y. Analyzing the Response of Learners to Use Kahoot as Gamification of Learning Physics. *Gravity J. Ilm. Penelit. Dan Pembelajaran Fis.* **2020**, *6*, 72–76. [CrossRef]
44. Böckle, M.; Novak, J.; Bick, M. Towards Adaptive Gamification: A Synthesis of Current Developments. In Proceedings of the Twenty-Fifth European Conference on Information Systems (ECIS), Guimarães, Portugal, 5–10 June 2017; pp. 158–174.

45. Santos, W.O.D.; Bittencourt, I.I.; Vassileva, J. Design of Tailored Gamified Educational Systems Based on Gamer Types. In *Proceedings of the Congresso Brasileiro de Informática na Educação*; SBC: Porto Alegre, Brazil, 2018; pp. 42–51.
46. Rapp, A.; Hopfgartner, F.; Hamari, J.; Linehan, C.; Cena, F. Strengthening Gamification Studies: Current Trends and Future Opportunities of Gamification Research. *Int. J. Hum.-Comput. Stud.* **2019**, *127*, 1–6. [CrossRef]
47. Pelletier, K.; Brown, M.; Brooks, D.C.; Mccormack, M.; Reeves, J.; Arbino, N.; Bozkurt, A.; Crawford, S.; Czerniewicz, L.; Gibson, R.; et al. *2021 EDUCAUSE Horizon Report, Teaching and Learning Edition*; EDUCAUSE: Boulder, CO, USA, 2021.
48. Werbach, K.; Hunter, D.; Dixon, W. *For the Win: How Game Thinking Can Revolutionize Your Business*; Wharton Digital Press: Philadelphia, PA, USA, 2012; Volume 1.
49. Monterrat, B.; Desmarais, M.; Lavoué, É.; George, S. A Player Model for Adaptive Gamification in Learning Environments. In *Proceedings of the Artificial Intelligence in Education: 17th International Conference, Madrid, Spain, 22–26 June 2015*; Springer: Cham, Switzerland, 2015; Volume 9112, pp. 297–306.
50. Mora, A.; Tondello, G.F.; Nacke, L.E.; Arnedo-Moreno, J. Effect of Personalized Gameful Design on Student Engagement. In *Proceedings of the IEEE Global Engineering Education Conference, EDUCON, Santa Cruz de Tenerife, Spain, 17–20 April 2018*; IEEE Computer Society: Piscataway, NJ, USA, 2018; pp. 1925–1933.
51. Roosta, F.; Taghiyareh, F.; Mosharraf, M. Personalization of Gamification-Elements in an e-Learning Environment Based on Learners' Motivation. In *Proceedings of the 8th International Symposium on Telecommunications (IST), Tehran, Iran, 27–28 September 2016*; Institute of Electrical and Electronics Engineers Inc.: Piscataway, NJ, USA, 2017; pp. 637–642.
52. Nacke, L.E.; Bateman, C.; Mandryk, R.L. BrainHex: A Neurobiological Gamer Typology Survey. *Entertain. Comput.* **2014**, *5*, 55–62. [CrossRef]
53. Oliveira, W.; Bittencourt, I.I. *Tailored Gamification to Educational Technologies*; Springer: Singapore, 2019; ISBN 9789813298118.
54. Oliveira, W.; Isotani, S.; Toda, A.M.; Palomino, P.T.; Vassileva, J.; Shi, L.; Bittencourt, I.I. Does Tailoring Gamified Educational Systems Matter? The Impact on Students' Flow Experience. In Proceedings of the 53rd Hawaii International Conference on System Sciences, Maui, HI, USA, 7–10 January 2020; pp. 1226–1235.
55. Hallifax, S.; Lavoué, E.; Serna, A. To Tailor or Not to Tailor Gamification? An Analysis of the Impact of Tailored Game Elements on Learners' Behaviours and Motivation. In *Proceedings of the International Conference on Artificial Intelligence in Education, Ifrane, Morocco, 6–10 July 2020*; Springer: Cham, Switzerland, 2020; Volume 12163, pp. 216–227.
56. Seaborn, K.; Fels, D.I. Gamification in Theory and Action: A Survey. *Int. J. Hum.-Comput. Stud.* **2015**, *74*, 14–31. [CrossRef]
57. Nand, K.; Baghaei, N.; Casey, J.; Barmada, B.; Mehdipour, F.; Liang, H.-N. Engaging Children with Educational Content via Gamification. *Smart Learn. Environ.* **2019**, *6*, 6. [CrossRef]
58. Fitz-Walter, Z.; Johnson, D.; Wyeth, P.; Tjondronegoro, D.; Scott-Parker, B. Driven to Drive? Investigating the Effect of Gamification on Learner Driver Behavior, Perceived Motivation and User Experience. *Comput. Hum. Behav.* **2017**, *71*, 586–595. [CrossRef]
59. Hallifax, S.; Serna, A.; Marty, J.C.; Lavoué, G.; Lavoué, E. Factors to Consider for Tailored Gamification. In *Proceedings of the Annual Symposium on Computer-Human Interaction in Play, Barcelona, Spain, 22–25 October 2019*; Association for Computing Machinery, Inc.: New York, NY, USA, 2019; pp. 559–572.
60. Scheiner, C.; Haas, P.; Bretschneider, U.; Blohm, I.; Leimeister, J.M. Obstacles and Challenges in the Use of Gamification for Virtual Idea Communities. In *Gamification*; Springer: Cham, Switzerland, 2017; pp. 65–76.
61. Majchrzak, A.; Malhotra, A. Towards an Information Systems Perspective and Research Agenda on Crowdsourcing for Innovation. *J. Strateg. Inf. Syst.* **2013**, *22*, 257–268. [CrossRef]
62. Scheiner, C.W. The Motivational Fabric of Gamified Idea Competitions: The Evaluation of Game Mechanics from a Longitudinal Perspective. *Creat. Innov. Manag.* **2015**, *24*, 341–352. [CrossRef]
63. Mora, A.; Riera, D.; González, C.; Arnedo-Moreno, J. Gamification: A Systematic Review of Design Frameworks. *J. Comput. High. Educ.* **2017**, *29*, 516–548. [CrossRef]
64. Tondello, G.F.; Wehbe, R.R.; Diamond, L.; Busch, M.; Marczewski, A.; Nacke, L.E. The Gamification User Types Hexad Scale. In *Proceedings of the 2016 Annual Symposium on Computer-Human Interaction in Play, Austin, TX, USA, 16 October 2016*; Association for Computing Machinery, Inc.: New York, NY, USA, 2016; pp. 229–243.
65. Jimoyiannis, A. Designing and Implementing an Integrated Technological Pedagogical Science Knowledge Framework for Science Teachers Professional Development. *Comput. Educ.* **2010**, *55*, 1259–1269. [CrossRef]
66. Petousi, V.; Sifaki, E. Contextualising Harm in the Framework of Research Misconduct. Findings from Discourse Analysis of Scientific Publications. *Int. J. Sustain. Dev.* **2020**, *23*, 149–174. [CrossRef]
67. Rajendran, T.; Bin Naaim, N.A.; Yunus, M.M. Pupils Motivation And Perceptions Towards Learning English Using Quizvaganza. *Int. J. Sci. Res. Publ.* **2019**, *9*, 220–227. [CrossRef]
68. Halim, M.S.A.A.; Hashim, H.; Yunus, M.M. Pupils' Motivation and Perceptions on ESL Lessons through Online Quiz-Games. *J. Educ. e-Learn. Res.* **2020**, *7*, 229–234. [CrossRef]
69. Melkersson, J.; Lundin, J. *Gamification Badges vs Leaderboards as a Motivational Tool for University Students Learning a Second Language*; Malmö University: Malmö, Sweden, 2022.
70. Taggart, J.; Hood, N. Determinants of Autonomy in Multinational Corporation Subsidiaries. *Eur. Manag. J.* **1999**, *17*, 226–236. [CrossRef]

71. Rodrigues, L.; Palomino, P.T.; Toda, A.M.; Klock, A.C.T.; Oliveira, W.; Avila-Santos, A.P.; Gasparini, I.; Isotani, S. Personalization Improves Gamification: Evidence from a Mixed-Methods Study. In *Proceedings of the ACM on Human-Computer Interaction, Salamanca, Spain, 6 October 2021*; ACM: New York, NY, USA; Volume 5, pp. 1–25.
72. Papadakis, S.; Zourmpakis, A.I.; Kalogiannakis, M. Analyzing the Impact of a Gamification Approach on Primary Students' Motivation and Learning in Science Education. In *Proceedings of the International Conference on Interactive Collaborative Learning, Madrid, Spain, 26–29 September 2023*; Springer: Cham, Switzerland, 2023; Volume 633, pp. 701–711.
73. Aldemir, T.; Celik, B.; Kaplan, G. A Qualitative Investigation of Student Perceptions of Game Elements in a Gamified Course. *Comput. Hum. Behav.* **2018**, *78*, 235–254. [CrossRef]
74. Carlier, S.; Van der Paelt, S.; Ongenae, F.; De Backere, F.; De Turck, F. Empowering Children with ASD and Their Parents: Design of a Serious Game for Anxiety and Stress Reduction. *Sensors* **2020**, *20*, 966. [CrossRef]
75. Reyssier, S.; Hallifax, S.; Serna, A.; Marty, J.C.; Simonian, S.; Lavoue, E. The Impact of Game Elements on Learner Motivation: Influence of Initial Motivation and Player Profile. *IEEE Trans. Learn. Technol.* **2022**, *15*, 42–54. [CrossRef]
76. Klock, A.C.T.; Gasparini, I.; Pimenta, M.S.; Hamari, J. Tailored Gamification: A Review of Literature. *Int. J. Hum.-Comput. Stud.* **2020**, *144*, 102495. [CrossRef]
77. Bozkurt, A.; Sharma, R.C. Education in Normal, New Normal, and Next Normal: Observations from the Past, Insights from the Present and Projections for the Future. *Asian J. Distance Educ.* **2020**, *15*, i–x. [CrossRef]
78. Ferdig, R.E.; Baumgartner, E.; Hartshorne, R.; Kaplan-Rakowski, R.; Mouza, C. *Teaching, Technology, and Teacher Education during the COVID-19 Pandemic: Stories from the Field*; AACE: Waynesville, NC, USA, 2019; ISBN 9781939797490.
79. Teo, T. Pre-Service Teachers' Attitudes towards Computer Use: A Singapore Survey. *Australas. J. Educ. Technol.* **2008**, *24*, 413–424. [CrossRef]
80. Mekler, E.D.; Brühlmann, F.; Tuch, A.N.; Opwis, K. Towards Understanding the Effects of Individual Gamification Elements on Intrinsic Motivation and Performance. *Comput. Hum. Behav.* **2017**, *71*, 525–534. [CrossRef]
81. Xezonaki, A. Gamification in Preschool Science Education. *Adv. Mob. Learn. Educ. Res.* **2022**, *2*, 308–320. [CrossRef]

Disclaimer/Publisher's Note: The statements, opinions and data contained in all publications are solely those of the individual author(s) and contributor(s) and not of MDPI and/or the editor(s). MDPI and/or the editor(s) disclaim responsibility for any injury to people or property resulting from any ideas, methods, instructions or products referred to in the content.

MDPI AG
Grosspeteranlage 5
4052 Basel
Switzerland
Tel.: +41 61 683 77 34

Computers Editorial Office
E-mail: computers@mdpi.com
www.mdpi.com/journal/computers

Disclaimer/Publisher's Note: The title and front matter of this reprint are at the discretion of the Guest Editors. The publisher is not responsible for their content or any associated concerns. The statements, opinions and data contained in all individual articles are solely those of the individual Editors and contributors and not of MDPI. MDPI disclaims responsibility for any injury to people or property resulting from any ideas, methods, instructions or products referred to in the content.

www.ingramcontent.com/pod-product-compliance
Lightning Source LLC
LaVergne TN
LVHW072329090526
838202LV00019B/2375